Algorithms and Computation in Mathematics • Volume 8

Editors

E. Becker   M. Bronstein   H. Cohen
D. Eisenbud   R. Gilman

Springer
Berlin
Heidelberg
New York
Barcelona
Hong Kong
London
Milan
Paris
Tokyo

David Eisenbud   Daniel R. Grayson
Michael Stillman   Bernd Sturmfels
(Eds.)

Computations in Algebraic Geometry with Macaulay 2

 Springer

Editors

David Eisenbud
Mathematical Sciences
Research Institute
1000 Centennial Drive
Berkeley, CA 94720, USA
e-mail: de@msri.org

Daniel R. Grayson
University of Illinois
at Urbana-Champaign
Department of Mathematics
Urbana, IL 61801, USA
e-mail: drg@uiuc.edu

Michael Stillman
Cornell University
Department of Mathematics
Ithaca, NY 14853, USA
e-mail: mike@math.cornell.edu

Bernd Sturmfels
University of California
Department of Mathematics
Berkeley, CA 94720, USA
e-mail: bernd@math.berkeley.edu

Mathematics Subject Classification (2000): 13-04, 13P, 14-04, 14Q, 16Z05, 68W30, 11Y, 12Y05, 14P, 12D10, 14M15, 14N15

Library of Congress Cataloging-in-Publication Data applied for

Die Deutsche Bibliothek – CIP-Einheitsaufnahme

Computations in algebraic geometry with macaulay 2 / David Eisenbud ...
(ed.). - Berlin; Heidelberg; New York; Barcelona; Hong Kong; London;
Milan; Paris; Tokyo: Springer, 2002
(Algorithms and computation in mathematics; Vol. 8)
ISBN 3-540-42230-7

ISSN 1431-1550

ISBN 3-540-42230-7   Springer-Verlag Berlin Heidelberg New York

Springer-Verlag Berlin Heidelberg New York
a member of BertelsmannSpringer Science+Business Media GmbH

http://www.springer.de

© Springer-Verlag Berlin Heidelberg 2002
Printed in Germany

Typeset by the authors using a Springer LaTeX-package
Cover design: *design & production GmbH*, Heidelberg

SPIN: 10842488   46/3142LK - 5 4 3 2 1 0 – Printed on acid-free paper

Preface

Systems of polynomial equations arise throughout mathematics, science, and engineering. Algebraic geometry provides powerful theoretical techniques for studying the qualitative and quantitative features of their solution sets. Recently developed algorithms have made theoretical aspects of the subject accessible to a broad range of mathematicians and scientists. The algorithmic approach to the subject has two principal aims: developing new tools for research within mathematics, and providing new tools for modeling and solving problems that arise in the sciences and engineering. A healthy synergy emerges, as new theorems yield new algorithms and emerging applications lead to new theoretical questions.

This book presents algorithmic tools for algebraic geometry and experimental applications of them. It also introduces a software system in which the tools have been implemented and with which the experiments can be carried out. *Macaulay 2* is a computer algebra system devoted to supporting research in algebraic geometry, commutative algebra, and their applications. The reader of this book will encounter *Macaulay 2* in the context of concrete applications and practical computations in algebraic geometry.

The expositions of the algorithmic tools presented here are designed to serve as a useful guide for those wishing to bring such tools to bear on their own problems. A wide range of mathematical scientists should find these expositions valuable. This includes both the users of other programs similar to *Macaulay 2* (for example, Singular and CoCoA) and those who are not interested in explicit machine computations at all.

The chapters are ordered roughly by increasing mathematical difficulty. The first part of the book is meant to be accessible to graduate students and computer algebra users from across the mathematical sciences and is primarily concerned with introducing *Macaulay 2*. The second part emphasizes the mathematics: each chapter exposes some domain of mathematics at an accessible level, presents the relevant algorithms, sometimes with proofs, and illustrates the use of the program. In both parts, each chapter comes with its own abstract and its own bibliography; the index at the back of the book covers all of them.

One of the first computer algebra packages aimed at algebraic geometry was *Macaulay*, the predecessor of *Macaulay 2*, written during the years 1983-1993 by Dave Bayer and Mike Stillman. Worst-case estimates suggested that trying to compute Gröbner bases might be a hopeless approach to solving problems. But from the first prototype, *Macaulay* was successful surprisingly often, perhaps because of the geometrical origin of the problems attacked. *Macaulay* improved steadily during its first decade. It helped transform the theoretical notion of a projective resolution into an exciting new practical research tool, and became widely used for research and teaching in commutative algebra and algebraic geometry. It was possible to write routines in the top-level language, and many important algorithms were added by David Eisenbud and other users, enhancing the system and broadening its usefulness.

There were certain practical drawbacks for the researcher who wanted to use *Macaulay* effectively. A minor annoyance was that only finite prime fields were available as coefficient rings. The major problem was that the language made available to users was primitive and barely supported high-level development of new algorithms; it had few basic data types and didn't support the addition of new ones.

Macaulay 2 is based on experience gained from writing and using its predecessor *Macaulay*, but is otherwise a fresh start. It was written by Dan Grayson and Mike Stillman with the generous financial support of the U.S. National Science Foundation, with the work starting in 1993[1]. It also incorporates some code from other authors: the package SINGULAR-FACTORY[2] provides for factorization of polynomials; SINGULAR-LIBFAC[3] uses FACTORY to enable the computation of characteristic sets and thus the decomposition of subvarieties into their irreducible components; and GNU MP[4] by Torbjörn Granlund and others provides for multiple precision arithmetic.

Macaulay 2 aims to support efficient computation associated with a wide variety of high level mathematical objects, including Galois fields, number fields, polynomial rings, exterior algebras, Weyl algebras, quotient rings, ideals, modules, homomorphisms of rings and modules, graded modules, maps between graded modules, chain complexes, maps between chain complexes, free resolutions, algebraic varieties, and coherent sheaves. To make the system easily accessible, standard mathematical notation is followed closely.

As with *Macaulay*, it was hoped that users would join in the further development of new algorithms for *Macaulay 2*, so the developers tried to make the language available to the users as powerful as possible, yet easy to use. Indeed, much of the high-level part of the system is written in the same language available to the user. This ensures that the user will find it just as

---

[1] NSF grants DMS 92-10805, 92-10807, 96-23232, 96-22608, 99-70085, and 99-70348.

[2] SINGULAR-FACTORY, a subroutine library for factorization, by G.-M. Greuel, R. Stobbe, G. Pfister, H. Schoenemann, and J. Schmidt; available at ftp://helios.mathematik.uni-kl.de/pub/Math/Singular/Factory/.

[3] SINGULAR-LIBFAC, a subroutine library for characteristic sets and irreducible decomposition, by M. Messollen; available at ftp://helios.mathematik.uni-kl.de/pub/Math/Singular/Libfac/.

[4] GMP, a library for arbitrary precision arithmetic, by Torbjörn Granlund, John Amanatides, Paul Zimmermann, Ken Weber, Bennet Yee, Andreas Schwab, Robert Harley, Linus Nordberg, Kent Boortz, Kevin Ryde, and Guillaume Hanrot; available at ftp://ftp.gnu.org/gnu/gmp/.

easy as the developers did to implement a new type of mathematical object
or to modify the high-level aspects of the current algorithms.

The language available to the user is interpreted. The interpreter itself
is written in a convenient language designed to be mostly type-safe and to
handle memory allocation and initialization automatically. For maximum efficiency, the core mathematical algorithms are written in C++ and compiled,
not interpreted. This includes the arithmetic operations of rings, modules,
and matrices, the Gröbner basis algorithm (in several enhanced versions, tai-
lored for various situations), several algorithms for computing free resolutions
of modules, the algorithm for computing the Hilbert series of a graded ring or
module, the algorithms for computing determinants and Pfaffians, the basis
reduction algorithm, factoring, etc.

In one way *Macaulay 2* is like a standard computer algebra system, such
as *Mathematica* or *Maple*: the user enters mathematical expressions at the
keyboard, and the program computes the value of the expression and displays
the answer.

Here is the first input prompt offered to the user.

i1 :

In response to the prompt, the user may enter, for example, a simple arithmetic expression.

```
i1 : 3/5 + 7/11

      68
o1 = --
      55

o1 : QQ
```

The answer itself is displayed to the right of the output label

o1 =

and its type (or class) is displayed to the right of the following label.

o1 :

The symbol QQ appearing in this example denotes the class of all rational
numbers, and is meant to be reminiscent of the notation $\mathbb{Q}$.

Macaulay 2 often finds itself being run in a window with horizontal scroll
bars, so by default it does not wrap output lines, but instead lets them grow
without bound. This book was produced by an automated mechanism that
submits code provided by the authors to *Macaulay 2* and incorporates the
result into the text. Output lines that exceed the width of the pages of this
book are indicated with ellipses, as in the following example.

```
i2 : 100!

o2 = 9332621544394415268169923885626670049071596826438162146859296389 5 · · ·
```

Next we describe an important difference between general computer algebra systems (such as *Maple* and *Mathematica*) and *Macaulay 2*. Before entering an expression involving variables (such as x+y) into *Macaulay 2* the user must first create a ring containing those variables. Rings are important objects of study in algebraic geometry; quotient rings of polynomial rings, for example, encapsulate the essential information about a system of polynomial equations, including, for example, the field from which the coefficients are drawn. Often one has several rings under consideration at once, along with ring homomorphisms between them, so it is important to treat them as first-class objects in the computer, capable of being named and manipulated the same way numbers and characters can be manipulated in simpler programming languages.

Let's give a hint of the breadth of types of mathematical objects available in *Macaulay 2* with some examples. In *Macaulay 2* one defines a quotient ring of a polynomial ring R over the rational numbers by entering a command such as the one below.

```
i3 : R = QQ[x,y,z]/(x^3-y^3-z^3)

o3 = R

o3 : QuotientRing
```

Having done that, we can compute in the ring.

```
i4 : (x+y+z)^3

         2        2     3      2                  2         2        2     3
o4 = 3x y + 3x*y  + 2y  + 3x z + 6x*y*z + 3y z + 3x*z  + 3y*z  + 2z

o4 : R
```

We can make matrices over the ring.

```
i5 : b = vars R

o5 = | x y z |

              1      3
o5 : Matrix R  <--- R

i6 : c = matrix {{x^2,y^2,z^2}}

o6 = | x2 y2 z2 |

              1      3
o6 : Matrix R  <--- R
```

We can make modules over the ring.

```
i7 : M = coker b

o7 = cokernel | x y z |

                               1
o7 : R-module, quotient of R

i8 : N = ker c

o8 = image {2} | x    0   -y2 -z2 |
           {2} | -y  -z2  x2   0  |
           {2} | -z  y2    0   x2 |
```

```
                        3
o8 : R-module, submodule of R
```

We can make projective resolutions of modules.

```
i9 : res M

      1     3     4     4     4
o9 = R  <-- R  <-- R  <-- R  <-- R

      0     1     2     3     4

o9 : ChainComplex
```

We can make projective varieties.

```
i10 : X = Proj R

o10 = X

o10 : ProjectiveVariety
```

We can make coherent sheaves and compute their cohomology.

```
i11 : HH^1 cotangentSheaf X

      1
o11 = QQ

o11 : QQ-module, free
```

At this writing, *Macaulay 2* is available for GNU/Linux and other flavors of Unix, and also for Microsoft Windows and the Macintosh operating system. Although it can be used as a free-standing program, it is most convenient to use it in an editor's buffer; Emacs (on Unix or Windows systems) or MPW on Macintosh systems are currently the editors of choice. To obtain *Macaulay 2*, download it from the website[5] and unpack the file. Among the resulting files will be a file called `Macaulay2/README.txt`, which you should read. It will tell you how to run the `setup` script, and how to install a few lines of code in your `emacs` init file to enable you to run M2 in an emacs buffer and to edit *Macaulay 2* code. A system administrator of a Unix system may optionally arrange for those lines of code to be available to every emacs user.

The editors thank the authors of the chapters for their valuable contributions and hard work, and the National Science Foundation for funding the development of *Macaulay 2* and for partial funding of the authors who have contributed to this volume.

May, 2001

David Eisenbud
Daniel R. Grayson
Michael E. Stillman
Bernd Sturmfels

---

[5] *Macaulay 2*, a software system for research in algebraic geometry, by Daniel R. Grayson and Michael E. Stillman; available online in source code form and compiled for various architectures at `http://www.math.uiuc.edu/Macaulay2/`.

Table of Contents

Part II   Mathematical Computations

List of Contributors

Luchezar L. Avramov
Department of Mathematics
Purdue University
West Lafayette, IN 47907, USA
avramov@math.purdue.edu
www.math.purdue.edu/~avramov/

Wolfram Decker
FB Mathematik
Universität des Saarlandes
66041 Saarbrücken, Germany
decker@math.uni-sb.de
loge.math.uni-sb.de/~agdecker/

David Eisenbud
Mathematical Sciences Research Institute
1000 Centennial Drive
Berkeley, CA 94720, USA,
and
Department of Mathematics
University of California at Berkeley
Berkeley, CA 94720, USA
de@msri.org
www.msri.org/people/staff/de/

Daniel R. Grayson
Department of Mathematics
University of Illinois at Urbana-Champaign
Urbana, IL 61801, USA
dan@math.uiuc.edu
www.math.uiuc.edu/~dan/

Serkan Hoşten
Department of Mathematics
San Francisco State University
San Francisco, CA 94132, USA
serkan@math.sfsu.edu

Frank-Olaf Schreyer
Fakultät für Mathematik und Physik
Universität Bayreuth
95440 Bayreuth, Germany
frank.schreyer@uni-bayreuth.de

Gregory G. Smith
Department of Mathematics
Barnard College, Columbia University
New York, NY 10027, USA
ggsmith@math.berkeley.edu

Frank Sottile
Department of Mathematics and Statistics
University of Massachusetts, Amherst
Amherst, MA 01003, USA
sottile@math.umass.edu
www.math.umass.edu/~sottile/

Michael Stillman
Department of Mathematics
Cornell University
Ithaca, NY 14853, USA
mike@math.cornell.edu

Bernd Sturmfels
Department of Mathematics
University of California
Berkeley, California 94720, USA
bernd@math.berkeley.edu
www.math.berkeley.edu/~bernd/

Rekha Thomas
Department of Mathematics
University of Washington
Seattle, WA 98195, USA
thomas@math.washington.edu
www.math.washington.edu/~thomas/

Fabio Tonoli
Mathematisches Institut
Georg–August Universität
37073 Göttingen, Germany
tonoli@uni-math.gwdg.de

Uli Walther
Department of Mathematics
Purdue University
West Lafayette, IN 47907, USA
walther@math.purdue.edu
www.math.purdue.edu/~walther/

Part I

Introducing *Macaulay 2*

Ideals, Varieties and *Macaulay 2*

Bernd Sturmfels*

This chapter introduces *Macaulay 2* commands for some elementary computations in algebraic geometry. Familiarity with Gröbner bases is assumed.

Many students and researchers alike have their first encounter with Gröbner bases through the delightful text books [1] and [2] by David Cox, John Little and Donal O'Shea. This chapter illustrates the use of *Macaulay 2* for some computations discussed in these books. It can be used as a supplement for an advanced undergraduate course or first-year graduate course in computational algebraic geometry. The mathematically advanced reader will find this chapter a useful summary of some basic *Macaulay 2* commands.

1   A Curve in Affine Three-Space

Our first example concerns geometric objects in (complex) affine 3-space. We start by setting up the ring of polynomial functions with rational coefficients.

```
i1 : R = QQ[x,y,z]

o1 = R

o1 : PolynomialRing
```

Various monomial orderings are available in *Macaulay 2*; since we did not specify one explicitly, the monomials in the ring R will be sorted in graded reverse lexicographic order [1, §I.2, Definition 6]. We define an ideal generated by two polynomials in this ring and assign it to the variable named curve.

```
i2 : curve = ideal( x^4-y^5, x^3-y^7 )

            5    4    7    3
o2 = ideal (- y  + x , - y  + x )

o2 : Ideal of R
```

We compute the reduced Gröbner basis of our ideal:

```
i3 : gb curve

o3 = | y5-x4 x4y2-x3 x8-x3y3 |

o3 : GroebnerBasis
```

By inspecting leading terms (and using [1, §9.3, Theorem 8]), we see that our ideal curve does indeed define a one-dimensional affine variety. This can be tested directly with the following commands in *Macaulay 2*:

```
i4 : dim curve

o4 = 1
```

* Partially supported by the National Science Foundation (DMS-9970254).

```
i5 : codim curve
```

```
o5 = 2
```

The *degree* of a curve in complex affine 3-space is the number of intersection points with a general plane. It coincides with the degree [2, §6.4] of the projective closure [1, §8.4] of our curve, which we compute as follows:

```
i6 : degree curve
```

```
o6 = 28
```

The Gröbner basis in o3 contains two polynomials which are not irreducible: they contain a factor of x^3. This shows that our curve is not irreducible over **Q**. We first extract the components which are transverse to the plane $x = 0$:

```
i7 : curve1 = saturate(curve,ideal(x))

              2      5    4   5    3
o7 = ideal (x*y  - 1, y  - x , x  - y )

o7 : Ideal of R
```

And next we extract the component which lies in the plane $x = 0$:

```
i8 : curve2 = saturate(curve,curve1)

            3   5
o8 = ideal (x , y )

o8 : Ideal of R
```

The second component is a multiple line. Hence our input ideal was not radical. To test equality of ideals we use the command == .

```
i9 : curve == radical curve
```

```
o9 = false
```

We now replace our curve by its first component:

```
i10 : curve = curve1

               2      5    4   5    3
o10 = ideal (x*y  - 1, y  - x , x  - y )

o10 : Ideal of R

i11 : degree curve
```

```
o11 = 13
```

The ideal of this curve is radical:

```
i12 : curve == radical curve
```

```
o12 = true
```

Notice that the variable **z** does not appear among the generators of the ideal. Our curve consists of 13 straight lines (over **C**) parallel to the z-axis.

2   Intersecting Our Curve With a Surface

In this section we explore basic operations on ideals, starting with those described in [1, §4.3]. Consider the following surface in affine 3-space:

```
i13 : surface = ideal( x^5 + y^5 + z^5 - 1)

             5   5   5
o13 = ideal(x + y + z  - 1)

o13 : Ideal of R
```

The union of the curve and the surface is represented by the intersection of their ideals:

```
i14 : theirunion = intersect(curve,surface)

             6 2        7      2 5    5    5    5     2       5 5    1 · · ·
o14 = ideal (x y  + x*y  + x*y z  - x  - y  - z  - x*y  + 1, x y  + y   · · ·

o14 : Ideal of R
```

In this example this coincides with the product of the two ideals:

```
i15 : curve*surface == theirunion

o15 = true
```

The intersection of the curve and the surface is represented by the sum of their ideals. We get a finite set of points:

```
i16 : ourpoints = curve + surface

             2      5    4    5    3    5    5    5
o16 = ideal (x*y  - 1, y  - x , x  - y , x  + y  + z  - 1)

o16 : Ideal of R

i17 : dim ourpoints

o17 = 0
```

The number of points is sixty five:

```
i18 : degree ourpoints

o18 = 65
```

Each of the points is multiplicity-free:

```
i19 : degree radical ourpoints

o19 = 65
```

The number of points coincides with the number of monomials not in the initial ideal [2, §2.2]. These are called the *standard monomials*.

```
i20 : staircase = ideal leadTerm ourpoints

             2    5    5    5
o20 = ideal (x*y , z , y , x )

o20 : Ideal of R
```

The basis command can be used to list all the standard monomials

```
i21 : T = R/staircase;

i22 : basis T

o22 = | 1 x x2 x3 x4 x4y x4yz x4yz2 x4yz3 x4yz4 x4z x4z2 x4z3 x4z4 x3y · · ·

             1        65
o22 : Matrix T  <--- T
```

The assignment of the quotient ring to the global variable T had a side effect: the variables x, y, and z now have values in that ring. To bring the variables of R to the fore again, we must say:

```
i23 : use R;
```

Every polynomial function on our 65 points can be written uniquely as a linear combination of these standard monomials. This representation can be computed using the normal form command %.

```
i24 : anyOldPolynomial = y^5*x^5-x^9-y^8+y^3*x^5

        5 5    9    5 3    8
o24 = x y  - x  + x y  - y

o24 : R

i25 : anyOldPolynomial % ourpoints

        4    3
o25 = x y - x y

o25 : R
```

Clearly, the normal form is zero if and only the polynomial is in the ideal.

```
i26 : anotherPolynomial = y^5*x^5-x^9-y^8+y^3*x^4

        5 5    9    8    4 3
o26 = x y  - x  - y  + x y

o26 : R

i27 : anotherPolynomial % ourpoints

o27 = 0

o27 : R
```

3   Changing the Ambient Polynomial Ring

During a *Macaulay 2* session it sometimes becomes necessary to change the ambient ring in which the computations takes place. Our original ring, defined in i1, is the polynomial ring in three variables over the field **Q** of rational numbers with the graded reverse lexicographic order. In this section two modifications are made: first we replace the field of coefficients by a finite field, and later we replace the monomial order by an elimination order.

An important operation in algebraic geometry is the decomposition of algebraic varieties into irreducible components [1, §4.6]. Algebraic algorithms for this purpose are based on the *primary decomposition* of ideals [1, §4.7]. A future version of *Macaulay 2* will have an implementation of primary decomposition over any polynomial ring. The current version of *Macaulay 2* has a command decompose for finding all the minimal primes of an ideal, but, as it stands, this works only over a finite field.

Let us change our coefficient field to the field with 101 elements:

```
i28 : R' = ZZ/101[x,y,z];
```

We next move our ideal from the previous section into the new ring (fortunately, none of the coefficients of its generators have 101 in the denominator):

```
i29 : ourpoints' = substitute(ourpoints,R')

            2      5    4   5    3   5    5   5
o29 = ideal (x*y  - 1, y  - x , x  - y , x  + y  + z  - 1)

o29 : Ideal of R'

i30 : decompose ourpoints'

                                                                     . . .
o30 = {ideal (z + 36, y - 1, x - 1), ideal (z + 1, y - 1, x - 1), idea · · ·

o30 : List
```

Oops, that didn't fit on the display, so let's print them out one per line.

```
i31 : oo / print @@ print;
ideal (z + 36, y - 1, x - 1)

ideal (z + 1, y - 1, x - 1)

ideal (z - 6, y - 1, x - 1)

ideal (z - 14, y - 1, x - 1)

ideal (z - 17, y - 1, x - 1)

            3      2               2                   3     2    2    . . .
ideal (x  - 46x  + 28x*y - 27y  + 46x + y + 27, - 16x  + x y + x  - 15 · · ·

              2                                       2           . . .
ideal (- 32x  - 16x*y + x*z - 16x - 27y - 30z - 14, - 34x  - 14x*y + y · · ·

            2                                       2          2   . . .
ideal (44x  + 22x*y + x*z + 22x - 26y - 30z - 6, 18x  + 12x*y + y  + 1 · · ·

              2                                       2           2 . . .
ideal (- 41x  + 30x*y + x*z + 30x + 38y - 30z + 1, - 26x  - 10x*y + y  · · ·

            2                                        2            2 . . .
ideal (39x  - 31x*y + x*z - 31x - 46y - 30z + 36, - 32x  - 13x*y + y  · · ·

              2                                       2           2 . . .
ideal (- 10x  - 5x*y + x*z - 5x - 40y - 30z - 17, - 37x  + 35x*y + y  · · ·
```

If we just want to see the degrees of the irreducible components, then we say:

```
i32 : ooo / degree

o32 = {1, 1, 1, 1, 1, 30, 6, 6, 6, 6, 6}

o32 : List
```

Note that the expressions oo and ooo refer to the previous and prior-to-previous output lines respectively.

Suppose we wish to compute the x-coordinates of our sixty five points. Then we must use an elimination order, for instance, the one described in [1, §3.2, Exercise 6.a]. We define a new polynomial ring with the elimination order for $\{y, z\} > \{x\}$ as follows:

```
i33 : S = QQ[z,y,x, MonomialOrder => Eliminate 2]

o33 = S

o33 : PolynomialRing
```

We move our ideal into the new ring,

```
i34 : ourpoints'' = substitute(ourpoints,S)

            2       5   4     3     5   5     5     5
o34 = ideal (y x - 1, y  - x , - y + x , z + y + x  - 1)

o34 : Ideal of S
```

and we compute the reduced Gröbner basis in this new order:

```
i35 : G = gens gb ourpoints''

o35 = | x13-1 y-x6 z5+x5+x4-1 |

                 1      3
o35 : Matrix S  <--- S
```

To compute the elimination ideal we use the following command:

```
i36 : ideal selectInSubring(1,G)

             13
o36 = ideal(x   - 1)

o36 : Ideal of S
```

4   Monomials Under the Staircase

Invariants of an algebraic variety, such as its dimension and degree, are computed from an initial monomial ideal. This computation amounts to the combinatorial task of analyzing the collection of standard monomials, that is, the monomials under the staircase [1, Chapter 9]. In this section we demonstrate some basic operations on monomial ideals in *Macaulay 2*.

Let us create a non-trivial staircase in three dimensions by taking the third power of the initial monomial from line i20.

```
i37 : M = staircase^3

              3 6    2 4 5   2 9   7 4    2 10    7 5   6 2 5     12  ···
o37 = ideal (x y , x y z , x y , x y , x*y z , x*y z , x y z , x*y   , ···

o37 : Ideal of R
```

The number of current generators of this ideal equals

```
i38 : numgens M

o38 = 20
```

To see all generators we can transpose the matrix of minimal generators:

```
i39 : transpose gens M
```

```
o39 = {-9}  | x3y6   |
       {-11} | x2y4z5 |
       {-11} | x2y9   |
       {-11} | x7y4   |
       {-13} | xy2z10 |
       {-13} | xy7z5  |
       {-13} | x6y2z5 |
       {-13} | xy12   |
       {-13} | x6y7   |
       {-13} | x11y2  |
       {-15} | z15    |
       {-15} | y5z10  |
       {-15} | x5z10  |
       {-15} | y10z5  |
       {-15} | x5y5z5 |
       {-15} | x10z5  |
       {-15} | y15    |
       {-15} | x5y10  |
       {-15} | x10y5  |
       {-15} | x15    |

                20        1
o39 : Matrix R    <--- R
```

Note that this generating set is not minimal; see o48 below. The number of standard monomials equals

```
i40 : degree M

o40 = 690
```

To list all the standard monomials we first create the residue ring

```
i41 : S = R/M

o41 = S

o41 : QuotientRing
```

and then we ask for a vector space basis of the residue ring:

```
i42 : basis S

o42 = | 1 x x2 x3 x4 x5 x6 x7 x8 x9 x10 x11 x12 x13 x14 x14y x14yz x14 ···

                1          690
o42 : Matrix S  <--- S
```

Let us count how many standard monomials there are of a given degree. The following table represents the Hilbert function of the residue ring.

```
i43 : tally apply(flatten entries basis(S),degree)

o43 = Tally{{0}  => 1   }
                {1}  => 3
                {10} => 63
                {11} => 69
                {12} => 73
                {13} => 71
                {14} => 66
                {15} => 53
                {16} => 38
                {17} => 23
                {18} => 12
```

```
{19} => 3
{2} => 6
{3} => 10
{4} => 15
{5} => 21
{6} => 28
{7} => 36
{8} => 45
{9} => 54
```

o43 : Tally

Thus the largest degree of a standard monomial is nineteen, and there are three standard monomials of that degree:

i44 : basis(19,S)

o44 = | x14yz4 x9yz9 x4yz14 |

$$\begin{array}{cc} & 1 \qquad 3 \\ \text{o44 : Matrix S} & \texttt{<---} \text{ S} \end{array}$$

The most recently defined ring involving x, y, and z was S, so all computations involving those variables are done in the residue ring S. For instance, we can also obtain the standard monomials of degree nineteen as follows:

i45 : (x+y+z)^19

```
             14   4         9  9        4   14
o45 = 58140x  y*z  + 923780x y*z  + 58140x y*z
```

o45 : S

An operation on ideals which will occur frequently throughout this book is the computation of minimal free resolutions. This is done as follows:

i46 : C = res M

```
       1       16      27       12
o46 = R   <-- R   <-- R   <-- R   <-- 0

       0       1       2        3       4
```

o46 : ChainComplex

This shows that our ideal M has sixteen minimal generators. They are the entries in the leftmost matrix of the chain complex C:

i47 : C.dd_1

o47 = | x3y6 x7y4 x2y9 x2y4z5 x11y2 xy12 x6y2z5 xy7z5 xy2z10 x15 y15 x ···

```
            1      16
o47 : Matrix R   <--- R
```

This means that four of the twenty generators in o39 were redundant. We construct the set consisting of the four redundant generators as follows:

i48 : set flatten entries gens M - set flatten entries C.dd_1

```
            6 7    10 5   5 10   5 5 5
o48 = Set {x y , x  y , x y  , x y z }
```

o48 : Set

Here `flatten entries` turns the matrix M into a single list. The command `set` turns that list into a set, to which we can apply the difference operation for sets.

Let us now take a look at the first syzygies (or *minimal S-pairs* [1, §2.9]) among the sixteen minimal generators. They correspond to the columns of the second matrix in our resolution C:

```
i49 : C.dd_2

o49 = {9}  | -y3 -x4 0   -z5 0   0   0   0   0   0   0   0   0   0   0  ···
      {11} | 0   y2  0   0   0   -x4 0   0   -z5 0   0   0   0   0   0  ···
      {11} | x   0   -y3 0   0   0   0   0   0   -z5 0   0   0   0   0  ···
      {11} | 0   0   0   xy2 -y3 0   -x4 0   x5  y5  0   -z5 0   0   0  ···
      {13} | 0   0   0   0   0   y2  0   0   0   0   0   0   0   -x4 0  ···
      {13} | 0   0   x   0   0   0   0   -y3 0   0   0   0   0   0   0  ···
      {13} | 0   0   0   0   0   0   y2  0   0   0   0   0   0   0   -  ···
      {13} | 0   0   0   0   x   0   0   0   0   0   -y3 0   0   0   0  ···
      {13} | 0   0   0   0   0   0   0   0   0   0   0   xy2 -y3 0   0  ···
      {15} | 0   0   0   0   0   0   0   0   0   0   0   0   0   0   y2 0 ···
      {15} | 0   0   0   0   0   0   0   0   x   0   0   0   0   0   0  ···
      {15} | 0   0   0   0   0   0   0   0   0   0   0   0   0   0   y  ···
      {15} | 0   0   0   0   0   0   0   0   0   0   x   0   0   0   0  ···
      {15} | 0   0   0   0   0   0   0   0   0   0   0   0   0   x   0  ···
      {15} | 0   0   0   0   0   0   0   0   0   0   0   0   0   0   0  ···

              16        27
o49 : Matrix R   <--- R
```

The first column represents the S-pair between the first generator x^3y^6 and the third generator x^2y^9. It is natural to form the *S-pair graph* with 16 vertices and 27 edges represented by this matrix. According to the general theory described in [3], this is a planar graph with 12 regions. The regions correspond to the 12 second syzygies, that is, to the columns of the matrix

```
i50 : C.dd_3

o50 = {12} | z5  0   0   0   0   0   0   0   0   0   0   0   |
      {13} | 0   z5  0   0   0   0   0   0   0   0   0   0   |
      {14} | 0   0   z5  0   0   0   0   0   0   0   0   0   |
      {14} | -y3 -x4 0   0   0   0   0   0   0   0   0   0   |
      {14} | 0   0   -y5 z5  0   0   0   0   0   0   0   0   |
      {15} | 0   0   0   0   z5  0   0   0   0   0   0   0   |
      {15} | 0   0   0   0   -x5 z5  0   0   0   0   0   0   |
      {16} | 0   0   0   0   0   0   z5  0   0   0   0   0   |
      {16} | 0   y2  0   0   -x4 0   0   0   0   0   0   0   |
      {16} | x   0   -y3 0   0   0   0   0   0   0   0   0   |
      {16} | 0   0   0   0   0   0   -y5 z5  0   0   0   0   |
      {16} | 0   0   0   -y3 0   -x4 0   0   0   0   0   0   |
      {16} | 0   0   0   0   0   0   0   -y5 z5  0   0   0   |
      {17} | 0   0   0   0   0   0   0   0   0   z5  0   0   |
      {17} | 0   0   0   0   0   0   0   0   0   -x5 z5  0   |
      {17} | 0   0   0   0   0   0   0   0   0   0   -x5 z5  |
      {18} | 0   0   0   0   y2  0   0   0   0   -x4 0   0   |
      {18} | 0   0   x   0   0   0   -y3 0   0   0   0   0   |
      {18} | 0   0   0   0   0   y2  0   0   0   0   -x4 0   |
      {18} | 0   0   0   x   0   0   0   -y3 0   0   0   0   |
      {18} | 0   0   0   0   0   0   0   0   -y3 0   0   -x4 |
      {20} | 0   0   0   0   0   0   0   0   0   y2  0   0   |
```

```
{20} | 0   0   0   0   0   0   x   0   0   0   0   0   |
{20} | 0   0   0   0   0   0   0   0   0   0   y2  0   |
{20} | 0   0   0   0   0   0   0   x   0   0   0   0   |
{20} | 0   0   0   0   0   0   0   0   0   0   0   y2  |
{20} | 0   0   0   0   0   0   0   0   x   0   0   0   |
              27        12
o50 : Matrix R    <--- R
```

But we are getting ahead of ourselves. Homological algebra and resolutions will be covered in the next chapter, and monomial ideals will appear in the chapter of Hoşten and Smith. Let us return to Cox, Little and O'Shea [2].

5   Pennies, Nickels, Dimes and Quarters

We now come to an application of Gröbner bases which appears in [2, Section 8.1]: *Integer Programming*. This is the problem of minimizing a linear objective function over the set of non-negative integer solutions of a system of linear equations. We demonstrate some techniques for doing this in *Macaulay 2*. Along the way, we learn about multigraded polynomial rings and how to compute Gröbner bases with respect to monomial orders defined by weights. Our running example is the linear system defined by the matrix:

```
i51 : A = {{1, 1, 1, 1},
          {1, 5,10,25}}

o51 = {{1, 1, 1, 1}, {1, 5, 10, 25}}

o51 : List
```

For the algebraic study of integer programming problems, a good starting point is to work in a multigraded polynomial ring, here in four variables:

```
i52 : R = QQ[p,n,d,q, Degrees => transpose A]

o52 = R

o52 : PolynomialRing
```

The degree of each variable is the corresponding column vector of the matrix Each variable represents one of the four coins in the U.S. currency system:

```
i53 : degree d

o53 = {1, 10}

o53 : List

i54 : degree q

o54 = {1, 25}

o54 : List
```

Each monomial represents a collection of coins. For instance, suppose you own four pennies, eight nickels, ten dimes, and three quarters:

```
i55 : degree(p^4*n^8*d^10*q^3)

o55 = {25, 219}

o55 : List
```

Then you have a total of 25 coins worth two dollars and nineteen cents. There are nine other possible ways of having 25 coins of the same value:

```
i56 : h = basis({25,219}, R)
```

```
o56 = | p14n2d2q7 p9n8d2q6 p9n5d6q5 p9n2d10q4 p4n14d2q5 p4n11d6q4 p4n8 ···
```

```
                1       9
o56 : Matrix R  <--- R
```

For just counting the number of columns of this matrix we can use the command

```
i57 : rank source h
```

```
o57 = 9
```

How many ways can you make change for ten dollars using 100 coins?

```
i58 : rank source basis({100,1000}, R)
```

```
o58 = 182
```

A typical integer programming problem is this: among all 182 ways of expressing ten dollars using 100 coins, which one uses the fewest dimes? We set up the Conti-Traverso algorithm [2, §8.1] for answering this question. We use the following ring with the lexicographic order and with the variable order: dimes (d) before pennies (p) before nickels (n) before quarters (q).

```
i59 : S = QQ[x, y, d, p, n, q,
            MonomialOrder => Lex, MonomialSize => 16]
```

```
o59 = S
```

```
o59 : PolynomialRing
```

The option `MonomialSize` advises *Macaulay 2* to use more space to store the exponents of monomials, thereby avoiding a potential overflow.

We define an ideal with one generator for each column of the matrix A.

```
i60 : I = ideal( p - x*y, n - x*y^5, d - x*y^10, q - x*y^25)
```

```
                         5          10          25
o60 = ideal (- x*y + p, - x*y  + n, - x*y   + d, - x*y   + q)
```

```
o60 : Ideal of S
```

The integer program is solved by normal form reduction with respect to the following Gröbner basis consisting of binomials.

```
i61 : transpose gens gb I
```

```
o61 = {-6} | p5q-n6     |
      {-4} | d4-n3q     |
      {-3} | yn2-dp     |
      {-6} | yp4q-dn4   |
      {-4} | yd3-pnq    |
      {-6} | y2p3q-d2n2 |
      {-5} | y2d2n-p2q  |
      {-7} | y2d2p3-n5  |
      {-6} | y3p2q-d3   |
      {-6} | y3dp2-n3   |
      {-5} | y4p-n      |
      {-6} | y5n-d      |
```

```
{-8}  | y6d2-pq   |
{-16} | y15d-q    |
{-7}  | xq-y5d2   |
{-5}  | xn-y3p2   |
{-2}  | xd-n2     |
{-2}  | xy-p      |

         18      1
o61 : Matrix S   <--- S
```

We fix the quotient ring, so the reduction to normal form will happen automatically.

```
i62 : S' = S/I

o62 = S'

o62 : QuotientRing
```

You need at least two dimes to express one dollar with ten coins.

```
i63 : x^10 * y^100

        2 6 2
o63 = d n q

o63 : S'
```

But you can express ten dollars with a hundred coins none of which is a dime.

```
i64 : x^100 * y^1000

        75 25
o64 = n  q

o64 : S'
```

The integer program is infeasible if and only if the normal form still contains the variable x or the variable y. For instance, you cannot express ten dollars with less than forty coins:

```
i65 : x^39 * y^1000

        25 39
o65 = y  q

o65 : S'
```

We now introduce a new term order on the polynomial ring, defined by assigning a weight to each variable. Specifically, we assign weights for each of the coins. For instance, let pennies have weight 5, nickels weight 7, dimes weight 13 and quarters weight 17.

```
i66 : weight = (5,7,13,17)

o66 = (5, 7, 13, 17)

o66 : Sequence
```

We set up a new ring with the resulting weight term order, and work modulo the same ideal as before in this new ring.

```
i67 : T = QQ[x, y, p, n, d, q,
            Weights => {{1,1,0,0,0,0},{0,0,weight}},
            MonomialSize => 16]/
            (p - x*y, n - x*y^5, d - x*y^10, q - x*y^25);
```

One dollar with ten coins:

```
i68 : x^10 * y^100

        5 2 3
o68 = p d q

o68 : T
```

Ten dollars with one hundred coins:

```
i69 : x^100 * y^1000

        60 3 37
o69 = p   n q

o69 : T
```

Here is an optimal solution which involves all four types of coins:

```
i70 : x^234 * y^5677

        2 4 3 225
o70 = p n d q

o70 : T
```

References

1. David Cox, John Little, and Donal O'Shea: *Ideals, varieties, and algorithms.* Springer-Verlag, New York, second edition, 1997. An introduction to computational algebraic geometry and commutative algebra.
2. David Cox, John Little, and Donal O'Shea: *Using algebraic geometry.* Springer-Verlag, New York, 1998.
3. Ezra Miller and Bernd Sturmfels: Monomial ideals and planar graphs. In S. Lin M. Fossorier, H. Imai and A. Poli, editors, *Applied Algebra, Algebraic Algorithms and Error-Correcting Codes*, volume 1719 of *Springer Lecture Notes in Computer Science*, pages 19–28, 1999.

Projective Geometry and Homological Algebra

David Eisenbud*

We provide an introduction to many of the homological commands in *Macaulay 2* (modules, free resolutions, Ext and Tor...) by means of examples showing how to use homological tools to study projective varieties.

In this chapter we will illustrate how one can manipulate projective varieties and sheaves, using the rich collection of tools *Macaulay 2* provides. One of our goals is to show how homological methods can be effective in solving concrete geometric problems.

The first four sections can be read by anyone who knows about projective varieties at the level of a first graduate course and knows the definitions of Ext and Tor. The last section assumes that the reader is familiar with the theory of curves and surfaces roughly at the level of the books of Hartshorne [7] and Harris [6].

We will work with projective schemes over a field kk. *Macaulay 2* can work over any finite field of characteristic at most 32749, and also a variety of fields in characteristic 0 (except for the primary decomposition commands, which at this writing are still restricted to positive characteristics). Our main interest is in geometry over an algebraically closed field of characteristic 0. Nevertheless, it is most convenient to work over a large prime field. It is known that the intermediate results in Gröbner basis computations (as in the Euclidean Algorithm computations they generalize) often involve coefficients far larger than those in the input data, so that work in characteristic zero essentially requires infinite precision arithmetic, a significant additional overhead. If we work over a finite field where the scalars can be represented in one machine word, we avoid this coefficient explosion. Experience with the sort of computations we will be doing shows that working over $\mathbb{Z}/p$, where p is a moderately large prime, gives results identical to the results we would get in characteristic 0. Of course one still has to be careful about the fact that our fields are not algebraically closed, especially when using primary decomposition. The largest prime p we can work with being 32749, we choose the field **Z**/32749. The name of the *Macaulay 2* constant representing the integers is ZZ, and by analogy we will call our field kk:

```
i1 : kk = ZZ/32749
```

```
o1 = kk
```

```
o1 : QuotientRing
```

In *Macaulay 2* we will represent projective space $\mathbb{P}^n$ by its homogeneous coordinate ring ringPn = $kk[x_0, \ldots, x_n]$. A projective scheme X in $\mathbb{P}^n$ may

---

* Supported by the NSF.

be most conveniently represented, depending on the situation, by its homogeneous ideal idealX or its homogeneous coordinate ring, represented either as a ring ringPn/idealX or as a module OX over ringPn. Coherent sheaves on the projective space, or on its subvarieties, will be represented by finitely generated graded modules over ringPn, using the Serre correspondence. For example, the structure sheaf $\mathcal{O}_X$ of the subvariety X would be represented by the module ringPn^1/idealX; here ringPn^1 denotes the free module of rank one over the ring ringPn.

1   The Twisted Cubic

As a first illustration, we give three constructions of the twisted cubic curve in $\mathbb{P}^3$. We represent $\mathbb{P}^3$ by

```
i2 : ringP3 = kk[x_0..x_3]

o2 = ringP3

o2 : PolynomialRing
```

The twisted cubic is the image of the map $\mathbb{P}^1 \to \mathbb{P}^3$ sending a point with homogeneous coordinates (s, t) to the point with homogeneous coordinates (s^3, s^2t, st^2, t^3). We can compute its relations directly with

```
i3 : ringP1 = kk[s,t]

o3 = ringP1

o3 : PolynomialRing

i4 : cubicMap = map(ringP1,ringP3,{s^3, s^2*t, s*t^2, t^3})

                          3   2      2   3
o4 = map(ringP1,ringP3,{s , s t, s*t , t })

o4 : RingMap ringP1 <--- ringP3


i5 : idealCubic = kernel cubicMap

              2                      2
o5 = ideal (x  - x x , x x  - x x , x  - x x )
             2    1 3   1 2    0 3   1    0 2

o5 : Ideal of ringP3
```

We could also use *Macaulay 2*'s built-in facility, and say

```
i6 : idealCubic2 = monomialCurveIdeal(ringP3,{1,2,3})

                      2                  2
o6 = ideal (x x  - x x , x  - x x , x  - x x )
             1 2    0 3   2    1 3   1    0 2

o6 : Ideal of ringP3
```

which uses precisely the same method.

Of course we might remember that the ideal of the twisted cubic is generated by the 2×2 minors of the matrix

$$\begin{pmatrix} x_0 & x_1 & x_2 \\ x_1 & x_2 & x_3 \end{pmatrix},$$

which we can realize with the commands

```
i7 : M = matrix{{x_0,x_1,x_2},{x_1,x_2,x_3}}

o7 = | x_0 x_1 x_2 |
     | x_1 x_2 x_3 |

                 2            3
o7 : Matrix ringP3  <--- ringP3


i8 : idealCubic3 = minors(2, M)

            2                          2
o8 = ideal (- x  + x x ,  - x x  + x x ,  - x  + x x )
              1    0 2       1 2    0 3      2    1 3

o8 : Ideal of ringP3
```

We can get some useful information about the ideal `idealCubic` with

```
i9 : codim idealCubic

o9 = 2


i10 : degree idealCubic

o10 = 3
```

This shows that we do indeed have a cubic curve. Note that the command

```
i11 : dim idealCubic

o11 = 2
```

gives 2, not 1; it represents the dimension of the ideal in `ringP3`, the dimension of the affine cone over the curve.

We can easily assure ourselves that these ideals are the same. For example, to see whether the ideal `idealCubic` is contained in the ideal of minors of M, we can reduce the former modulo the latter, and see whether we get zero. The reduction operator `%` takes two maps with the same target as its arguments, so we must replace each ideal by a matrix whose entries generate it. This is done by the function `gens` as in

```
i12 : gens idealCubic

o12 = | x_2^2-x_1x_3 x_1x_2-x_0x_3 x_1^2-x_0x_2 |

                  1            3
o12 : Matrix ringP3  <--- ringP3
```

Thus for one of the inclusions we check

```
i13 : 0 == (gens idealCubic)%(gens idealCubic3)

o13 = true
```

Both inclusions can be checked automatically in this way with

```
i14 : idealCubic == idealCubic3

o14 = true
```

2 The Cotangent Bundle of $\mathbb{P}^3$

Many invariants of varieties are defined in terms of their tangent and cotangent bundles. We identify a bundle with its sheaf of sections, which is locally free. Any coherent locally free sheaf arises this way. (One can also regard a bundle as a variety in its own right, but this view is used in algebraic geometry more rarely.) In this section and the next we construct the cotangent bundle $\Omega_{\mathbb{P}^3}$ of $\mathbb{P}^3$ and its restriction to the twisted cubic above.

Consulting Hartshorne [7, Theorem II.8.13], we find that the cotangent bundle to $\mathbb{P}^n$ can be described by the *cotangent sequence*:

$$0 \longrightarrow \Omega_{\mathbb{P}^n} \longrightarrow \mathcal{O}_{\mathbb{P}^n}(-1)^{n+1} \xrightarrow{\ f\ } \mathcal{O}_{\mathbb{P}^n} \longrightarrow 0$$

where f is defined by the matrix of variables $(x_0, \ldots, x_n)$. We can translate this description directly into the language of *Macaulay 2*, here in the case $n = 3$:

```
i15 : f = vars ringP3

o15 = | x_0 x_1 x_2 x_3 |

                1            4
o15 : Matrix ringP3  <--- ringP3


i16 : OmegaP3 = kernel f

o16 = image {1} | 0    0    0    -x_1 -x_2 -x_3 |
            {1} | 0    -x_2 -x_3 x_0  0    0    |
            {1} | -x_3 x_1  0    0    x_0  0    |
            {1} | x_2  0    x_1  0    0    x_0  |

                                4
o16 : ringP3-module, submodule of ringP3
```

Note that the module which we specified as a kernel is now given as the image of a matrix. We can recover this matrix with

```
i17 : g=generators OmegaP3

o17 = {1} | 0    0    0    -x_1 -x_2 -x_3 |
      {1} | 0    -x_2 -x_3 x_0  0    0    |
      {1} | -x_3 x_1  0    0    x_0  0    |
      {1} | x_2  0    x_1  0    0    x_0  |

                4            6
o17 : Matrix ringP3  <--- ringP3
```

and we could correspondingly write

```
i18 : OmegaP3=image g
```

```
o18 = image {1} | 0    0    0    -x_1 -x_2 -x_3 |
            {1} | 0    -x_2 -x_3 x_0  0    0    |
            {1} | -x_3 x_1  0    0    x_0  0    |
            {1} | x_2  0    x_1  0    0    x_0  |
```

```
                                4
o18 : ringP3-module, submodule of ringP3
```

An even more elementary way to give a module is by generators and relations, and we can see this "free presentation" too with

```
i19 : presentation OmegaP3
```

```
o19 = {2} | x_1  0    0    x_0  |
      {2} | x_3  x_0  0    0    |
      {2} | -x_2 0    x_0  0    |
      {2} | 0    x_2  x_3  0    |
      {2} | 0    -x_1 0    x_3  |
      {2} | 0    0    -x_1 -x_2 |
```

```
                  6          4
o19 : Matrix ringP3  <--- ringP3
```

The astute reader will have noticed that we have just been computing the first few terms in the free resolution of the cokernel of the map of free modules corresponding to f. We could see the whole resolution at once with

```
i20 : G = res coker f
```

```
           1          4          6          4          1
o20 = ringP3  <-- ringP3  <-- ringP3  <-- ringP3  <-- ringP3  <-- 0

        0          1          2          3          4          5

o20 : ChainComplex
```

and then see all the matrices in the resolution with

```
i21 : G.dd
```

```
              1                                4
o21 = 0 : ringP3  <---------------------- ringP3  : 1
                      | x_0 x_1 x_2 x_3 |
```

```
              4                                                6
      1 : ringP3  <------------------------------------- ringP3  : 2
                      {1} | -x_1 -x_2 0    -x_3 0    0    |
                      {1} | x_0  0    -x_2 0    -x_3 0    |
                      {1} | 0    x_0  x_1  0    0    -x_3 |
                      {1} | 0    0    0    x_0  x_1  x_2  |
```

```
              6                                4
      2 : ringP3  <---------------------- ringP3  : 3
                      {2} | x_2  x_3  0    0    |
                      {2} | -x_1 0    x_3  0    |
                      {2} | x_0  0    0    x_3  |
                      {2} | 0    -x_1 -x_2 0    |
                      {2} | 0    x_0  0    -x_2 |
                      {2} | 0    0    x_0  x_1  |
```

```
                4                              1
    3 : ringP3   <---------------- ringP3   : 4
                    {3} | -x_3 |
                    {3} | x_2  |
                    {3} | -x_1 |
                    {3} | x_0  |

                1
    4 : ringP3   <----- 0 : 5
                0
```

```
o21 : ChainComplexMap
```

or just one of them, say the second, with

```
i22 : G.dd_2
```

```
o22 = {1} | -x_1 -x_2 0    -x_3 0    0    |
      {1} | x_0  0    -x_2 0    -x_3 0    |
      {1} | 0    x_0  x_1  0    0    -x_3 |
      {1} | 0    0    0    x_0  x_1  x_2  |

                    4              6
o22 : Matrix ringP3   <--- ringP3
```

Note that this matrix does not look exactly the same as the matrix produced by computing the kernel of f. This is because when *Macaulay 2* is asked to compute a whole resolution, it does not do the "obvious" thing and compute kernels over and over; it defaults to a more efficient algorithm, first proposed by Frank Schreyer [10, Appendix].

Any graded map of free modules, such as a map in a graded free resolution of a graded module, comes with some numerical data: the degrees of the generators of the source and target free modules. We can extract this information one module at a time with the command **degrees**, as in

```
i23 : degrees source G.dd_2
```

```
o23 = {{2}, {2}, {2}, {2}, {2}, {2}}
```

```
o23 : List
```

```
i24 : degrees target G.dd_2
```

```
o24 = {{1}, {1}, {1}, {1}}
```

```
o24 : List
```

Macaulay 2 has a more convenient mechanism for examining this numerical data, which we take time out to explain. First, for the resolution just computed, we can call

```
i25 : betti G
```

```
o25 = total: 1 4 6 4 1
           0: 1 4 6 4 1
```

The diagram shows the degrees of the generators of each free module in the resolution in coded form. To understand the code, it may be helpful to look at a less symmetric example, say the free resolution of `ringP3^1/I` where I is the ideal generated by the minors of the following 2×4 matrix.

```
i26 : m = matrix{{x_0^3, x_1^2, x_2,x_3},{x_1^3,x_2^2,x_3,0}}

o26 = | x_0^3 x_1^2 x_2 x_3 |
      | x_1^3 x_2^2 x_3 0  |

                  2            4
o26 : Matrix ringP3 <--- ringP3
```

We do this with

```
i27 : I = minors(2,m)

             5     3 2     3       3     3     2     3     2     2
o27 = ideal (- x  + x x , - x x  + x x , - x  + x x , -x x , -x x , -x )
             1     0 2     1 2     0 3     2     1 3    1 3    2 3    3

o27 : Ideal of ringP3

i28 : F = res(ringP3^1/I)

            1           6           8           3
o28 = ringP3 <-- ringP3 <-- ringP3 <-- ringP3 <-- 0

         0           1           2           3           4

o28 : ChainComplex

i29 : betti F

o29 = total: 1 6 8 3
          0: 1 . . .
          1: . 1 . .
          2: . 2 2 .
          3: . 2 2 .
          4: . 1 4 3
```

The resulting Betti diagram should be interpreted as follows. First, the maps go from right to left, so the beginning of the resolution is on the left. The given Betti diagram thus corresponds to an exact sequence of graded free modules

$$F_0 \longleftarrow F_1 \longleftarrow F_2 \longleftarrow F_3 \longleftarrow 0.$$

The top row of the diagram, 1,6,8,3, shows the ranks of the free modules F_i in the resolution. For example the 1 on the left means that F_0 has rank 1 (and, indeed, the module ringP3^1/I we are resolving is cyclic). The 6 shows that the rank of F_1 is 6, or equivalently that the ideal I is minimally generated by 6 elements—in this case the $6 = \binom{4}{2}$ minors of size 2 of the 2×4 matrix m.

The first column of the diagram shows degrees. The successive columns indicate how many generators of each degree occur in the successive F_i. The free module F_0 has a single generator in degree 0, and this is the significance of the second column. Note that F_1 could not have any generators of degree less than or equal to zero, because the resolution is minimal! Thus for compactness, the diagram is skewed: in each successive column the places correspond to larger degrees. More precisely, a number a occurring opposite the degree indication "i:" in the column corresponding to F_j signifies that F_j has a generators in degree $i + j$. Thus for example the 1 in the third column opposite the one on the left corresponds to a generator of degree 2 in the free module F_1; and altogether F_1 has one generator of degree 2, two generators of degree 3, two of degree 4 and one of degree 5.

Returning to the diagram

```
i30 : betti G

o30 = total: 1 4 6 4 1
          0: 1 4 6 4 1
```

we see that the successive free modules of G are each generated in degree 1 higher than the previous one; that is, the matrices in G.dd all have linear entries, as we have already seen.

3 The Cotangent Bundle of a Projective Variety

It is easy to construct the cotangent bundle Ω_X of a projective variety X starting from the cotangent bundle of the ambient projective space. We use the *conormal sequence* (Hartshorne [7, Proposition II.8.12] or Eisenbud [4, Proposition 16.3]). Writing I for the ideal of a variety X in $\mathbb{P}^n$ there is an exact sequence of sheaves

$$I \xrightarrow{\delta} \Omega_{\mathbb{P}^n} \otimes \mathcal{O}_X \longrightarrow \Omega_X \longrightarrow 0$$

where the map δ takes a function f to the element $df \otimes 1$. If I is generated by forms $f_1, \ldots, f_m$ then δ is represented by the Jacobian matrix (df_i/dx_j).

First of all, we must compute a module corresponding to $\Omega_{\mathbb{P}^n} \otimes \mathcal{O}_X$, the restriction of the sheaf $\Omega_{\mathbb{P}^n}$ to X. The simplest approach would be to take the tensor product of graded modules representing $\Omega_{\mathbb{P}^n}$ and $\mathcal{O}_X$. The result would represent the right sheaf, but would not be the module of twisted global sections of $\Omega_{\mathbb{P}^n} \otimes \mathcal{O}_X$ (the unique module of depth two representing the sheaf). This would make further computations less efficient.

Thus we take a different approach: since the cotangent sequence given in the previous section is a sequence of locally free sheaves, it is locally split, and thus remains exact when tensored by $\mathcal{O}_X$. Consequently $\Omega_{\mathbb{P}^n} \otimes \mathcal{O}_X$ is also represented by the kernel of the map $f \otimes \mathcal{O}_X$, where f is the map used in the definition of the cotangent bundle of $\mathbb{P}^n$. In *Macaulay 2*, working on $\mathbb{P}^3$, with X the twisted cubic, we can translate this into

```
i31 : OmegaP3res = kernel (f ** (ringP3^1/idealCubic))

o31 = subquotient ({1} | -x_3  0     0     -x_2 -x_3  0     -x_1 -x_2 -x_3  · · ·
                   {1} |  x_2 -x_3   0      x_1  0    -x_3  x_0  0     0     · · ·
                   {1} |  0    x_2  -x_3    0    x_1   0     0    x_0   0     · · ·
                   {1} |  0    0     x_2    0    0     x_1   0    0     x_0   · · ·
                                              4
o31 : ringP3-module, subquotient of ringP3
```

(The operator ** is *Macaulay 2*'s symbol for tensor product.) Since the map is a map between free modules over ringP3/idealCubic, the kernel has depth (at least) two.

Next, we form the Jacobian matrix of the generators of idealCubic, which represents a map from this ideal to the free module ringP3^4.

```
i32 : delta1 = jacobian idealCubic
```

```
o32 = {1} | 0     -x_3 -x_2 |
      {1} | -x_3 x_2  2x_1 |
      {1} | 2x_2 x_1  -x_0 |
      {1} | -x_1 -x_0 0    |
```

```
                    4            3
o32 : Matrix ringP3  <--- ringP3
```

We need to make this into a map to OmegaP3res, which as defined is a subquotient of ringP3^4. To this end we must first express the image of delta1 in terms of the generators of OmegaP3res. The division command // does this with

```
i33 : delta2 = delta1 // (gens OmegaP3res)
```

```
o33 = {2} | 0  1  0  |
      {2} | 2  0  0  |
      {2} | 0  0  0  |
      {2} | 0  0  2  |
      {2} | 0  1  0  |
      {2} | -1 0  0  |
      {2} | 0  0  0  |
      {2} | 0  0  -1 |
      {2} | 0  -1 0  |
```

```
                    9            3
o33 : Matrix ringP3  <--- ringP3
```

Once this is done we can use this matrix to form the necessary map $\delta : I \to \Omega_{\mathbb{P}^3} \otimes \mathcal{O}_X$:

```
i34 : delta = map(OmegaP3res, module idealCubic, delta2)
```

```
o34 = {2} | 0  1  0  |
      {2} | 2  0  0  |
      {2} | 0  0  0  |
      {2} | 0  0  2  |
      {2} | 0  1  0  |
      {2} | -1 0  0  |
      {2} | 0  0  0  |
      {2} | 0  0  -1 |
      {2} | 0  -1 0  |
```

```
o34 : Matrix
```

A minimal free presentation of Ω_X — or rather of one module over ringP3 that represents it — can be obtained with

```
i35 : OmegaCubic = prune coker delta
```

```
o35 = cokernel {2} | -10917x_3 0    -10917x_3 x_2  0         0         ...
               {2} | 0         0    x_2       0         16374x_3 0      ...
               {2} | 0         -x_3 0         16373x_3 0         x_2     ...
               {2} | x_3       x_2  0         0         0         0      ...
               {2} | 0         0    0         0         0         -2x_3  ...
               {2} | 0         0    0         0         x_2       0      ...
```

```
                                  6
o35 : ringP3-module, quotient of ringP3
```

We have used the function **prune** to compute minimal presentation matrices; these often make subsequent computations faster, and also allow us to inspect the final answer more easily.

The module **OmegaCubic** represents the sheaf Ω_X, where X is the cubic, but it is not the simplest possibility. A better representative is the graded module $\oplus_{d\in\mathbb{Z}}H^0(\Omega_X(d))$. We can at least find a minimal presentation of the submodule $\oplus_{d\geq 0}H^0(\Omega_X(d))$ with

```
i36 : prune HH^0((sheaf OmegaCubic)(>=0))

o36 = cokernel {1} | 16374x_3 16374x_2 16374x_1 |
               {1} | x_2      x_1      x_0      |

                                          2
o36 : ringP3-module, quotient of ringP3
```

The large coefficients appearing in the matrix arise in finite characteristic as the result of chance division by small integers. We see from the degrees labeling the rows of the matrix in the output of this command that the generators of the submodule are in degree 1, so in particular $H^0(\Omega_X) = 0$. It follows that that $H^0(\Omega_X(d)) = 0$ for all $d \leq 0$, so the submodule we computed was actually the whole module that we wanted! (If this had not been the case we could have tried HH^0((sheaf OmegaCubic)(>=d)) to compute the cohomology of all the twists greater than a given negative integer d, or simply used the submodule we had already computed, since it also represents the sheaf Ω_X.)

The sequence of commands we have used to construct the cotangent sheaf can be obtained also with the following built-in commands.

```
i37 : Cubic = Proj(ringP3/idealCubic)

o37 = Cubic

o37 : ProjectiveVariety

i38 : cotangentSheaf Cubic

o38 = cokernel {1} | x_2  x_1  x_0  |
               {1} | -x_3 -x_2 -x_1 |

                                              2
o38 : coherent sheaf on Cubic, quotient of OO    (-1)
                                            Cubic
```

Since X is a smooth curve, its cotangent bundle is equal to its *canonical bundle*, and also to its *dualizing sheaf* (see Hartshorne [7, sections II.8 and III.7] for definitions). We will see another (generally more efficient) method of computing this dualizing sheaf by using **Ext** and duality theory.

4   Intersections by Serre's Method

To introduce homological algebra in a simple geometric context, consider the problem of computing the intersection multiplicities of two varieties X and Y in $\mathbb{P}^n$, assuming for simplicity that $\dim X + \dim Y = n$ and that the two meet in a zero-dimensional scheme. Beginning in the 19th century, many people struggled to make a definition of local intersection multiplicity that would make *Bézout's Theorem* true: the product of the degrees of X and Y should be the number of points of intersection, each counted with its local intersection multiplicity (multiplied by the degree of the point, if the point is not rational over the ground field). In the simplest case, where the two varieties are Cohen-Macaulay, the right answer is that a point p should count with multiplicity equal to the length of the local ring $\mathcal{O}_{X,p} \otimes_{\mathcal{O}_{\mathbb{P}^n,p}} \mathcal{O}_{Y,p}$, and at first it was naively assumed that this would be the right answer in general.

Here is a famous example in $\mathbf{P}^4$ showing that the naive value can be wrong: in it, the scheme X is a 2-plane and the scheme $Y = L_1 \cup L_2$ is the union of two 2-planes. The planes L_1 and L_2 meet at just one point p, and we assume that X passes through p as well, and is general enough so that it meets Y only in p. Since $\mathrm{degree}(X) = 1, \mathrm{degree}(Y) = 2$, Bézout's Theorem requires that the multiplicity of the intersection at p should be 2. However, we have:

```
i39 : ringP4 = kk[x_0..x_4]

o39 = ringP4

o39 : PolynomialRing

i40 : idealX = ideal(x_1+x_3, x_2+x_4)

o40 = ideal (x  + x , x  + x )
              1    3   2    4

o40 : Ideal of ringP4

i41 : idealL1 = ideal(x_1,x_2)

o41 = ideal (x , x )
              1   2

o41 : Ideal of ringP4

i42 : idealL2 = ideal(x_3,x_4)

o42 = ideal (x , x )
              3   4

o42 : Ideal of ringP4

i43 : idealY = intersect(idealL1,idealL2)

o43 = ideal (x x , x x , x x , x x )
              2 4   1 4   2 3   1 3

o43 : Ideal of ringP4

i44 : degree(idealX+idealY)

o44 = 3
```

That is, the length of $\mathcal{O}_{X,p} \otimes_{\mathcal{O}_{\mathbb{P}^n,p}} \mathcal{O}_{Y,p}$ is 3 rather than 2. (We can do this computation without first passing to local rings because there is only one point of intersection, and because all the constructions we are using commute with localization.)

It was the happy discovery of Jean-Pierre Serre [11, V.B.3] that the naive measure of intersection multiplicity can be fixed in a simple way that works for all intersections in smooth varieties. One simply replaces the length of the tensor product

$$\mathcal{O}_{X,p} \otimes_{\mathcal{O}_{\mathbb{P}^n,p}} \mathcal{O}_{Y,p} = \mathrm{Tor}_0^{\mathcal{O}_{\mathbb{P}^n,p}}(\mathcal{O}_{X,p}, \mathcal{O}_{Y,p})$$

with the alternating sum of the Tor functors

$$\sum_i (-1)^i \, \mathrm{length} \, \mathrm{Tor}_i^{\mathcal{O}_{\mathbb{P}^n,p}}(\mathcal{O}_{X,p}, \mathcal{O}_{Y,p}).$$

In *Macaulay 2* we can proceed as follows:

```
i45 : degree Tor_0(ringP4^1/idealX, ringP4^1/idealY)

o45 = 3

i46 : degree Tor_1(ringP4^1/idealX, ringP4^1/idealY)

o46 = 1

i47 : degree Tor_2(ringP4^1/idealX, ringP4^1/idealY)

o47 = 0
```

The other Tor's are 0 because the projective dimension of `ringP4^1/idealX` is only two, as we see from

```
i48 : res (ringP4^1/idealX)

            1          2          1
o48 = ringP4  <-- ringP4  <-- ringP4  <-- 0

            0          1          2          3

o48 : ChainComplex
```

Thus, indeed, the alternating sum is 2, and Bézout's Theorem is upheld.

5   A Mystery Variety in $\mathbb{P}^3$

In the file `mystery.m2` is a function called `mystery` that will compute the ideal of a subvariety X of $\mathbb{P}^3$. We'll reveal what it does at the end of the chapter. Let's run it.

```
i49 : ringP3 = kk[x_0..x_3];

i50 : load "mystery.m2"

i51 : idealX = mystery ringP3

            4        2        2        2 2    2 2              2           ...
o51 = ideal (x  - 2x x x  - x x x  + x x , x x  - 10915x x x  - 10917x ...
             1     0 1 3    1 2 3    0 3   0 1         0 1 2          ...

o51 : Ideal of ringP3
```

We can't see all the generators of the ideal; the same file contains a function prettyPrint which will display the generators visibly.

```
i52 : prettyPrint gens idealX
x_1^4-2*x_0*x_1^2*x_3-x_1^2*x_2*x_3+x_0^2*x_3^2,
x_0^2*x_1^2-10915*x_0*x_1^2*x_2-10917*x_0^3*x_3+10916*x_0^2*x_2*x_3-
    10916*x_0*x_2^2*x_3-10916*x_1*x_3^3,
x_0*x_1^2*x_2^2+11909*x_0^4*x_3+5954*x_0^3*x_2*x_3+2977*x_0^2*x_2^2*x_3+
    11910*x_0*x_2^3*x_3-2978*x_1^3*x_3^2+14887*x_0*x_1*x_3^3+
    11910*x_1*x_2*x_3^3,
x_0*x_1^3*x_2-13099*x_1^3*x_2^2-6550*x_0^3*x_1*x_3-
    13100*x_0^2*x_1*x_2*x_3-6550*x_0*x_1*x_2^2*x_3+13099*x_1*x_2^3*x_3+
    13100*x_1^2*x_3*x_3^3+13099*x_0*x_3^4,
x_0^5+5*x_0^2*x_2^3+5*x_0*x_2^4-3*x_0*x_1^3*x_3-4*x_1^3*x_2*x_3+
    4*x_0^2*x_1*x_3^2+10*x_0*x_1*x_2*x_3^2+5*x_1*x_2^2*x_3^2,
x_1^2*x_2^4-8932*x_0^4*x_2*x_3+11909*x_0^3*x_2^2*x_3+5954*x_0^2*x_2^3*x_3-
    8934*x_0*x_2^4*x_3-x_2^5*x_3+2*x_0*x_1^3*x_3^2-5952*x_1^3*x_2*x_3^2-
    x_0^2*x_1*x_3^3-2979*x_0*x_1*x_2*x_3^3-8934*x_1*x_2^2*x_3^3+x_3^6
```

Imagine that you found yourself looking at the scheme X in $\mathbb{P}^3$ defined by the 6 equations above.

```
i53 : X = variety idealX

o53 = X

o53 : ProjectiveVariety
```

How would you analyze the scheme X? We will illustrate one approach.

In outline, we will first look at the topological invariants: the number and dimensions of the irreducible components, and how they meet if there is more than one; the topological type of each component; and the degree of each component in $\mathbb{P}^3$. We will then see what we can say about the analytic invariants of X using adjunction theory (we give some references at the end).

Since we are interested in the projective scheme defined by idealX we could work with any ideal having the same saturation. It is usually the case that working with the saturation itself greatly eases subsequent computation so, as a matter of good practice, we begin by checking whether the ideal is saturated. If not, we should replace it with its saturation.

```
i54 : idealX == saturate idealX

o54 = true
```

Thus we see that idealX is already saturated. Perhaps the most basic invariant of X is its dimension:

```
i55 : dim X

o55 = 1
```

This shows that X consists of a curve, and possibly some zero-dimensional components. The command

```
i56 : idealXtop = top idealX

              4      2       2      2 2     2 2           2          ...
o56 = ideal (x  - 2x x x  - x x x + x x , x x  - 10915x x x  - 10917x ...
              1      0 1 3   1 2 3   0 3   0 1         0 1 2         ...

o56 : Ideal of ringP3
```

returns the ideal of the largest dimensional components of X. If there were 0-dimensional components (or if idealX were not saturated) then `idealXtop` would be larger than `idealX`. To test this we reduce `idealXtop` modulo `idealX` and see whether we get 0:

```
i57 : (gens idealXtop)%(gens idealX) == 0

o57 = true
```

Thus X is a purely one-dimensional scheme.

Is X singular?

```
i58 : codim singularLocus idealX

o58 = 4
```

A variety of codimension 4 in $\mathbb{P}^3$ must be empty, so X is a nonsingular curve.

A nonsingular curve in $\mathbb{P}^3$ could still be reducible, but since the intersection of two components would be a singular point, the curve would then be disconnected. A straightforward way to decide is to use the command `decompose`, which returns a list of irreducible components defined over kk. The length of this list,

```
i59 : # decompose idealX

o59 = 1
```

is thus the number of irreducible components that are defined over kk, and we see there is only one. (Warning: at this writing (December 2000), the command "decompose" works only in positive characteristic).

Often what we really want to know is whether X is *absolutely irreducible* (that is, irreducible over the algebraic closure of kk). The property of being smooth transfers to the algebraic closure, so again the question is the number of connected components we would get over the algebraic closure. For any reduced scheme X over a perfect field (such as our finite field kk) this number is $h^0 \mathcal{O}_X := \dim_{kk} H^0 \mathcal{O}_X$. We compute it with

```
i60 : HH^0 OO_X

          1
o60 = kk

o60 : kk-module, free

i61 : rank oo

o61 = 1
```

This command works much faster than the decompose command. (You can compute the time by adding the command `time` to the beginning of the line where the command to be timed starts.) Since we already know that `idealX` is saturated, this also shows that `idealX` is prime.

We next ask for the genus of the curve X. Here the *genus* may be defined as the dimension of the space $H^1 \mathcal{O}_X$. We can get this space with

```
i62 : HH^1 OO_X

          6
o62 = kk
```

```
o62 : kk-module, free
```

The genus of the curve is the dimension of this space, which we can see to be 6. Next, the cohomology class of X in $\mathbb{P}^3$ is determined by the degree of X:

```
i63 : degree idealX

o63 = 10
```

In sum: X is a smooth, absolutely irreducible curve of genus 6 and degree 10.

We next ask for analytic information about the curve and the embedding. A reasonable place to start is with the relation between the line bundle defining the embedding and the canonical sheaf ω_X. Notice first that the degree of the hyperplane divisor (the degree of the curve) is 10 = 2g-2, the same as the canonical bundle. By Riemann-Roch the embedding line bundle either is the canonical bundle or has first cohomology 0, which we can check with

```
i64 : P3 = Proj ringP3

o64 = P3

o64 : ProjectiveVariety

i65 : HH^1((OO_P3(1)/idealX)(>=0))

o65 = cokernel | x_3 x_2 x_1 x_0 |

                                1

o65 : ringP3-module, quotient of ringP3
```

Let's examine the degree of the generator of that module.

```
i66 : degrees oo

o66 = {{0}}

o66 : List
```

From that and the presentation matrix above we see that this cohomology module is the residue class field $\text{ringP3}/(x_0, x_1, x_2, x_3)$, concentrated in degree 0. Thus the embedding line bundle $\mathcal{O}_X(1)$ is isomorphic to ω_X. On the other hand the dimension of the space of sections of this line bundle has already been computed; it is $g = 6$. The curve is embedded in $\mathbb{P}^3$, so only 4 of these sections were used—the embedding is a projection of the same curve, embedded in $\mathbb{P}^6$ by the *canonical map*.

We next ask more about the curve itself. After the genus, the gonality and the Clifford index are among the most interesting invariants. Recall that the *gonality* of X is the smallest degree of a mapping from X to $\mathbb{P}^1$. To define the Clifford index of X we first define the *Clifford index of a line bundle* L on X to be degree$(L) - 2(\text{h}^0(L) - 1)$. For example, the Clifford indices of the structure sheaf $\mathcal{O}_X$ and the canonical sheaf ω_X are both equal to 0. The *Clifford index of the curve* X is defined to be the minimum value of the Clifford index of a line bundle L on X for which both $\text{h}^0(L) \geq 2$ and $\text{h}^1(L) \geq 2$. The Clifford index of a curve of genus g lies between 0 (for a hyperelliptic curve) and $\lfloor (g-1)/2 \rfloor$ (for a general curve). The Clifford index of any curve is bounded above by the gonality minus 2.

For a curve of genus 6 such as X, the gonality is either 2 (the hyperelliptic case), 3 (the trigonal case) or 4 (the value for general curves). The Clifford index, on the other hand is either 0 (the hyperelliptic case) or 1 (the case of a trigonal curve OR a smooth plane quintic curve—which is necessarily of gonality 4) or 2 (the case of a general curve). Thus for most curves (and this is true in any genus) the Clifford index is equal to the gonality minus 2.

We can make a start on distinguishing these cases already: since our curve is embedded in $\mathbb{P}^3$ by a subseries of the canonical series, X cannot be hyperelliptic (for hyperelliptic curves, the canonical series maps the curve two-to-one onto a rational curve.)

To make further progress we use an idea of Mark Green (see Green and Lazarsfeld [5]). Green conjectured a formula for the Clifford index that depends only on numerical data about the free resolution of the curve in its complete canonical embedding (where the hyperplanes cut out all the canonical divisors). The conjecture is known for genus 6 and in many other cases; see for example Schreyer [9].

We therefore begin by computing the canonical embedding of X. We could proceed to find the canonical bundle as in the computation for $\mathbb{P}^3$ above, or indeed as $\mathcal{O}_X(1)$, but instead we describe the general method that is most efficient: duality, as described (for example) in the book of Altman and Kleiman [1]. The module $\oplus_{d \in \mathbb{Z}} H^0(\omega_X(d))$ can be computed as

```
i67 : omegaX = Ext^(codim idealX)(ringP3^1/idealX, ringP3^{-4})

o67 = cokernel {0}  | 9359x_3         -4677x_3          -10105x_1    ...
               {0}  | 12014x_1        2552x_1           2626x_0      ...
               {-1} | x_0x_3-2553x_2x_3 x_1^2-1702x_2x_3 x_0x_1-8086x_ ...

                                      3
o67 : ringP3-module, quotient of ringP3
```

To find the equations of the canonical embedding of X, we first compute a basis of $H^0(\omega_X)$, which is the degree 0 part of the module omegaX. The desired equations are computed as the algebraic relations among the images of this basis under any monomorphism $\omega_X \to \mathcal{O}_X$.

As the ring ringP3/idealX is a domain, and ω_X is the module corresponding to a line bundle, any nonzero map from ω_X to ringP3/idealX will be an embedding. We can compute the module of such maps with

```
i68 : dualModule = Hom(omegaX, ringP3^1/idealX)

o68 = subquotient ({0}  | x_0^3x_2^2+10915x_0^2x_2^3+807x_0x_2^4+4043x_ ...
                   {0}  | 10105x_0x_1x_2^3+6063x_1x_2^4+11820x_0x_1^2x_ ...
                   {1}  | 10105x_0^2x_2_2^2-11322x_0x_2^3+11322x_2^4+8396 ...

                                      3
o68 : ringP3-module, subquotient of ringP3
```

and examine it with

```
i69 : betti prune dualModule

o69 = relations : total: 10 26
                      3:  3  2
```

```
4:   6 14
5:   1  9
6:   .  1
```

For want of a better idea we take the first generator, dualModule_{0}, which we can turn into an actual homomorphism with

```
i70 : f = homomorphism dualModule_{0}

o70 =  | x_0^3x_2^2+10915x_0^2x_2^3+807x_0x_2^4+4043x_2^5+7655x_0x_1x_2 · · ·

o70 : Matrix
```

The image of a basis of ω_X is given by the columns of the matrix

```
i71 : canGens = f*basis(0,omegaX)

o71 =  | x_0^3x_2^2+10915x_0^2x_2^3+807x_0x_2^4+4043x_2^5+7655x_0x_1x_2 · · ·

o71 : Matrix
```

regarded as elements of

```
i72 : ringX = ringP3/idealX

o72 = ringX

o72 : QuotientRing
```

Because of the particular homomorphism we chose, they have degree 5.

We can now compute the defining ideal for X in its canonical embedding as the relations on these elements. We first define a ring with 6 variables corresponding to the columns of canGens

```
i73 : ringP5 = kk[x_0..x_5]

o73 = ringP5

o73 : PolynomialRing
```

and then compute the canonical ideal as the kernel of the corresponding map from this ring to ringX with

```
i74 : idealXcan = trim kernel map(ringX, ringP5,
                                 substitute(matrix canGens,ringX),
                                 DegreeMap => i -> 5*i)

              2                                              · · ·
o74 = ideal (x  + 5040x x  - 8565x x  - 11589x x , x x  - 6048x x  - 1 · · ·
              3         0 5        2 5          4 5   1 3        0 5     · · ·

o74 : Ideal of ringP5
```

Here the command trim is used to extract a minimal set of generators of the desired ideal, and the command matrix replaces the map of (nonfree) modules canGens by the matrix that gives its action on the generators. The DegreeMap option specifies a function which transforms degrees (represented as lists of integers) as the ring homomorphism does; using it here makes the ring map homogeneous.

To get information about the Clifford index, we examine the free resolution with

```
i75 : betti res idealXcan

o75 = total: 1 9 16 9 1
          0: 1 .  . . .
          1: . 6  8 3 .
          2: . 3  8 6 .
          3: . .  . . 1
```

Quite generally, for a non-hyperelliptic curve of genus $g \geq 3$ the ideal of the canonical embedding requires $g - \binom{2}{2}$ quadratic generators, in our case 6. It is known that the curve is trigonal (Clifford index 1) if and only if the ideal also requires cubic generators, that is, the first term in the free resolution requires generators of degree $3 = 1 + 2$; and Green's conjecture says in general that the curve has Clifford index c if the $c - 1$ term in the resolution does not require generators of degree $(c - 1) + 2 = c + 1$ but the c term does require generators of degree $c + 2$. Thus from the Betti diagram above, and the truth of Green's conjecture in low genus, we see that our curve has Clifford index 1 and is thus either trigonal or a plane quintic.

If X is trigonal, that is, X has a map of degree 3 to $\mathbb{P}^1$, then the fibers of this map form a linear series whose elements are divisors of degree three. The geometric form of the Riemann-Roch theorem says that if

$$p_1, \ldots, p_d \in X \subset \mathbb{P}^g$$

are points on a canonically embedded curve X, then the dimension of the linear system in which the divisor $p_1 + \cdots + p_d$ moves is the amount by which the points fail to be linearly independent: $d - 1$ minus the dimension of the projective plane spanned by the points. In particular, the 3 points in the fiber of a three-to-one map to $\mathbb{P}^1$ are linearly dependent, that is, they span a projective line. This "explains" why the ideal of a trigonal curve requires cubic generators: the quadrics all contain three points of these lines and thus contain the whole lines! It is known (see St-Donat [8]) that, in the trigonal case, the 6 quadrics in the ideal of the canonical curve generate the defining ideal of the variety which is the union of these lines, and that variety is a rational normal scroll. In case X is a plane quintic, the *adjunction formula* (Hartshorne [7, II.8.20.3]) shows that the canonical embedding of X is obtained from the plane embedding by composing with the Veronese embedding of the plane in $\mathbb{P}^5$ as the Veronese surface; and the 6 quadrics in the ideal of the canonical curve generate the defining ideal of the Veronese surface.

Thus if we let S denote the variety defined by the quadrics in the ideal of X, we can decide whether X is a trigonal curve or a plane quintic by deciding whether S is a rational normal scroll or a Veronese surface. To compute the ideal of S we first ascertain which of the generators of the ideal of the canonical curve have degree 2 with

```
i76 : deg2places = positions(degrees idealXcan, i->i=={2})

o76 = {0, 1, 2, 3, 4, 5}

o76 : List
```

and then compute

```
i77 : idealS= ideal (gens idealXcan)_deg2places

            2                                                      ...
o77 = ideal (x  + 5040x x  - 8565x x  - 11589x x , x x  - 6048x x  - 1 ···
             3        0 5        2 5          4 5  1 3        0 5     ···

o77 : Ideal of ringP5
```

One of the scrolls that could appear is singular, the cone over the rational quartic in $\mathbb{P}^4$. We check for singularity first:

```
i78 : codim singularLocus idealS

o78 = 6
```

Since the codimension is 6, the surface S is nonsingular, and thus must be one of the nonsingular scrolls or the Veronese surface (which is by definition the image of $\mathbf{P}^2$, embedded in $\mathbf{P}^5$ by the linear series of conics.)

The ideals defining any rational normal scroll of codimension 3, and the ideal of a Veronese surface all have free resolutions with the same Betti diagrams, so we need a subtler method to determine the identity of S. The most powerful tool for such purposes is adjunction theory; we will use a simple version.

The idea is to compare the embedding bundle (the "hyperplane bundle") with the canonical bundle. On the Veronese surface, the canonical bundle is the bundle associated to -3 lines in $\mathbf{P}^2$, while the hyperplane bundle is associated to 2 lines in $\mathbf{P}^2$. Thus the inverse of the square of the canonical bundle is the cube of the hyperplane bundle, $\mathcal{O}_S(3)$. For a scroll on the other hand, these two bundles are different.

As before we follow the homological method for computing the canonical bundle:

```
i79 : omegaS = Ext^(codim idealS)(ringP5^1/idealS, ringP5^{-6})

o79 = cokernel {2} | 4032x_5 0        14811x_5 -4032x_3   6549x_3       ···
                {2} | x_3     x_2      x_1      -x_4       x_0-14291x_   ···
                {2} | -6852x_5 6549x_3 362x_5   x_1-6248x_3 0           ···

                                 3
o79 : ringP5-module, quotient of ringP5

i80 : OS = ringP5^1/idealS

o80 = cokernel | x_3^2+5040x_0x_5-8565x_2x_5-11589x_4x_5 x_1x_3-6048x_ ···

                                 1
o80 : ringP5-module, quotient of ringP5
```

We want the square of the canonical bundle, which we can compute as the tensor square

```
i81 : omegaS**omegaS

o81 = cokernel {4} | 4032x_5 0        14811x_5 -4032x_3   6549x_3       ···
                {4} | x_3     x_2      x_1      -x_4       x_0-14291x_   ···
                {4} | -6852x_5 6549x_3 362x_5   x_1-6248x_3 0           ···
                {4} | 0        0        0        0          0            ···
```

```
{4} | 0      0      0      0      0      · · ·
{4} | 0      0      0      0      0      · · ·
{4} | 0      0      0      0      0      · · ·
{4} | 0      0      0      0      0      · · ·
{4} | 0      0      0      0      0      · · ·

                        9
o81 : ringP5-module, quotient of ringP5
```

But while this module represents the correct sheaf, it is hard to interpret, since it may not be (is not, in this case) the module of all twisted global sections of the square of the line bundle. Since the free resolution of OS (visible inside the Betti diagram of the resolution of idealXcan) has length 3, the module OS has depth 2. Thus we can find the module of all twisted global sections of omega2S by taking the double dual

```
i82 : omega2S = Hom(Hom(omegaS**omegaS, OS),OS)

o82 = cokernel {3} | x_3^2+5040x_0x_5-8565x_2x_5-11589x_4x_5 x_1x_3-60 · · ·

                        1
o82 : ringP5-module, quotient of ringP5
```

We see from the output that this module is generated by 1 element of degree 3. It follows that $\omega_S^2 \cong \mathcal{O}_S(-3)$. This in turn shows that S is the Veronese surface.

We now know that the canonical embedding of the curve X is the Veronese map applied to a planar embedding of X of degree 5, and we can ask to see the plane embedding. Since the *anticanonical bundle* ω_S^{-1} on S corresponds to 3 lines in the plane and the hyperplane bundle to 2 lines, we can recover the line bundle corresponding to 1 line, giving the isomorphism of X to the plane, as the quotient

```
i83 : L = Hom(omegaS, OS**(ringP5^{-1}))

o83 = subquotient ({-1} | 14401x_2+16185x_4    x_0-14291x_4 -5359x_1+1 · · ·
                   {-1} | -1488x_1-10598x_3     -6549x_3        -11789x_5 · · ·
                   {-1} | x_0+7742x_2-15779x_4 x_2             x_1+6551x_ · · ·

                        3
o83 : ringP5-module, subquotient of ringP5
```

and the line bundle on Xcan that gives the embedding in $\mathbb{P}^2$ will be the restriction of L to Xcan. To realize the map from X to $\mathbb{P}^2$, we proceed as before:

```
i84 : dualModule = Hom(L, OS)

o84 = subquotient (| x_0+7742x_2-15779x_4 14401x_2+16185x_4 x_1-301x_3 · · ·
                   | x_2                  x_0-14291x_4       4032x_3   · · ·
                   | x_1+6551x_3          -5359x_1+14409x_3 -9874x_5   · · ·

                        3
o84 : ringP5-module, subquotient of ringP5

i85 : betti generators dualModule

o85 = total: 3 3
          0: 3 3
```

Again, we may choose any homomorphism from L to OS, for example

```
i86 : g = homomorphism dualModule_{0}

o86 = | x_0+7742x_2-15779x_4 x_2 x_1+6551x_3 |

o86 : Matrix

i87 : toP2 = g*basis(0,L)

o87 = | x_0+7742x_2-15779x_4 x_2 x_1+6551x_3 |

o87 : Matrix

i88 : ringXcan = ringP5/idealXcan

o88 = ringXcan

o88 : QuotientRing

i89 : ringP2 = kk[x_0..x_2]

o89 = ringP2

o89 : PolynomialRing

i90 : idealXplane = trim kernel map(ringXcan, ringP2,
                              substitute(matrix toP2,ringXcan))

                5         4        3 2       2 3         4        5 ...
o90 = ideal(x   + 13394x x  - 13014x x  + 9232x x  + 12418x x  - 2746x  ...
             0         0 1        0 1        0 1        0 1        1 ...

o90 : Ideal of ringP2
```

We have effectively computed the square root of the line bundle embedding X in $\mathbb{P}^3$ with which we started, and exchanged a messy set of defining equations of an unknown scheme for a single equation defining a smooth plane curve whose properties are easy to deduce. The same curve may also be defined by a much simpler plane equation (see Appendix A below). I do not know any general method for choosing a coordinate transformation to simplify a given equation! Can the reader find one that will work at least in this case?

There is not yet a textbook-level exposition of the sort of methods we have used (although an introduction will be contained in a forthcoming elementary book of Decker and Schreyer). The reader who would like to go further into such ideas can find a high-level survey of how adjunction theory is used in the paper of Decker and Schreyer [3]. For a group of powerful methods with a different flavor, see Aure, Decker, Hulek, Popescu, and Ranestad [2].

Appendix A. How the "Mystery Variety" was Made

For those who would like to try out the computations above over a different field (perhaps the field of rational numbers QQ), and for the curious, we include the code used to produce the equations of the variety X above.

Start with the Fermat quintic in the plane

```
i91 : ringP2 = kk[x_0..x_2]

o91 = ringP2

o91 : PolynomialRing

i92 : idealC2 = ideal(x_0^5+x_1^5+x_2^5)

              5   5   5
o92 = ideal(x  + x  + x )
              0   1   2

o92 : Ideal of ringP2
```

Embed it by the Veronese map in $\mathbb{P}^5$:

```
i93 : ringC2 = ringP2/idealC2

o93 = ringC2

o93 : QuotientRing

i94 : ringP5 = kk[x_0..x_5]

o94 = ringP5

o94 : PolynomialRing

i95 : idealC5 = trim kernel map(ringC2, ringP5,
              gens (ideal vars ringC2)^2)

              2                                  2                  ...
o95 = ideal (x  - x x , x x  - x x , x x  - x x , x  - x x , x x  - x  ...
              4    3 5   2 4    1 5   2 3    1 4   2    0 5   1 2    0 ...

o95 : Ideal of ringP5
```

Finally, choose a projection into $\mathbb{P}^3$, from a line not meeting C5, which is an isomorphism onto its image. (This requires the image to be a smooth curve of degree 10).

```
i96 : ringC5 = ringP5/idealC5

o96 = ringC5

o96 : QuotientRing

i97 : use ringC5

o97 = ringC5

o97 : QuotientRing

i98 : idealC = trim kernel map(ringC5, ringP3,
              matrix{{x_0+x_1,x_2,x_3,x_5}})

              4       2       2     2 2   2 2        2            ...
o98 = ideal (x  - 2x x x  - x x x  + x x , x x  - 10915x x x  - 10917x ...
              1     0 1 3   1 2 3   0 3   0 1        0 1 2         ...

o98 : Ideal of ringP3
```

Let's check that this is the same ideal as that of the mystery variety.

```
i99 : idealC == idealX

o99 = true
```

Here is the code of the function `mystery`, which does the steps above.

```
i100 : code mystery

o100 = -- mystery.m2:1-13
       mystery = ringP3 -> (
           kk := coefficientRing ringP3;
           x := local x;
           ringP2 := kk[x_0..x_2];
           idealC2 := ideal(x_0^5+x_1^5+x_2^5);
           ringC2 := ringP2/idealC2;
           ringP5 := kk[x_0..x_5];
           idealC5 := trim kernel map(ringC2, ringP5,
               gens (ideal vars ringC2)^2);
           ringC5 := ringP5/idealC5;
           use ringC5;
           trim kernel map(ringC5, ringP3,
               matrix{{x_0+x_1,x_2,x_3,x_5}}))
```

And here is the code of the function `prettyPrint`.

```
i101 : code prettyPrint

o101 = -- mystery.m2:15-51
       prettyPrint = f -> (
           -- accept a matrix f and print its entries prettily,
           -- separated by commas
           wid := 74;
           -- page width
           post := (c,s) -> (
               -- This function concatenates string c to end of each
               -- string in list s except the last one
               concatenate \ pack_2 between_c s);
           strings := post_"," (toString \ flatten entries f);
           -- list of strings, one for each polynomial, with commas
           istate := ("",0);
           -- initial state = (out : output string, col : column number)
           strings = apply(
               strings,
               poly -> first fold(
                   -- break each poly into lines
                   (state,term) -> (
                       (out,col) -> (
                           if col + #term > wid -- too wide?
                           then (
                               out = out | "\n   ";
                               col = 3;
                               -- insert line break
                               );
                           (out | term, col + #term) -- new state
                           )
                       ) state,
                   istate,
                   fold( -- separate poly into terms
                       {"+","-"},
                       {poly},
                       (delimiter,poly) -> flatten(
                           post_delimiter \ separate_delimiter \ poly
                           )))); 
           print stack strings;  -- stack them vertically, then print
           )
```

References

1. Allen Altman and Steven Kleiman: *Introduction to Grothendieck duality theory.* Springer-Verlag, Berlin, 1970. Lecture Notes in Mathematics, Vol. 146.

2. Alf Aure, Wolfram Decker, Klaus Hulek, Sorin Popescu, and Kristian Ranestad: Syzygies of abelian and bielliptic surfaces in $\mathbf{p}^4$. *Internat. J. Math.*, 8(7):849–919, 1997.

3. Wolfram Decker and Frank-Olaf Schreyer: Non-general type surfaces in $\mathbf{p}^4$: some remarks on bounds and constructions. *J. Symbolic Comput.*, 29(4-5):545–582, 2000. Symbolic computation in algebra, analysis, and geometry (Berkeley, CA, 1998).

4. David Eisenbud: *Commutative algebra.* Springer-Verlag, New York, 1995. With a view toward algebraic geometry.

5. Mark Green and Robert Lazarsfeld: On the projective normality of complete linear series on an algebraic curve. *Invent. Math.*, 83(1):73–90, 1985.

6. Joe Harris: *Algebraic geometry.* Springer-Verlag, New York, 1995. A first course, Corrected reprint of the 1992 original.

7. Robin Hartshorne: *Algebraic geometry.* Springer-Verlag, New York, 1977. Graduate Texts in Mathematics, No. 52.

8. B. Saint-Donat: On Petri's analysis of the linear system of quadrics through a canonical curve. *Math. Ann.*, 206:157–175, 1973.

9. Frank-Olaf Schreyer: Syzygies of canonical curves and special linear series. *Math. Ann.*, 275(1):105–137, 1986.

10. Frank-Olaf Schreyer: A standard basis approach to syzygies of canonical curves. *J. Reine Angew. Math.*, 421:83–123, 1991.

11. Jean-Pierre Serre: *Algèbre locale. Multiplicités.* Springer-Verlag, Berlin, 1965. Cours au Collège de France, 1957–1958, rédigé par Pierre Gabriel. Seconde édition, 1965. Lecture Notes in Mathematics, 11.

Data Types, Functions, and Programming

Daniel R. Grayson* and Michael E. Stillman**

In this chapter we present an introduction to the structure of *Macaulay 2* commands and the writing of functions in the *Macaulay 2* language. For further details see the *Macaulay 2* manual distributed with the program [1].

1 Basic Data Types

The basic data types of *Macaulay 2* include numbers of various types (integers, rational numbers, floating point numbers, complex numbers), lists (basic lists, and three types of visible lists, depending on the delimiter used), hash tables, strings of characters (both 1-dimensional and 2-dimensional), Boolean values (true and false), symbols, and functions. Higher level types useful in mathematics are derived from these basic types using facilities provided in the *Macaulay 2* language. Except for the simplest types (integers and Boolean values), *Macaulay 2* normally displays the type of the output value on a second labeled output line.

Symbols have a name which consists of letters, digits, or apostrophes, the first of which is a letter. Values can be assigned to symbols and recalled later.

```
i1 : w

o1 = w

o1 : Symbol

i2 : w = 2^100

o2 = 1267650600228229401496703205376

i3 : w

o3 = 1267650600228229401496703205376
```

Multiple values can be assigned in parallel.

```
i4 : (w,w') = (33,44)

o4 = (33, 44)

o4 : Sequence

i5 : w

o5 = 33

i6 : w'

o6 = 44
```

* Supported by NSF grant DMS 99-70085.
** Supported by NSF grant 99-70348.

Comments are initiated by `--` and extend to the end of the line.

```
i7 : (w,w') = (33,    -- this is a comment
                44)

o7 = (33, 44)

o7 : Sequence
```

Strings of characters are delimited by quotation marks.

```
i8 : w = "abcdefghij"

o8 = abcdefghij
```

They may be joined horizontally to make longer strings, or vertically to make a two-dimensional version called a *net*.

```
i9 : w | w

o9 = abcdefghijabcdefghij

i10 : w || w

o10 = abcdefghij
      abcdefghij
```

Nets are used in the preparation of two dimensional output for polynomials.

Floating point numbers are distinguished from integers by the presence of a decimal point, and rational numbers are entered as fractions.

```
i11 : 2^100

o11 = 1267650600228229401496703205376

i12 : 2.^100

o12 = 1.26765 10^30

o12 : RR

i13 : (36 + 1/8)^6

      582622237229761
o13 = ---------------
           262144

o13 : QQ
```

Parentheses, braces, and brackets are used as delimiters for the three types of *visible lists*: lists, sequences, and arrays.

```
i14 : x1 = {1,a}

o14 = {1, a}

o14 : List

i15 : x2 = (2,b)

o15 = (2, b)

o15 : Sequence

i16 : x3 = [3,c,d,e]

o16 = [3, c, d, e]

o16 : Array
```

Even though they use braces, lists should not be confused with sets, which will be treated later. A double period can be used to construct a sequence of consecutive elements in various contexts.

```
i17 : 1 .. 6

o17 = (1, 2, 3, 4, 5, 6)

o17 : Sequence

i18 : a .. f

o18 = (a, b, c, d, e, f)

o18 : Sequence
```

Lists can be nested.

```
i19 : xx = {x1,x2,x3}

o19 = {{1, a}, (2, b), [3, c, d, e]}

o19 : List
```

The number of entries in a list is provided by #.

```
i20 : #xx

o20 = 3
```

The entries in a list are numbered starting with 0, and can be recovered with # used as a binary operator.

```
i21 : xx#0

o21 = {1, a}

o21 : List

i22 : xx#0#1

o22 = a

o22 : Symbol
```

We can join visible lists and *append* or *prepend* an element to a visible list. The output will be the same type of visible list that was provided in the input: a list, a sequence, or an array; if the arguments are various types of lists, the output will be same type as the first argument.

```
i23 : join(x1,x2,x3)

o23 = {1, a, 2, b, 3, c, d, e}

o23 : List


i24 : append(x3,f)

o24 = [3, c, d, e, f]

o24 : Array
```

```
i25 : prepend(f,x3)

o25 = [f, 3, c, d, e]

o25 : Array
```

Use sum or product to produce the sum or product of all the elements in a list.

```
i26 : sum {1,2,3,4}

o26 = 10


i27 : product {1,2,3,4}

o27 = 24
```

2   Control Structures

Commands for later execution are encapsulated in *functions*. A function is created using the operator -> to separate the parameter or sequence of parameters from the code to be executed later. Let's try an elementary example of a function with two arguments.

```
i28 : f = (x,y) -> 1000 * x + y

o28 = f

o28 : Function
```

The parameters x and y are symbols that will acquire a value later when the function is executed. They are *local* in the sense that they are completely different from any symbols with the same name that occur elsewhere. Additional local variables for use within the body of a function can be created by assigning a value to them with := (first time only). We illustrate this by rewriting the function above.

```
i29 : f = (x,y) -> (z := 1000 * x; z + y)

o29 = f

o29 : Function
```

Let's apply the function to some arguments.

```
i30 : f(3,7)

o30 = 3007
```

The sequence of arguments can be assembled first, and then passed to the function.

```
i31 : s = (3,7)

o31 = (3, 7)

o31 : Sequence
```

```
i32 : f s

o32 = 3007
```

As above, functions receiving one argument may be called without parentheses.

```
i33 : sin 2.1

o33 = 0.863209

o33 : RR
```

A compact notation for functions makes it convenient to apply them without naming them first. For example, we may use apply to apply a function to every element of a list and to collect the results into a list.

```
i34 : apply(1 .. 10, i -> i^3)

o34 = (1, 8, 27, 64, 125, 216, 343, 512, 729, 1000)

o34 : Sequence
```

The function scan will do the same thing, but discard the results.

```
i35 : scan(1 .. 5, print)
1
2
3
4
5
```

Use if ... then ... else ... to perform alternative actions based on the truth of a condition.

```
i36 : apply(1 .. 10, i -> if even i then 1000*i else i)

o36 = (1, 2000, 3, 4000, 5, 6000, 7, 8000, 9, 10000)

o36 : Sequence
```

A function can be terminated prematurely with return.

```
i37 : apply(1 .. 10, i -> (if even i then return 1000*i; -i))

o37 = (-1, 2000, -3, 4000, -5, 6000, -7, 8000, -9, 10000)

o37 : Sequence
```

Loops in a program can be implemented with while ... do ....

```
i38 : i = 1; while i < 50 do (print i; i = 2*i)
1
2
4
8
16
32
```

Another way to implement loops is with for and do or list, with optional clauses introduced by the keywords from, to, and when.

```
i40 : for i from 1 to 10 list i^3

o40 = {1, 8, 27, 64, 125, 216, 343, 512, 729, 1000}

o40 : List
```

```
i41 : for i from 1 to 4 do print i
1
2
3
4
```

A loop can be terminated prematurely with **break**, which accepts an optional value to return as the value of the loop expression.

```
i42 : for i from 2 to 100 do if not isPrime i then break i

o42 = 4
```

If no value needs to be returned, the condition for continuing can be provided with the keyword **when**; iteration continues only as long as the predicate following the keyword returns **true**.

```
i43 : for i from 2 to 100 when isPrime i do print i
2
3
```

3   Input and Output

The function **print** can be used to display something on the screen.

```
i44 : print 2^100
1267650600228229401496703205376
```

For example, it could be used to display the elements of a list on separate lines.

```
i45 : (1 .. 5) / print;
1
2
3
4
5
```

The operator **<<** can be used to display something on the screen, without the newline character.

```
i46 : << 2^100
1267650600228229401496703205376
o46 = stdio

o46 : File

    -- the standard input output file
```

Notice the value returned is a *file*. A *file* in *Macaulay 2* is a data type that represents a channel through which data can be passed, as input, as output, or in both directions. The file **stdio** encountered above corresponds to your shell window or terminal, and is used for two-way communication between the program and the user. A file may correspond to what one usually calls a file, i.e., a sequence of data bytes associated with a given name and stored on your disk drive. A file may also correspond to a *socket*, a channel for communication with other programs over the network.

Files can be used with the binary form of the operator **<<** to display something else on the same line.

```
i47 : << "the value is : " << 2^100
the value is :  1267650600228229401496703205376
o47 = stdio

o47 : File

    --  the standard input output file
```

Using **endl** to represent the new line character or character sequence, we can produce multiple lines of output.

```
i48 : << "A = " << 2^100 << endl << "B = " << 2^200 << endl;
A = 1267650600228229401496703205376
B = 1606938044258990275541962092341162602522202993782792835301376
```

We can send the same output to a disk file named **foo**, but we must remember to close it with **close**.

```
i49 : "foo" << "A = " << 2^100 << endl << close

o49 = foo

o49 : File
```

The contents of the file can be recovered as a string with **get**.

```
i50 : get "foo"

o50 = A = 1267650600228229401496703205376
```

If the file contains valid *Macaulay 2* commands, as it does in this case, we can execute those commands with **load**.

```
i51 : load "foo"
```

We can verify that the command took effect by evaluating **A**.

```
i52 : A

o52 = 1267650600228229401496703205376
```

Alternatively, if we want to see those commands and the output they produce, we may use **input**.

```
i53 : input "foo"

i54 : A = 1267650600228229401496703205376

o54 = 1267650600228229401496703205376

i55 :
```

Let's set up a ring for computation in *Macaulay 2*.

```
i56 : R = QQ[x,y,z]

o56 = R

o56 : PolynomialRing


i57 : f = (x+y)^3

        3     2       2    3
o57 = x  + 3x y + 3x*y  + y

o57 : R
```

Printing, and printing to files, works for polynomials, too.

```
i58 : "foo" << f << close;
```

The two-dimensional output is readable by humans, but is not easy to convert back into a polynomial.

```
i59 : get "foo"

o59 = 3     2        2    3
      x  + 3x y + 3x*y  + y
```

Use `toString` to create a 1-dimensional form of the polynomial that can be stored in a file in a format readable by *Macaulay 2* and by other symbolic algebra programs, such as *Mathematica* or *Maple*.

```
i60 : toString f

o60 = x^3+3*x^2*y+3*x*y^2+y^3
```

Send it to the file.

```
i61 : "foo" << toString f << close;
```

Get it back.

```
i62 : get "foo"

o62 = x^3+3*x^2*y+3*x*y^2+y^3
```

Convert the string back to a polynomial with `value`, using oo to recover the value of the expression on the previous line.

```
i63 : value oo

        3     2        2    3
o63 = x  + 3x y + 3x*y  + y

o63 : R
```

The same thing works for matrices, and a little more detail is provided by `toExternalString`, if needed.

```
i64 : vars R

o64 = | x y z |

                1      3
o64 : Matrix R  <--- R

i65 : toString vars R

o65 = matrix {{x, y, z}}

i66 : toExternalString vars R

o66 = map(R^{{0}}, R^{{-1}, {-1}, {-1}}, {{x, y, z}})
```

4   Hash Tables

Recall how one sets up a quotient ring for computation in *Macaulay 2*.

```
i67 : R = QQ[x,y,z]/(x^3-y)

o67 = R

o67 : QuotientRing

i68 : (x+y)^4

            2 2       3    4         2
o68 = 6x y  + 4x*y  + y  + x*y + 4y

o68 : R
```

How does *Macaulay 2* represent a ring like R in the computer? To answer that, first think about what sort of information needs to be retained about R. We may need to remember the coefficient ring of R, the names of the variables in R, the monoid of monomials in the variables, the degrees of the variables, the characteristic of the ring, whether the ring is commutative, the ideal modulo which we are working, and so on. We also may need to remember various bits of code: the code for performing the basic arithmetic operations, such as addition and multiplication, on elements of R; the code for preparing a readable representation of an element of R, either 2-dimensional (with superscripts above the line and subscripts below), or 1-dimensional. Finally, we may want to remember certain things that take a lot of time to compute, such as the Gröbner basis of the ideal.

A *hash table* is, by definition, a way of representing (in the computer) a function whose domain is a finite set. In *Macaulay 2*, hash tables are extremely flexible: the elements of the domain (or *keys*) and the elements of the range (or *values*) of the function may be any of the other objects represented in the computer. It's easy to come up with uses for functions whose domain is finite: for example, a monomial can be represented by the function that associates to a variable its nonzero exponent; a polynomial can be represented by a function that associates to a monomial its nonzero coefficient; a set can be represented by any function with that set as its domain; a (sparse) matrix can be represented as a function from pairs of natural numbers to the corresponding nonzero entry.

Let's create a hash table and name it.

```
i69 : f = new HashTable from { a=>444, Daniel=>555, {c,d}=>{1,2,3,4}}

o69 = HashTable{{c, d} => {1, 2, 3, 4}}
                    a => 444
                    Daniel => 555

o69 : HashTable
```

The operator => is used to represent a key-value pair. We can use the operator # to recover the value from the key.

```
i70 : f#Daniel

o70 = 555

i71 : f#{c,d}

o71 = {1, 2, 3, 4}
```

```
o71 : List
```

If the key is a symbol, we can use the operator . instead; this is convenient
if the symbol has a value that we want to ignore.

```
i72 : Daniel = a

o72 = a

o72 : Symbol

i73 : f.Daniel

o73 = 555
```

We can use #? to test whether a given key occurs in the hash table.

```
i74 : f#?a

o74 = true

i75 : f#?c

o75 = false
```

Finite sets are implemented in *Macaulay 2* as hash tables: the elements of the
set are stored as the keys in the hash table, with the accompanying values all
being 1. (Multisets are implemented by using values larger than 1, and are
called *tallies*.)

```
i76 : x = set{1,a,{4,5},a}

o76 = Set {{4, 5}, 1, a}

o76 : Set


i77 : x#?a

o77 = true

i78 : peek x

o78 = Set{{4, 5} => 1}
            1 => 1
            a => 1

i79 : y = tally{1,a,{4,5},a}

o79 = Tally{{4, 5} => 1}
              1 => 1
              a => 2

o79 : Tally

i80 : y#a

o80 = 2
```

We might use `tally` to tally how often a function attains its various possible
values. For example, how often does an integer have 3 prime factors? Or 4?
Use `factor` to factor an integer.

```
i81 : factor 60

          2
o81 = 2 3*5

o81 : Product
```

Then use # to get the number of factors.

```
i82 : # factor 60

o82 = 3
```

Use apply to list some values of the function.

```
i83 : apply(2 .. 1000, i -> # factor i)

o83 = (1, 1, 1, 1, 2, 1, 1, 1, 2, 1, 2, 1, 2, 2, 1, 1, 2, 1, 2, 2, 2,  ···

o83 : Sequence
```

Finally, use tally to summarize the results.

```
i84 : tally oo

o84 = Tally{1 => 193}
            2 => 508
            3 => 275
            4 => 23

o84 : Tally
```

Hash tables turn out to be convenient entities for storing odd bits and pieces of information about something in a way that's easy to think about and use. In *Macaulay 2*, rings are represented as hash tables, as are ideals, matrices, modules, chain complexes, and so on. For example, although it isn't a documented feature, the key ideal is used to preserve the ideal that was used above to define the quotient ring R, as part of the information stored in R.

```
i85 : R.ideal

            3
o85 = ideal(x  - y)

o85 : Ideal of QQ [x, y, z]
```

The preferred and documented way for a user to recover this information is with the function ideal.

```
i86 : ideal R

            3
o86 = ideal(x  - y)

o86 : Ideal of QQ [x, y, z]
```

Users who want to introduce a new high-level mathematical concept to *Macaulay 2* may learn about hash tables by referring to the *Macaulay 2* manual [1].

5   Methods

You may use the `code` command to locate the source code for a given function, at least if it is one of those functions written in the *Macaulay 2* language. For example, here is the code for `demark`, which may be used to put commas between strings in a list.

```
i87 : code demark

o87 = -- ../../../m2/fold.m2:23
      demark = (s,v) -> concatenate between(s,v)
```

The code for tensoring a ring map with a module can be displayed in this way.

```
i88 : code(symbol **, RingMap, Module)

o88 = -- ../../../m2/ringmap.m2:294-298
      RingMap ** Module := Module => (f,M) -> (
          R := source f;
          S := target f;
          if R =!= ring M then error "expected module over source ring";
          cokernel f(presentation M));
```

The code implementing the `ideal` function when applied to a quotient ring can be displayed as follows.

```
i89 : code(ideal, QuotientRing)

o89 = -- ../../../m2/quotring.m2:7
      ideal QuotientRing := R -> R.ideal
```

Notice that it uses the key `ideal` to extract the information from the ring's hash table, as you might have guessed from the previous discussion. The bit of code displayed above may be called a *method* as a way of indicating that several methods for dealing with various types of arguments are attached to the function named `ideal`. New such *method functions* may be created with the function `method`. Let's illustrate that with an example: we'll write a function called `denom` which should produce the denominator of a rational number. When applied to an integer, it should return 1. First we create the method function.

```
i90 : denom = method();
```

Then we tell it what to do with an argument from the class `QQ` of rational numbers.

```
i91 : denom QQ := x -> denominator x;
```

And also what to do with an argument from the class `ZZ` of integers.

```
i92 : denom ZZ := x -> 1;
```

Let's test it.

```
i93 : denom(5/3)

o93 = 3

i94 : denom 5

o94 = 1
```

6   Pointers to the Source Code

A substantial part of *Macaulay 2* is written in the same language provided to the users. A good way to learn more about the *Macaulay 2* language is to peruse the source code that comes with the system in the directory Macaulay2/m2. Use the code function, as described in the previous section, for locating the bit of code you wish to view.

The source code for the interpreter of the *Macaulay 2* language is in the directory Macaulay2/d. It is written in another language designed to be mostly type-safe, which is translated into C by the translator whose own C source code is in the directory Macaulay2/c. Here is a sample line of code from the file Macaulay2/d/tokens.d, which shows how the translator provides for allocation and initialization of dynamic data structures.

```
globalFrame := Frame(dummyFrame,globalScope.seqno,Sequence(nullE));
```

And here is the C code produced by the translator.

```
tokens_Frame tokens_globalFrame;
tokens_Frame tmp__23;
Sequence tmp__24;
tmp__24 = (Sequence) GC_MALLOC(sizeof(struct S259_)+(1-1)*sizeof(Expr));
if (0 == tmp__24) outofmem();
tmp__24->len_ = 1;
tmp__24->array_[0] = tokens_nullE;
tmp__23 = (tokens_Frame) GC_MALLOC(sizeof(struct S260_));
if (0 == tmp__23) outofmem();
tmp__23->next = tokens_dummyFrame;
tmp__23->scopenum = tokens_globalScope->seqno;
tmp__23->values = tmp__24;
tokens_globalFrame = tmp__23;
```

The core algebraic algorithms constitute the *engine* of *Macaulay 2* and are written in C++, with the source files in the directory Macaulay2/e. In the current version of the program, the interface between the interpreter and the core algorithms consists of a single two-directional stream of bytes. The manual that comes with the system [1] describes the engine communication protocol used in that interface.

References

1. Daniel R. Grayson and Michael E. Stillman: *Macaulay 2*, a software system for research in algebraic geometry and commutative algebra. Available in source code form and compiled for various architectures, with documentation, at http://www.math.uiuc.edu/Macaulay2/.

Teaching the Geometry of Schemes

Gregory G. Smith and Bernd Sturmfels

This chapter presents a collection of graduate level problems in algebraic geometry illustrating the power of *Macaulay 2* as an educational tool.

When teaching an advanced subject, like the language of schemes, we think it is important to provide plenty of concrete instances of the theory. Computer algebra systems, such as *Macaulay 2*, provide students with an invaluable tool for studying complicated examples. Furthermore, we believe that the explicit nature of a computational approach leads to a better understanding of the objects being examined. This chapter presents some problems which we feel illustrate this point of view.

Our examples are selected from the homework of an algebraic geometry class given at the University of California at Berkeley in the fall of 1999. This graduate course was taught by the second author with assistance from the first author. Our choice of problems, as the title suggests, follows the material in David Eisenbud and Joe Harris' textbook *The Geometry of Schemes* [5].

1 Distinguished Open Sets

We begin with a simple example involving the Zariski topology of an affine scheme. This example also indicates some of the subtleties involved in working with arithmetic schemes.

Problem. *Let $S = \mathbb{Z}[x, y, z]$ and $X = \mathrm{Spec}(S)$. If $f = x$ and X_f is the corresponding basic open subset in X, then establish the following:*

(1) *If $e_1 = x + y + z$, $e_2 = xy + xz + yz$ and $e_3 = xyz$ are the elementary symmetric functions then the set $\{X_{e_i}\}_{1 \leq i \leq 3}$ is an open cover of X_f.*

(2) *If $p_1 = x + y + z$, $p_2 = x^2 + y^2 + z^2$ and $p_3 = x^3 + y^3 + z^3$ are the power sum symmetric functions then $\{X_{p_i}\}_{1 \leq i \leq 3}$ is not an open cover of X_f.*

Solution. (1) To prove that $\{X_{e_i}\}_{1 \leq i \leq 3}$ is an open cover of X_f, it suffices to show that e_1, e_2 and e_3 generate the unit ideal in S_f; see Lemma I-16 in Eisenbud and Harris [5]. This is equivalent to showing that x^m belongs to the S-ideal $\langle e_1, e_2, e_3 \rangle$ for some $m \in \mathbb{N}$. In other words, the saturation $(\langle e_1, e_2, e_3 \rangle : x^\infty)$ is the unit ideal if and only if $\{X_{e_i}\}_{1 \leq i \leq 3}$ is an open cover of X_f. We verify this in *Macaulay 2* as follows:

```
i1 : S = ZZ[x, y, z];

i2 : elementaryBasis = ideal(x+y+z, x*y+x*z+y*z, x*y*z);

o2 : Ideal of S
```

```
i3 : saturate(elementaryBasis, x)

o3 = ideal 1

o3 : Ideal of S
```

(2) Similarly, to show that $\{X_{p_i}\}_{1 \leq i \leq 3}$ is not an open cover of X_f, we prove that $\big(\langle p_1, p_2, p_3 \rangle : x^\infty\big)$ is not the unit ideal. Calculating this saturation, we find

```
i4 : powerSumBasis = ideal(x+y+z, x^2+y^2+z^2, x^3+y^3+z^3);

o4 : Ideal of S

i5 : saturate(powerSumBasis, x)

                  2              2
o5 = ideal (6, x + y + z, 2y + 2y*z + 2z , 3y*z)

o5 : Ideal of S

i6 : clearAll
```

which is not the unit ideal.   □

The fact that 6 is a generator of the ideal $\big(\langle p_1, p_2, p_3 \rangle : x^\infty\big)$ indicates that $\{X_{p_i}\}_{1 \leq i \leq 3}$ does not contain the points in X lying over the points $\langle 2 \rangle$ and $\langle 3 \rangle$ in $\mathrm{Spec}(\mathbb{Z})$. If we work over a base ring in which 6 is a unit, then $\{X_{p_i}\}_{1 \leq i \leq 3}$ would, in fact, be an open cover of X_f.

2 Irreducibility

The study of complex semisimple Lie algebras gives rise to an important family of algebraic varieties called nilpotent orbits. The next problem examines the irreducibility of a particular nilpotent orbit.

Problem. *Let X be the set of nilpotent complex 3×3 matrices. Show that X is an irreducible algebraic variety.*

Solution. A 3×3 matrix M is nilpotent if and only if its minimal polynomial $p(\mathsf{T})$ equals T^k, for some $k \in \mathbb{N}$. Since each irreducible factor of the characteristic polynomial of M is also a factor of $p(\mathsf{T})$, it follows that the characteristic polynomial of M is T^3. We conclude that the coefficients of the characteristic polynomial of a generic 3×3 matrix define the algebraic variety X.

To prove that X is irreducible over $\mathbb{C}$, we construct a rational parameterization. First, observe that $\mathrm{GL}_3(\mathbb{C})$ acts on X by conjugation. Jordan's canonical form theorem implies that there are exactly three orbits; one for each of the following matrices:

$$N_{(1,1,1)} = \begin{bmatrix} 0 & 0 & 0 \\ 0 & 0 & 0 \\ 0 & 0 & 0 \end{bmatrix}, \quad N_{(2,1)} = \begin{bmatrix} 0 & 1 & 0 \\ 0 & 0 & 0 \\ 0 & 0 & 0 \end{bmatrix} \text{ and } N_{(3)} = \begin{bmatrix} 0 & 1 & 0 \\ 0 & 0 & 1 \\ 0 & 0 & 0 \end{bmatrix} .$$

Each orbit is defined by a rational parameterization, so it suffices to show that the closure of the orbit containing $N_{(3)}$ is the entire variety X. We demonstrate this as follows:

```
i7 : S = QQ[t, y_0 .. y_8, a..i, MonomialOrder => Eliminate 10];

i8 : N3 = (matrix {{0,1,0},{0,0,1},{0,0,0}}) ** S

o8 = | 0 1 0 |
     | 0 0 1 |
     | 0 0 0 |

             3       3
o8 : Matrix S  <--- S

i9 : G = genericMatrix(S, y_0, 3, 3)

o9 = | y_0 y_3 y_6 |
     | y_1 y_4 y_7 |
     | y_2 y_5 y_8 |

             3       3
o9 : Matrix S  <--- S
```

To determine the entries in $G \cdot N_{(3)} \cdot G^{-1}$, we use the classical adjoint to construct the matrix $\det(G) \cdot G^{-1}$.

```
i10 : classicalAdjoint = (G) -> (
            n := degree target G;
            m := degree source G;
            matrix table(n, n, (i, j) -> (-1)^(i+j) * det(
                    submatrix(G, {0..j-1, j+1..n-1},
                              {0..i-1, i+1..m-1})))));

i11 : num = G * N3 * classicalAdjoint(G);

             3       3
o11 : Matrix S  <--- S

i12 : D = det(G);

i13 : M = genericMatrix(S, a, 3, 3);

             3       3
o13 : Matrix S  <--- S
```

The entries in $G \cdot N_{(3)} \cdot G^{-1}$ give a rational parameterization of the orbit generated by $N_{(3)}$. Using elimination theory — see section 3.3 in Cox, Little and O'Shea [2] — we give an "implicit representation" of this variety.

```
i14 : elimIdeal = minors(1, (D*id_(S^3))*M - num) + ideal(1-D*t);

o14 : Ideal of S

i15 : closureOfOrbit = ideal selectInSubring(1, gens gb elimIdeal);

o15 : Ideal of S
```

Finally, we verify that this orbit closure equals X scheme-theoretically. Recall that X is defined by the coefficients of the characteristic polynomial of a generic 3×3 matrix M.

```
i16 : X = ideal substitute(
              contract(matrix{{t^2,t,1}}, det(t-M)),
              {t => 0_S})

o16 = ideal (- a - e - i, - b*d + a*e - c*g - f*h + a*i + e*i, c*e*g - ...

o16 : Ideal of S
```

```
i17 : closureOfOrbit == X

o17 = true

i18 : clearAll
```

This completes our solution.   □

More generally, Kostant shows that the set of all nilpotent elements in a complex semisimple Lie algebra form an irreducible variety. We refer the reader to Chriss and Ginzburg [1] for a proof of this result (Corollary 3.2.8) and a discussion of its applications in representation theory.

3   Singular Points

In our third question, we study the singular locus of a family of elliptic curves.

Problem. *Consider a general form of degree 3 in* $\mathbb{Q}[x, y, z]$:

$$F = ax^3 + bx^2y + cx^2z + dxy^2 + exyz + fxz^2 + gy^3 + hy^2z + iyz^2 + jz^3 \ .$$

Give necessary and sufficient conditions in terms of $a, \ldots, j$ *for the cubic curve* $\mathrm{Proj}\left(\mathbb{Q}[x, y, z]/\langle F\rangle\right)$ *to have a singular point.*

Solution. The singular locus of F is defined by a polynomial of degree 12 in the 10 variables $a, \ldots, j$. We calculate this polynomial in two different ways.

Our first method is an elementary but time consuming elimination. Carrying it out in *Macaulay 2*, we have

```
i19 : S = QQ[x, y, z, a..j, MonomialOrder => Eliminate 2];

i20 : F = a*x^3+b*x^2*y+c*x^2*z+d*x*y^2+e*x*y*z+f*x*z^2+g*y^3+h*y^2*z+
          i*y*z^2+j*z^3;

i21 : partials = submatrix(jacobian matrix{{F}}, {0..2}, {0})

o21 = {1} | 3x2a+2xyb+y2d+2xzc+yze+z2f |
      {1} | x2b+2xyd+3y2g+xze+2yzh+z2i |
      {1} | x2c+xye+y2h+2xzf+2yzi+3z2j |

                   3       1
o21 : Matrix S  <--- S

i22 : singularities = ideal(partials) + ideal(F);

o22 : Ideal of S

i23 : elimDiscr = time ideal selectInSubring(1,gens gb singularities);
      -- used 64.27 seconds

o23 : Ideal of S

i24 : elimDiscr = substitute(elimDiscr, {z => 1});

o24 : Ideal of S
```

On the other hand, there is also an elegant and more useful determinantal formula for this discriminant; it is a specialization of the formula (2.8) in section 3.2 of Cox, Little and O'Shea [3]. To apply this determinantal formula, we first create the coefficient matrix A of the partial derivatives of F.

```
i25 : A = contract(matrix{{x^2,x*y,y^2,x*z,y*z,z^2}},
            diff(transpose matrix{{x,y,z}},F))

o25 = {1} | 3a 2b d  2c e  f |
      {1} | b  2d 3g e  2h i |
      {1} | c  e  h  2f 2i 3j |

            3      6
o25 : Matrix S <--- S
```

We also construct the coefficient matrix B of the partial derivatives of the Hessian of F.

```
i26 : hess = det submatrix(jacobian ideal partials, {0..2}, {0..2});

i27 : B = contract(matrix{{x^2,x*y,y^2,x*z,y*z,z^2}},
            diff(transpose matrix{{x,y,z}},hess))

o27 = {1} | -24c2d+24bce-18ae2-24b2f+72adf        4be2-16bdf-48 · · ·
      {1} | 2be2-8bdf-24c2g+72afg+16bch-24aeh-8b2i+24adi 4de2-16d2f-48 · · ·
      {1} | 2ce2-8cdf-8c2h+24afh+16bci-24aei-24b2j+72adj 2e3-8def-24cf · · ·

            3      6
o27 : Matrix S <--- S
```

To obtain the discriminant, we combine these two matrices and take the determinant.

```
i28 : detDiscr = ideal det (A || B);

o28 : Ideal of S
```

Finally, we check that our two discriminants are equal

```
i29 : detDiscr == elimDiscr

o29 = true
```

and examine the generator.

```
                2    4 3 2          5 3 2        6 3 2        2 2 2 · · ·
o30 = 13824c d*e f g  - 13824b*c*e f g  + 13824a*e f g  - 110592c d e  · · ·

i30 : detDiscr_0

o30 : S

i31 : numgens detDiscr

o31 = 1

i32 : # terms detDiscr_0

o32 = 2040

i33 : clearAll
```

Hence, the singular locus is given by a single polynomial of degree 12 with 2040 terms.   □

For a further discussion of singularities and discriminants see Section V.3 in Eisenbud and Harris [5]. For information on resultants and discriminants see Chapter 2 in Cox, Little and O'Shea [3].

4   Fields of Definition

Schemes over non-algebraically closed fields arise in number theory. Our fourth problem looks at one technique for working with number fields in *Macaulay 2.*

Problem (Exercise II-6 in [5]). *An inclusion of fields* $K \hookrightarrow L$ *induces a map* $\mathbb{A}^n_L \to \mathbb{A}^n_K$. *Find the images in* $\mathbb{A}^2_{\mathbb{Q}}$ *of the following points of* $\mathbb{A}^2_{\overline{\mathbb{Q}}}$ *under this map.*

(1) $\langle x - \sqrt{2}, y - \sqrt{2} \rangle$;
(2) $\langle x - \sqrt{2}, y - \sqrt{3} \rangle$;
(3) $\langle x - \zeta, y - \zeta^{-1} \rangle$ *where* ζ *is a 5-th root of unity* ;
(4) $\langle \sqrt{2}x - \sqrt{3}y \rangle$;
(5) $\langle \sqrt{2}x - \sqrt{3}y - 1 \rangle$.

Solution. The images can be determined by using the following three step algorithm: (1) replace the coefficients not contained in K with indeterminates, (2) add the minimal polynomials of these coefficients to the given ideal in $\mathbb{A}^2_L$, and (3) eliminate the new indeterminates. Here are the five examples:

```
i34 : S = QQ[a,b,x,y, MonomialOrder => Eliminate 2];

i35 : I1 = ideal(x-a, y-a, a^2-2);

o35 : Ideal of S

i36 : ideal selectInSubring(1, gens gb I1)

                     2
o36 = ideal (x - y, y  - 2)

o36 : Ideal of S

i37 : I2 = ideal(x-a, y-b, a^2-2, b^2-3);

o37 : Ideal of S

i38 : ideal selectInSubring(1, gens gb I2)

              2      2
o38 = ideal (y  - 3, x  - 2)

o38 : Ideal of S

i39 : I3 = ideal(x-a, y-a^4, a^4+a^3+a^2+a+1);

o39 : Ideal of S
```

```
i40 : ideal selectInSubring(1, gens gb I3)

              2   2              3   2
o40 = ideal (x*y - 1, x  + y  + x + y + 1, y  + y  + x + y + 1)

o40 : Ideal of S

i41 : I4 = ideal(a*x+b*y, a^2-2, b^2-3);

o41 : Ideal of S

i42 : ideal selectInSubring(1, gens gb I4)

            2   3 2
o42 = ideal(x  - -*y )
                 2

o42 : Ideal of S

i43 : I5 = ideal(a*x+b*y-1, a^2-2, b^2-3);

o43 : Ideal of S

i44 : ideal selectInSubring(1, gens gb I5)

            4        2 2   9 4    2   3 2   1
o44 = ideal(x  - 3x y + -*y  - x - -*y + -)
                        4            2     4

o44 : Ideal of S

i45 : clearAll
```

□

It is worth noting that the points in $A_{\overline{\mathbb{Q}}}^n$ correspond to orbits of the action of $\mathrm{Gal}(\overline{\mathbb{Q}}/\mathbb{Q})$ on the points of $A_{\mathbb{Q}}^n$. For more examples and information, see section II.2 in Eisenbud and Harris [5].

5 Multiplicity

The multiplicity of a zero-dimensional scheme X at a point $p \in X$ is defined to be the length of the local ring $\mathcal{O}_{X,p}$. Unfortunately, we cannot work directly in the local ring in *Macaulay 2*. What we can do, however, is to compute the multiplicity by computing the degree of the component of X supported at p; see page 66 in Eisenbud and Harris [5].

Problem. *What is the multiplicity of the origin as a zero of the polynomial equations* $x^5 + y^3 + z^3 = x^3 + y^5 + z^3 = x^3 + y^3 + z^5 = 0$?

Solution. If I is the ideal generated by $x^5+y^3+z^3$, $x^3+y^5+z^3$ and $x^3+y^3+z^5$ in $\mathbb{Q}[x, y, z]$, then the multiplicity of the origin is

$$\dim_{\mathbb{Q}} \frac{\mathbb{Q}[x, y, z]_{\langle x,y,z \rangle}}{I\mathbb{Q}[x, y, z]_{\langle x,y,z \rangle}}.$$

It follows that the multiplicity is the vector space dimension of the ring $\mathbb{Q}[x, y, z]/\varphi^{-1}(I\mathbb{Q}[x, y, z]_{\langle x,y,z\rangle})$ where $\varphi\colon \mathbb{Q}[x, y, z] \to \mathbb{Q}[x, y, z]_{\langle x,y,z\rangle}$ is the natural map. Moreover, we can express this using ideal quotients:

$$\varphi^{-1}(I\mathbb{Q}[x, y, z]_{\langle x,y,z\rangle}) = \left(I : (I : \langle x, y, z\rangle^{\infty})\right).$$

Carrying out this calculation in *Macaulay 2*, we obtain:

```
i46 : S = QQ[x, y, z];

i47 : I = ideal(x^5+y^3+z^3, x^3+y^5+z^3, x^3+y^3+z^5);

o47 : Ideal of S

i48 : multiplicity = degree(I : saturate(I))

o48 = 27

i49 : clearAll
```

Thus, we conclude that the multiplicity is 27.   □

There are algorithms (not yet implemented in *Macaulay 2*) for working directly in the local ring $\mathbb{Q}[x, y, z]_{\langle x,y,z\rangle}$. We refer the interested reader to Chapter 4 in Cox, Little and O'Shea [3].

6   Flat Families

Non-reduced schemes arise naturally as flat limits of a family of reduced schemes. Our next problem illustrates how a family of skew lines in $\mathbb{P}^3$ gives rise to a double line with an embedded point.

Problem (Exercise III-68 in [5]). *Let L and M be the lines in $\mathbb{P}^3_{k[t]}$ given by $x = y = 0$ and $x - tz = y + t^2w = 0$ respectively. Show that the flat limit as $t \to 0$ of the union $L \cup M$ is the double line $x^2 = y = 0$ with an embedded point of degree 1 located at the point $(0 : 0 : 0 : 1)$.*

Solution. We first find the flat limit by saturating the intersection ideal and setting $t = 0$.

```
i50 : PP3 = QQ[t, x, y, z, w];

i51 : L = ideal(x, y);

o51 : Ideal of PP3

i52 : M = ideal(x-t*z, y+t^2*w);

o52 : Ideal of PP3

i53 : X = intersect(L, M);

o53 : Ideal of PP3
```

```
i54 : Xzero = trim substitute(saturate(X, t), {t => 0})
```

$$o54 = \text{ideal } (y*z, \; y\;^2, \; x*y, \; x\;^2)$$

```
o54 : Ideal of PP3
```

Secondly, we verify that this is the union of a double line and an embedded point of degree 1.

```
i55 : Xzero == intersect(ideal(x^2, y), ideal(x, y^2, z))

o55 = true

i56 : degree(ideal(x^2, y ) / ideal(x, y^2, z))

o56 = 1

i57 : clearAll
```

□

Section III.3.4 in Eisenbud and Harris [5] contains several other interesting limits of various flat families.

7 Bézout's Theorem

Bézout's Theorem — Theorem III-78 in Eisenbud and Harris [5] — may fail without the Cohen-Macaulay hypothesis. Our seventh problem is to demonstrate this.

Problem (Exercise III-81 in [5]). *Find irreducible closed subvarieties X and Y in $\mathbb{P}^4$ such that*

$$\text{codim}(X \cap Y) = \text{codim}(X) + \text{codim}(Y)$$
$$\deg(X \cap Y) > \deg(X) \cdot \deg(Y).$$

Solution. We show that the assertion holds when X is the cone over the nonsingular rational quartic curve in $\mathbb{P}^3$ and Y is a two-plane passing through the vertex of the cone. First, recall that the rational quartic curve is given by the 2×2 minors of the matrix $\begin{bmatrix} a & b^2 & bd & c \\ b & ac & c^2 & d \end{bmatrix}$; see Exercise 18.8 in Eisenbud [4]. Thus, we have

```
i58 : S = QQ[a, b, c, d, e];

i59 : IX = trim minors(2, matrix{{a, b^2, b*d, c},{b, a*c, c^2, d}})
```

$$o59 = \text{ideal } (b*c - a*d, \; c\;^3 - b*d\;^2, \; a*c\;^2 - b\;^2 d, \; b\;^3 - a\;^2 c)$$

```
o59 : Ideal of S

i60 : IY = ideal(a, d);

o60 : Ideal of S
```

```
i61 : codim IX + codim IY == codim (IX + IY)

o61 = true

i62 : (degree IX) * (degree IY)

o62 = 4

i63 : degree (IX + IY)

o63 = 5
```

which establishes the assertion.   □

To understand how this example works, it is enlightening to express Y as the intersection of two hyperplanes; one given by $a = 0$ and the other given by $d = 0$. Intersecting X with the first hyperplane yields

```
i64 : J = ideal mingens (IX + ideal(a))

              3    2  2    3
o64 = ideal (a, b*c, c  - b*d , b d, b )

o64 : Ideal of S
```

However, this first intersection has an embedded point;

```
i65 : J == intersect(ideal(a, b*c, b^2, c^3-b*d^2),
               ideal(a, d, b*c, c^3, b^3)) -- embedded point

o65 = true

i66 : clearAll
```

The second hyperplane passes through this embedded point which explains the extra intersection.

8   Constructing Blow-ups

The blow-up of a scheme X along a subscheme Y can be constructed from the Rees algebra associated to the ideal sheaf of Y in X; see Theorem IV-22 in Eisenbud and Harris [5]. Gröbner basis techniques allow one to express the Rees algebra in terms of generators and relations. We illustrate this method in the next solution.

Problem (Exercises IV-43 & IV-44 in [5]). *Find the blow-up X of the affine plane $\mathbb{A}^2 = \mathrm{Spec}\left(\mathbb{Q}[x, y]\right)$ along the subscheme defined by $\langle x^3, xy, y^2 \rangle$. Show that X is nonsingular and its fiber over the origin is the union of two copies of $\mathbb{P}^1$ meeting at a point.*

Solution. We first provide a general function which returns the ideal of relations for the Rees algebra.

```
i67 : blowUpIdeal = (I) -> (
          r := numgens I;
          S := ring I;
          n := numgens S;
          K := coefficientRing S;
          tR := K[t, gens S, vars(0..r-1),
                    MonomialOrder => Eliminate 1];
          f := map(tR, S, submatrix(vars tR, {1..n}));
          F := f(gens I);
          J := ideal apply(1..r, j -> (gens tR)_(n+j)-t*F_(0,(j-1)));
          L := ideal selectInSubring(1, gens gb J);
          R := K[gens S, vars(0..r-1)];
          g := map(R, tR, 0 | vars R);
          trim g(L));
```

Now, applying the function to our specific case yields:

```
i68 : S = QQ[x, y];

i69 : I = ideal(x^3, x*y, y^2);

o69 : Ideal of S

i70 : J = blowUpIdeal(I)

                 2          2        3      2
o70 = ideal (y*b - x*c, x*b  - a*c, x b - y*a, x c - y a)

o70 : Ideal of QQ [x, y, a, b, c]
```

Therefore, the blow-up of the affine plane along the given subscheme is

$$X = \mathrm{Proj}\left(\frac{(\mathbb{Q}[x,y])[a,b,c]}{\langle yb - xc, xb^2 - ac, x^2b - ya, x^3c - y^2a\rangle}\right).$$

Using *Macaulay 2*, we can also verify that the scheme X is nonsingular;

```
i71 : J + ideal jacobian J == ideal gens ring J

o71 = true

i72 : clearAll
```

Since we have

$$\frac{(\mathbb{Q}[x,y])[a,b,c]}{\langle yb - xc, xb^2 - ac, x^2b - ya, x^3c - y^2a\rangle} \otimes \frac{\mathbb{Q}[x,y]}{\langle x,y\rangle} \simeq \frac{\mathbb{Q}[a,b,c]}{\langle ac\rangle},$$

the fiber over the origin $\langle x, y\rangle$ in $\mathbb{A}^2$ is clearly a union of two copies of $\mathbb{P}^1$ meeting at one point. In particular, the exceptional fiber is not a projective space. □

Many other interesting blow-ups can be found in section II.2 in Eisenbud and Harris [5].

9   A Classic Blow-up

We consider the blow-up of the projective plane $\mathbb{P}^2$ at a point.

Problem. *Show that the following varieties are isomorphic.*

(a) *the image of the rational map from* $\mathbb{P}^2$ *to* $\mathbb{P}^4$ *given by*

$$(r : s : t) \mapsto (r^2 : s^2 : rs : rt : st);$$

(b) *the blow-up of the plane* $\mathbb{P}^2$ *at the point* $(0 : 0 : 1)$;

(c) *the determinantal variety defined by the* 2×2 *minors of the matrix* $\begin{bmatrix} a & c & d \\ b & d & e \end{bmatrix}$ *where* $\mathbb{P}^4 = \mathrm{Proj}\left(k[a, b, c, d, e]\right)$.

This surface is called the cubic scroll *in* $\mathbb{P}^4$.

Solution. We find the ideal in part (a) by elimination theory.

```
i73 : PP4 = QQ[a..e];

i74 : S = QQ[r..t, A..E, MonomialOrder => Eliminate 3];

i75 : I = ideal(A - r^2, B - s^2, C - r*s, D - r*t, E - s*t);

o75 : Ideal of S

i76 : phi = map(PP4, S, matrix{{0_PP4, 0_PP4, 0_PP4}} | vars PP4)

o76 = map(PP4,S,{0, 0, 0, a, b, c, d, e})

o76 : RingMap PP4 <--- S

i77 : surfaceA = phi ideal selectInSubring(1, gens gb I)

                            2
o77 = ideal (c*d - a*e, b*d - c*e, a*b - c )

o77 : Ideal of PP4
```

Next, we determine the surface in part (b). We construct the ideal defining the blow-up of $\mathbb{P}^2$

```
i78 : R = QQ[t, x, y, z, u, v, MonomialOrder => Eliminate 1];

i79 : blowUpIdeal = ideal selectInSubring(1, gens gb ideal(u-t*x,
          v-t*y))

o79 = ideal(y*u - x*v)

o79 : Ideal of R
```

and embed it in $\mathbb{P}^2 \times \mathbb{P}^1$.

```
i80 : PP2xPP1 = QQ[x, y, z, u, v];

i81 : embed = map(PP2xPP1, R, 0 | vars PP2xPP1);

o81 : RingMap PP2xPP1 <--- R

i82 : blowUp = PP2xPP1 / embed(blowUpIdeal);
```

We then map this surface into $\mathbb{P}^5$ using the Segre embedding.

```
i83 : PP5 = QQ[A .. F];
```

```
i84 : segre = map(blowUp, PP5, matrix{{x*u,y*u,z*u,x*v,y*v,z*v}});

o84 : RingMap blowUp <--- PP5

i85 : ker segre

                        2
o85 = ideal (B - D, C*E - D*F, D  - A*E, C*D - A*F)

o85 : Ideal of PP5
```

Note that the image under the Segre map lies on a hyperplane in $\mathbb{P}^5$. To get the desired surface in $\mathbb{P}^4$, we project

```
i86 : projection = map(PP4, PP5, matrix{{a, c, d, c, b, e}})

o86 = map(PP4,PP5,{a, c, d, c, b, e})

o86 : RingMap PP4 <--- PP5

i87 : surfaceB = trim projection ker segre

                                    2
o87 = ideal (c*d - a*e, b*d - c*e, a*b - c )

o87 : Ideal of PP4
```

Finally, we compute the surface in part (c).

```
i88 : determinantal = minors(2, matrix{{a, c, d}, {b, d, e}})

                                    2
o88 = ideal (- b*c + a*d, - b*d + a*e, - d  + c*e)

o88 : Ideal of PP4

i89 : sigma = map( PP4, PP4, matrix{{d, e, a, c, b}});

o89 : RingMap PP4 <--- PP4

i90 : surfaceC = sigma determinantal

                                    2
o90 = ideal (c*d - a*e, b*d - c*e, a*b - c )

o90 : Ideal of PP4
```

By incorporating a permutation of the variables into definition of surfaceC, we obtain the desired isomorphisms

```
i91 : surfaceA == surfaceB

o91 = true

i92 : surfaceB == surfaceC

o92 = true

i93 : clearAll
```

which completes the solution.   □

For more information of the geometry of rational normal scrolls, see Lecture 8 in Harris [6].

10   Fano Schemes

Our final example concerns the family of Fano schemes associated to a flat family of quadrics. Recall that the k-th Fano scheme $F_k(X)$ of a scheme $X \subseteq \mathbb{P}^n$ is the subscheme of the Grassmannian parametrizing k-planes contained in X.

Problem (Exercise IV-69 in [5]). *Consider the one-parameter family of quadrics tending to a double plane with equation*

$$Q = V(tx^2 + ty^2 + tz^2 + w^2) \subseteq \mathbb{P}^3_{\mathbb{Q}[t]} = \mathrm{Proj}\left(\mathbb{Q}[t][x, y, z, w]\right) \ .$$

What is the flat limit of the Fano schemes $F_1(Q_t)$?

Solution. We first compute the ideal defining $F_1(Q_t)$, the scheme parametrizing lines in Q.

```
i94 : PP3 = QQ[t, x, y, z, w];

i95 : Q = ideal( t*x^2+t*y^2+t*z^2+w^2 );

o95 : Ideal of PP3
```

To parametrize a line in our projective space, we introduce indeterminates u, v and $A, \ldots, H$.

```
i96 : R = QQ[t, u, v, A .. H];
```

We then make a map phi from PP3 to R sending the variables to the coordinates of the general point on a line.

```
i97 : phi = map(R, PP3, matrix{{t}} |
              u*matrix{{A, B, C, D}} + v*matrix{{E, F, G, H}});

o97 : RingMap R <--- PP3

i98 : imageFamily = phi Q;

o98 : Ideal of R
```

For a line to belong to Q, the imageFamily must vanish identically. In other words, $F_1(Q)$ is defined by the coefficients of the generators of imageFamily.

```
i99 : coeffOfFamily = contract(matrix{{u^2,u*v,v^2}}, gens imageFamily)

o99 = | tA2+tB2+tC2+D2 2tAE+2tBF+2tCG+2DH tE2+tF2+tG2+H2 |

                    1       3
o99 : Matrix R  <--- R
```

Since we don't need the variables u and v, we get rid of them.

```
i100 : S = QQ[t, A..H];

i101 : coeffOfFamily = substitute(coeffOfFamily, S);

                     1       3
o101 : Matrix S  <--- S

i102 : Sbar = S / (ideal coeffOfFamily);
```

Next, we move to the Grassmannian $\mathbb{G}(1,3) \subset \mathbb{P}^5$. Recall the homogeneous coordinates on $\mathbb{P}^5$ correspond to the 2×2 minors of a 2×4 matrix. We obtain these minors using the `exteriorPower` function in *Macaulay 2*.

```
i103 : psi = matrix{{t}} | exteriorPower(2,
               matrix{{A, B, C, D}, {E, F, G, H}})

o103 = | t -BE+AF -CE+AG -CF+BG -DE+AH -DF+BH -DG+CH |

                1               7
o103 : Matrix Sbar  <--- Sbar

i104 : PP5 = QQ[t, a..f];

i105 : fanoOfFamily = trim ker map(Sbar, PP5, psi);

o105 : Ideal of PP5
```

Now, to answer the question, we determine the limit as t tends to 0.

```
i106 : zeroFibre = trim substitute(saturate(fanoOfFamily, t), {t=>0})

              2   2           2                  ...
o106 = ideal (e*f, d*f, e , f , d*e, a*e + b*f, d , c*d - b*e + a*f, b ...

o106 : Ideal of PP5
```

Let's transpose the matrix of generators so all of its elements are visible on the printed page.

```
i107 : transpose gens zeroFibre

o107 = {-2} | ef       |
       {-2} | df       |
       {-2} | e2       |
       {-2} | f2       |
       {-2} | de       |
       {-2} | ae+bf    |
       {-2} | d2       |
       {-2} | cd-be+af |
       {-2} | bd+ce    |
       {-2} | ad-cf    |
       {-2} | a2+b2+c2 |

                11          1
o107 : Matrix PP5   <--- PP5
```

We see that $F_1(Q_0)$ is supported on the plane conic $\langle d, e, f, a^2 + b^2 + c^2 \rangle$. However, $F_1(Q_0)$ is not reduced; it has multiplicity two. On the other hand, the generic fiber is

```
i108 : oneFibre = trim substitute(saturate(fanoOfFamily, t), {t => 1})

               2   2   2                               ...
o108 = ideal (a*e + b*f, d + e + f , c*d - b*e + a*f, b*d + c*e, a*d ...

o108 : Ideal of PP5

i109 : oneFibre == intersect(ideal(c-d, b+e, a-f, d^2+e^2+f^2),
            ideal(c+d, b-e, a+f, d^2+e^2+f^2))

o109 = true
```

Hence, for $t \neq 0$, $F_1(Q_t)$ is the union of two conics lying in complementary planes and $F_1(Q_0)$ is the double conic obtained when the two conics move together.   □

References

1. Neil Chriss and Victor Ginzburg: *Representation theory and complex geometry.* Birkhäuser Boston Inc., Boston, MA, 1997.
2. David Cox, John Little, and Donal O'Shea: *Ideals, varieties, and algorithms.* Springer-Verlag, New York, second edition, 1997. An introduction to computational algebraic geometry and commutative algebra.
3. David Cox, John Little, and Donal O'Shea: *Using algebraic geometry.* Springer-Verlag, New York, 1998.
4. David Eisenbud: *Commutative algebra with a view toward algebraic geometry.* Springer-Verlag, New York, 1995.
5. David Eisenbud and Joe Harris: *The geometry of schemes.* Springer-Verlag, New York, 2000.
6. Joe Harris: *Algebraic geometry, A first course.* Springer-Verlag, New York, 1995.

Part II

Mathematical Computations

Monomial Ideals

Serkan Hoşten and Gregory G. Smith

Monomial ideals form an important link between commutative algebra and combinatorics. In this chapter, we demonstrate how to implement algorithms in *Macaulay 2* for studying and using monomial ideals. We illustrate these methods with examples from combinatorics, integer programming, and algebraic geometry.

An ideal I in $S = \mathbb{Q}[x_1, \ldots, x_n]$ is called a monomial ideal if it satisfies any of the following equivalent conditions:

(a) I is generated by monomials,

(b) if $f = \sum_{\alpha \in \mathbb{N}^n} k_\alpha x^\alpha$ belongs to I then $x^\alpha \in I$ whenever $k_\alpha \neq 0$,

(c) I is torus-fixed; in other words, if $(c_1, \ldots, c_n) \in (\mathbb{Q}^*)^n$, then I is fixed under the action $x_i \mapsto c_i x_i$ for all i.

It follows that a monomial ideal is uniquely determined by the monomials it contains. Most operations are far simpler for a monomial ideal than for an ideal generated by arbitrary polynomials. In particular, many invariants can be effectively determined for monomial ideals. As a result, one can solve a broad collection of problems by reducing to or encoding data in a monomial ideal. The aim of this chapter is to develop the computational aspects of monomial ideals in *Macaulay 2* and demonstrate a range of applications.

This chapter is divided into five sections. Each section begins with a discussion of a computational procedure involving monomial ideals. Algorithms are presented as *Macaulay 2* functions. We illustrate these methods by solving problems from various areas of mathematics. In particular, we include the *Macaulay 2* code for generating interesting families of monomial ideals. The first section introduces the basic functions on monomial ideals in *Macaulay 2*. To demonstrate these functions, we use the Stanley-Reisner ideal associated to a simplicial complex to compute its f-vector. Next, we present two algorithms for finding a primary decomposition of a monomial ideal. In a related example, we use graph ideals to study the complexity of determining the codimension of a monomial ideal. The third section focuses on the standard pairs of a monomial ideal; two methods are given for finding the set of standard pairs. As an application, we use standard pairs to solve integer linear programming problems. The fourth section examines Borel-fixed ideals and generic initial ideals. Combining these constructions with distractions, we demonstrate that the Hilbert scheme $\mathrm{Hilb}^{4t+1}(\mathbb{P}^4)$ is connected. Finally, we look at the chains of associated primes in various families of monomial ideals.

1   The Basics of Monomial Ideals

Creating monomial ideals in *Macaulay 2* is analogous to creating general ideals. The monomial ideal generated by a sequence or list of monomials can be constructed with the function `monomialIdeal`.

```
i1 : S = QQ[a, b, c, d];

i2 : I = monomialIdeal(a^2, a*b, b^3, a*c)

                2         3
o2 = monomialIdeal (a , a*b, b , a*c)

o2 : MonomialIdeal of S

i3 : J = monomialIdeal{a^2, a*b, b^2}

                2         2
o3 = monomialIdeal (a , a*b, b )

o3 : MonomialIdeal of S
```

The type `MonomialIdeal` is the class of all monomial ideals. If an entry in the sequence or list is not a single monomial, then `monomialIdeal` takes only the leading monomial; recall that every polynomial ring in *Macaulay 2* is equipped with a monomial ordering.

```
i4 : monomialIdeal(a^2+a*b, a*b+3, b^2+d)

                2         2
o4 = monomialIdeal (a , a*b, b )

o4 : MonomialIdeal of S
```

There are also several methods of associating a monomial ideal to an arbitrary ideal in a polynomial ring. The most important of these is the initial ideal — the monomial ideal generated by the leading monomials of all elements in the given ideal. When applied to an `Ideal`, the function `monomialIdeal` returns the initial ideal.

```
i5 : K = ideal(a^2, b^2, a*b+b*c)

           2  2
o5 = ideal (a , b , a*b + b*c)

o5 : Ideal of S

i6 : monomialIdeal K

                2         2      2
o6 = monomialIdeal (a , a*b, b , b*c )

o6 : MonomialIdeal of S
```

This is equivalent to taking the leading monomials of a Gröbner basis for K. In our example, the given generators for K are not a Gröbner basis.

```
i7 : monomialIdeal gens K

                2         2
o7 = monomialIdeal (a , a*b, b )

o7 : MonomialIdeal of S
```

One can also test if a general ideal is generated by monomials with the function isMonomialIdeal.

```
i8 : isMonomialIdeal K

o8 = false

i9 : isMonomialIdeal ideal(a^5, b^2*c, d^11)

o9 = true
```

The usual algebraic operations on monomial ideals are the same as on general ideals. For example, we have

```
i10 : I+J

                 2         2
o10 = monomialIdeal (a , a*b, b , a*c)

o10 : MonomialIdeal of S
```

Example: Stanley-Reisner Ideals and f-vectors

Radical monomial ideals — ideals generated by squarefree monomials — have a beautiful combinatorial interpretation in terms of simplicial complexes. More explicitly, a simplicial complex Δ on the vertex set $\{x_1, \ldots, x_n\}$ corresponds to the ideal I_Δ in $S = \mathbb{Q}[x_1, \ldots, x_n]$ generated by all monomials $x_{i_1} \cdots x_{i_p}$ such that $\{x_{i_1}, \ldots, x_{i_p}\} \notin \Delta$. The ideal I_Δ is called the Stanley-Reisner ideal of Δ.

To illustrate the connections between Stanley-Reisner ideals and simplicial complexes, we consider the f-vector. Perhaps the most important invariant of a simplicial complex, the f-vector of a d-dimensional simplicial complex Δ is $(f_0, f_1, \ldots, f_d) \in \mathbb{N}^{d+1}$, where f_i denotes the number of i-dimensional faces in Δ. From the monomial ideal point of view, the f-vector is encoded in the Hilbert series of the quotient ring S/I_Δ as follows:

Theorem 1.1. *If Δ is a simplicial complex with f-vector $(f_0, \ldots, f_d)$, then the Hilbert series of S/I_Δ is*

$$H_{S/I_\Delta}(t) = \sum_{i=-1}^{d} \frac{f_i t^{i+1}}{(1-t)^{i+1}},$$

where $f_{-1} = 1$.

Proof. Following Stanley [24], we work with the fine grading and then specialize. The fine grading of S is the $\mathbb{Z}^n$-grading defined by $\deg x_i = \mathbf{e}_i \in \mathbb{Z}^n$, where $\mathbf{e}_i$ is the i-th standard basis vector. The support of a monomial x^α is defined to be the set $\mathrm{supp}(x^\alpha) = \{x_i : \alpha_i > 0\}$. Observe that $x^\alpha \neq 0$ in S/I_Δ if and only if $\mathrm{supp}(x^\alpha) \in \Delta$. Moreover, the nonzero monomials x^α form a

$\mathbb{Q}$-basis of S/I_Δ. By counting such monomials according to their support, we obtain the following expression for the Hilbert series with the fine grading:

$$H_{S/I_\Delta}(\mathbf{t}) = \sum_{F \in \Delta} \sum_{\substack{\alpha \in \mathbb{N}^n \\ \operatorname{supp}(x^\alpha)=F}} \mathbf{t}^\alpha = \sum_{F \in \Delta} \prod_{x_i \in F} \frac{t_i}{1-t_i}.$$

Finally, by replacing each t_i with t, we complete the proof.   □

Since $H_{S/I_\Delta}(t)$ is typically expressed in the form $\frac{h_0+h_1t+\cdots+h_dt^d}{(1-t)^{d+1}}$, we can obtain the f-vector by using the identity $\sum_i h_i t^i = \sum_{j=0}^d f_{j-1}t^j(1-t)^{d-j}$. In particular, we can compute f-vectors from Stanley-Reisner ideals as follows:

```
i11 : fvector = I -> (
          R := (ring I)/I;
          d := dim R;
          N := poincare R;
          t := first gens ring N;
          while 0 == substitute(N, t => 1) do N = N // (1-t);
          h := apply(reverse toList(0..d), i -> N_(t^i));
          f := j -> sum(0..j+1, i -> binomial(d-i, j+1-i)*h#(d-i));
          apply(toList(0..d-1), j -> f(j)));
```

For example, we can demonstrate that the f-vector of the octahedron is $(6, 12, 8)$.

```
i12 : S = QQ[x_1 .. x_6];

i13 : octahedron = monomialIdeal(x_1*x_2, x_3*x_4, x_5*x_6)

o13 = monomialIdeal (x x , x x , x x )
                      1 2   3 4   5 6

o13 : MonomialIdeal of S

i14 : fvector octahedron

o14 = {6, 12, 8}

o14 : List
```

More generally, we can recursively construct simplicial 2-spheres with $f_0 \geq 4$, starting with the tetrahedron, by pulling a point in the relative interior of a facet. This procedure leads to the following family:

```
i15 : simplicial2sphere = v -> (
          S := QQ[x_1..x_v];
          if v === 4 then monomialIdeal product gens S
          else (
              L := {};
              scan(1..v-4, i -> L = L | apply(v-i-3,
                      j -> x_i*x_(i+j+4)));
              scan(2..v-3, i -> L = L | {x_i*x_(i+1)*x_(i+2)});
              monomialIdeal L));

i16 : apply({4,5,6,7,8}, j -> fvector simplicial2sphere(j))

o16 = {{4, 6, 4}, {5, 9, 6}, {6, 12, 8}, {7, 15, 10}, {8, 18, 12}}

o16 : List
```

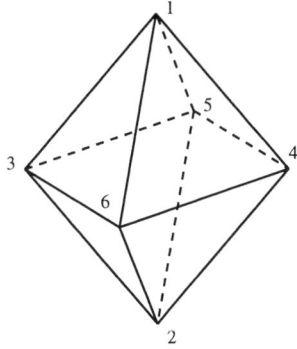

Fig. 1. The octahedron

In fact, it follows from Euler's formula that the f-vector of any simplicial 2-sphere has the form $(v, 3v - 6, 2v - 4)$ for $v \geq 4$. The problem of characterizing the f-vectors for triangulations of d-spheres is open for $d \geq 3$. One of the most important results in this direction is the upper bound theorem for simplicial spheres (Corollary 5.4.7 Bruns and Herzog [6]) which states that the cyclic polytope has the maximal number of i-faces for all i. We point out that Stanley's proof of this theorem depends heavily on these methods from commutative algebra.

On the other hand, the f-vectors for several major classes of simplicial complexes have been characterized. The Kruskal-Katona theorem (Theorem 8.32 in Ziegler [29]) gives necessary and sufficient conditions for a sequence of nonnegative integers to be an f-vector of a simplicial complex. Stanley [24] describes the f-vectors of pure shellable complexes and Cohen-Macaulay complexes. Given the Betti numbers of a simplicial complex, Björner and Kalai [5] specify the f-vectors. Finally, the g-theorem (Theorem 8.35 in Ziegler [29]) characterizes the f-vectors for boundary complexes of a simplicial convex polytope.

For a further study of Stanley-Reisner ideals see Bruns and Herzog [6] and Stanley [24]. For more information of f-vectors, see Ziegler [29] and Björner [4].

2 Primary Decomposition

A primary decomposition of an ideal I is an expression of I as a finite intersection of primary ideals; an ideal J is called primary if $r_1 r_2 \in J$ implies either $r_1 \in J$ or $r_2^\ell \in J$ for some $\ell > 0$. Providing an algorithm for computing the primary decomposition of an arbitrary ideal in a polynomial ring is quite difficult. However, for monomial ideals, there are two algorithms which are relatively simple to describe.

We first present a recursive method for generating an irreducible primary decomposition. It is based on the following two observations.

Lemma 2.1. *Let I be a monomial ideal in $S = \mathbb{Q}[x_1, \ldots, x_n]$.*

(1) *If I is generated by pure powers of a subset of the variables, then it is a primary ideal.*
(2) *If r is minimal generator of I such that $r = r_1 r_2$ where r_1 and r_2 are relatively prime, then $I = (I + \langle r_1 \rangle) \cap (I + \langle r_2 \rangle).$*

Proof. (1) This follows immediately from the definition of primary. (2) Since I is a monomial ideal, it is enough to show that I and $(I + \langle r_1 \rangle) \cap (I + \langle r_1 \rangle)$ contain the same monomials. A monomial r' belongs to $(I + \langle r_j \rangle)$ if and only if $r' \in I$ or r_j divides r'. Because r_1 and r_2 are relative prime, we have

$$r' \in (I + \langle r_1 \rangle) \cap (I + \langle r_1 \rangle) \Leftrightarrow r' \in I \text{ or } r_1 r_2 \text{ divides } r' \Leftrightarrow r' \in I. \qquad \square$$

The following is an implementation of the resulting algorithm:

```
i17 : supp = r -> select(gens ring r, e -> r % e == 0);

i18 : monomialDecompose = method();

i19 : monomialDecompose List := L -> (
          P := select(L, I -> all(first entries gens I,
                  r -> #supp(r) < 2) === false);
          if #P > 0 then (
              I := first P;
              m := first select(first entries gens I,
                  r -> #supp(r) > 1);
              E := first exponents m;
              i := position(E, e -> e =!= 0);
              r1 := product apply(E_{0..i}, (gens ring I)_{0..i},
                  (j, r) -> r^j);
              r2 := m // r1;
              monomialDecompose(delete(I, L) | {I+monomialIdeal(r1),
                  I+monomialIdeal(r2)}))
          else L);

i20 : monomialDecompose MonomialIdeal := I -> monomialDecompose {I};
```

Here is a small example illustrating this method.

```
i21 : S = QQ[a,b,c,d];

i22 : I = monomialIdeal(a^3*b, a^3*c, a*b^3, b^3*c, a*c^3, b*c^3)

                    3      3    3    3     3     3
o22 = monomialIdeal (a b, a*b , a c, b c, a*c , b*c )

o22 : MonomialIdeal of S

i23 : P = monomialDecompose I;

i24 : scan(P, J -> << endl << J << endl);

monomialIdeal (b, c)

monomialIdeal (a, c)
```

```
              3   3   3
monomialIdeal (a , b , c )

monomialIdeal (a, b)

              3       3
monomialIdeal (a , b, c )

monomialIdeal (a, b)

                  3   3
monomialIdeal (a, b , c )

i25 : I == intersect(P)

o25 = true
```

As we see from this example, this procedure doesn't necessarily yield an irredundant decomposition.

The second algorithm for finding a primary decomposition of a monomial ideal I is based on the Alexander dual of I. The Alexander dual was first introduced for squarefree monomial ideals. In this case, it is the monomial ideal of the dual of the simplicial complex Δ corresponding to I. By definition the dual complex of Δ is $\Delta^\vee = \{F : F^c \notin \Delta\}$, where $F^c = \{x_1, \ldots, x_n\} \setminus F$. The following general definition appears in Miller [16], [17]. If $I \subseteq \mathbb{Q}[x_1, \ldots, x_n]$ is a monomial ideal and x^λ is the least common multiple of the minimal generators of I, then the Alexander dual of I is

$$I^\vee = \left\langle \prod_{\beta_i > 0} x_i^{\lambda_i + 1 - \beta_i} : \begin{array}{l} \langle x_i^{\beta_i} : \beta_i \geq 1 \rangle \text{ is an irredundant} \\ \text{irreducible component of } I \end{array} \right\rangle.$$

In particular, the minimal generators of $I^\vee$ correspond to the irredundant irreducible components of I. The next proposition provides a useful way of computing $I^\vee$ given a set of generators for I.

Proposition 2.2. *If I is a monomial ideal and x^λ is the least common multiple of the minimal generators of I, then the generators for $I^\vee$ are those generators of the ideal $\left(\langle x_1^{\lambda_1+1}, \ldots, x_n^{\lambda_n+1} \rangle : I \right)$ that are not divisible by $x_i^{\lambda_i+1}$ for $1 \leq i \leq n$.*

Proof. See Theorem 2.1 in Miller [16]. □

Miller's definition of Alexander dual is even more general than the one above. The resulting algorithm for computing this general Alexander dual and primary decomposition are implemented in *Macaulay 2* as follows. For the Alexander dual we use, the list a that appears as an input argument for dual should be list of exponents of the least common multiple of the minimal generators of I.

```
i26 : code(dual, MonomialIdeal, List)
```

```
o26 = -- ../../../m2/monideal.m2:260-278
     dual(MonomialIdeal, List) := (I,a) -> ( -- Alexander dual
          R := ring I;
          X := gens R;
          aI := lcmOfGens I;
          if aI =!= a then (
               if #aI =!= #a
               then error (
                    "expected list of length ",
                    toString (#aI));
               scan(a, aI,
                    (b,c) -> (
                         if b<c then
                         error "exponent vector not large enough"
                    ));
          );
          S := R/(I + monomialIdeal apply(#X, i -> X#i^(a#i+1)));
          monomialIdeal contract(
               lift(syz transpose vars S, R),
               product(#X, i -> X#i^(a#i))))

i27 : code(primaryDecomposition, MonomialIdeal)

o27 = -- ../../../m2/monideal.m2:286-295
     primaryDecomposition MonomialIdeal := (I) -> (
          R := ring I;
          aI := lcmOfGens I;
          M := first entries gens dual I;
          L := unique apply(#M, i -> first exponents M_i);
          apply(L, i -> monomialIdeal apply(#i, j -> (
                         if i#j === 0 then 0_R
                         else R_j^(aI#j+1-i#j)
                    )))
          )
```

This direct algorithm is more efficient than our recursive algorithm. In particular, it gives an irredundant decomposition. For example, when we use it to determine a primary decomposition for the ideal I above, we obtain

```
i28 : L = primaryDecomposition I;

i29 : scan(L, J -> << endl << J << endl);

                 3   3   3
monomialIdeal (a , b , c )

monomialIdeal (b, c)

monomialIdeal (a, b)

monomialIdeal (a, c)

i30 : I == intersect L

o30 = true
```

For a family of larger examples, we consider the tree ideals:

$$\left\langle \left(\prod_{i \in F} x_i\right)^{n-|F|+1} : \emptyset \neq F \subseteq \{x_1, \ldots, x_n\}\right\rangle .$$

These ideals are so named because their standard monomials (the monomials not in the ideal) correspond to trees on $n + 1$ labeled vertices. We determine the number of irredundant irreducible components as follows:

```
i31 : treeIdeal = n -> (
          S = QQ[vars(0..n-1)];
          L := delete({}, subsets gens S);
          monomialIdeal apply(L, F -> (product F)^(n - #F +1)));

i32 : apply(2..6, i -> #primaryDecomposition treeIdeal i)

o32 = (2, 6, 24, 120, 720)

o32 : Sequence
```

Example: Graph Ideals and Complexity Theory

Monomial ideals also arise in graph theory. Given a graph G with vertices $\{x_1, \ldots, x_n\}$, we associate the ideal I_G in $\mathbb{Q}[x_1, \ldots, x_n]$ generated by the quadratic monomials $x_i x_j$ such that x_i is adjacent to x_j. The primary decomposition of I_G is related to the graph G as follows. Recall that a subset $F \subseteq \{x_1, \ldots, x_n\}$ is called a *vertex cover* of G if each edge in G is incident to at least one vertex in F.

Lemma 2.3. *If G is a graph and $\mathcal{C}$ is the set of minimal vertex covers of G then the irreducible irredundant primary decomposition of I_G is $\bigcap_{F \in \mathcal{C}} P_{F^c}$, where P_{F^c} is the prime ideal $\langle x_i : x_i \notin F^c \rangle = \langle x_i : x_i \in F \rangle$.*

Proof. Since each generator of I_G corresponds to an edge in G, it follows from the `monomialDecompose` algorithm that I_G has an irreducible primary decomposition of the form: $I_G = \bigcap P_{F^c}$, where F is a vertex cover. To obtain an irredundant decomposition, one clearly needs only the minimal vertex covers. □

As an application of graph ideals, we examine the complexity of determining the codimension of a monomial ideal. In fact, following Bayer and Stillman [3], we prove

Proposition 2.4. *The following decision problem is NP-complete:*

> *Given a monomial ideal $I \subseteq \mathbb{Q}[x_1, \ldots, x_n]$ and* $\qquad$ (CODIM)
> *$m \in \mathbb{N}$, is $\operatorname{codim} I \leq m$?*

By definition, a decision problem is NP-complete if all other problems in the class NP can be reduced to it. To prove that a particular problem is NP-complete, it suffices to show: (1) the problem belongs to the class NP; (2) some known NP-complete problem reduces to the given decision problem (see Lemma 2.3 in Garey and Johnson [8]). One of the "standard NP-complete" problems (see section 3.1 in Garey and Johnson [8]) is the following:

> Given a graph G and $m \in \mathbb{N}$, is there a vertex $\qquad$ (VERTEX COVER)
> cover F such that $|F| \leq m$?

Proof of Proposition. (1) Observe that a monomial ideal I has codimension at most m if and only if $I \subseteq P_{F^c}$ for some F with $|F| \leq m$. Now, if I has codimension at most m, then given an appropriate choice of F, one can verify in polynomial time that $I \subseteq P_{F^c}$ and $|F| \leq m$. Therefore, the CODIM problem belongs to the class NP.

(2) Lemma 2.3 implies that I_G has codimension m if and only if G has a vertex cover of size at most m. In particular, the VERTEX COVER problem reduces to the CODIM problem.   □

Thus, assuming P $\neq$ NP, there is no polynomial time algorithm for finding the codimension of a monomial ideal. Nevertheless, we can effectively compute the codimension for many interesting examples.

To illustrate this point, we consider the following family of examples. Let $S = \mathbb{Q}[X]$ denote the polynomial ring generated by the entries of a generic $m \times n$ matrix $X = [x_{i,j}]$. Let I_k be the ideal generated by the $k \times k$ minors of X. Since the Hilbert function of S/I_k equals the Hilbert function of $S/\operatorname{in}(I_k)$ (see Theorem 15.26 in Eisenbud [7]), we can determine the codimension I_k by working with the monomial ideal $\operatorname{in}(I_k)$. Because Sturmfels [25] shows that the set of $k \times k$-minors of X is the reduced Gröbner basis of I_k with respect to the lexicographic term order induced from the variable order

$$x_{1,n} > x_{1,n-1} > \cdots > x_{1,1} > x_{2,n} > \cdots > x_{2,1} > \cdots > x_{m,n} > \cdots > x_{m,1},$$

we can easily calculate $\operatorname{in}(I_k)$. In particular, in *Macaulay 2* we have

```
i33 : minorsIdeal = (m,n,k) -> (
            S := QQ[x_1..x_(m*n), MonomialOrder => Lex];
            I := minors(k, matrix table(m, n, (i,j) -> x_(i*n+n-j)));
            forceGB gens I;
            I);
```

```
i34 : apply(2..8, i -> time codim monomialIdeal minorsIdeal(i,2*i,2))
      -- used 0.02 seconds
      -- used 0.05 seconds
      -- used 0.1 seconds
      -- used 0.36 seconds
      -- used 1.41 seconds
      -- used 5.94 seconds
      -- used 25.51 seconds
```

```
o34 = (3, 10, 21, 36, 55, 78, 105)
```

```
o34 : Sequence
```

The properties of I_k are further developed in chapter 11 of Sturmfels [26] and chapter 7 of Bruns and Herzog [6]

For more on the relationships between a graph and its associated ideal, see Villarreal [28], Simis, Vasconcelos and Villarreal [23], and Ohsugi and Hibi [19].

```
i35 : erase symbol x;
```

3   Standard Pairs

In this section, we examine a combinatorial object associated to a monomial ideal. In particular, we present two algorithms for computing the standard pairs of a monomial ideal from its minimal generators. Before giving the definition of a standard pair, we consider an example.

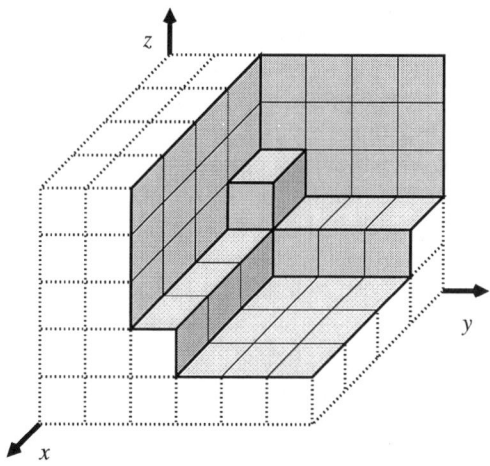

Fig. 2. Staircase diagram for $I = \langle xy^3z, xy^2z^2, y^3z^2, y^2z^3 \rangle$

Example 3.1. Let $I = \langle xy^3z, xy^2z^2, y^3z^2, y^2z^3 \rangle$ in $\mathbb{Q}[x, y, z]$. We identify the monomials in $\mathbb{Q}[x, y, z]$ with the lattice points in $\mathbb{N}^3$; see Figure 2. The standard monomials of I, those monomials which are not in I, can be enumerated as follows: (i) monomials corresponding to lattice points in the xy-plane, (ii) monomials corresponding to lattice points in the xz-plane, (iii) monomials corresponding to lattice points in the plane parallel to the xz-plane containing $(0, 1, 0)$, (iv) monomials corresponding to lattice points on the line parallel to the y-axis containing $(0, 0, 1)$, (v) monomials corresponding to lattices point on the line parallel to the x-axis containing $(0, 2, 1)$, and (vi) the monomial y^2z^2.

Following Sturmfels, Trung and Vogel [27], we make the following definitions. Given a monomial x^α and a subset $F \subseteq \{x_1, \dots, x_n\}$, we index the set of monomials of the form $x^\alpha \cdot x^\beta$ where $\mathrm{supp}(x^\beta) \subseteq F$ by the pair (x^α, F). A standard pair of a monomial ideal I is a pair (x^α, F) satisfying the following three conditions:

(1) $\mathrm{supp}(x^\alpha) \cap F = \emptyset$,

(2) all of the monomials represented by this pair are standard, and
(3) $(x^\alpha, F) \not\subseteq (x^\beta, G)$ for any other pair (x^β, G) satisfying the first two conditions.

Hence, the six standard pairs

$$(1, \{x, y\}), (1, \{x, z\}), (y, \{x, z\}), (z, \{y\}), (y^2 z, \{x\}), (y^2 z^2, \emptyset)$$

in Example 3.1 correspond to (i)–(vi).

Observe that the set of standard pairs of I gives an irreducible decomposition of $I = \bigcap \langle x_i^{\alpha_i + 1} : x_i \notin F \rangle$, where the intersection is over all standard pairs (x^α, F). Moreover, the prime ideal $P_F := \langle x_i : x_i \notin F \rangle$ is an associated prime of I if and only if there exists a standard pair of the form $(\bullet, F)$; see Sturmfels, Trung and Vogel [27] for details.

Our first algorithm for computing the set of standard pairs is taken from Hoşten and Thomas [15]. The ideas behind it are as follows: given a witness $w_1 = x^\alpha$ for the associated prime $P_F := \langle x_i : x_i \notin F \rangle$, that is $(I : x^\alpha) = P_F$, set $w_2 = \prod_{x_i \in \text{supp}(w_1) \cap F^c} x_i^{\alpha_i}$. It follows that (w_2, F) is a standard pair of I. Now, consider the standard pairs of the slightly larger ideal $I + \langle w_1 \rangle$. Clearly (w_2, F) is not a standard pair of this ideal because w_1 "destroys" it. This larger ideal might have standard pairs which cover standard monomials in (w_2, F) that are not in the pair (w_1, F). However, all other standard pairs are the same as the original ideal I. Thus, the problem of finding all standard pairs of I reduces to determining if a standard pair of $I + \langle w_1 \rangle$ is a standard pair for I. To decide if a pair (x^β, G) of $I + \langle w_1 \rangle$ is a standard pair of I, we first check that P_F is an associated prime of I. If this is true, we determine if (w_2, F) is covered by (x^β, G).

The *Macaulay 2* version of this algorithm takes the following form:

```
i36 : stdPairs = I -> (
          S := ring I;
          X := gens S;
          std := {};
          J := I;
          while J != S do (
              w1 := 1_S;
              F := X;
              K := J;
              while K != 0 do (
                  g1 := (ideal mingens ideal K)_0;
                  x := first supp g1;
                  w1 = w1 * g1 // x;
                  F = delete(x, F);
                  K = K : monomialIdeal(g1 // x);
                  L := select(first entries gens K,
                      r -> not member(x, supp r));
                  if #L > 0 then K = monomialIdeal L
                  else K = monomialIdeal 0_S;);
              w2 := w1;
              scan(X, r -> if not member(r, supp w1) or member(r, F)
                  then w2 = substitute(w2, {r => 1}));
              P := monomialIdeal select(X, r -> not member(r, F));
```

```
        if (I:(I:P) == P) and (all(std, p ->
                    (w2 % (first p) != 0) or not
                    isSubset(supp(w2 // first p) | F, last p)))
        then std = std | {{w2, F}};
        J = J + monomialIdeal(w1););
    std);
```

We can compute the standard pairs of Example 3.1 using this *Macaulay 2*
function:

```
i37 : S = QQ[x,y,z];

i38 : I = monomialIdeal(x*y^3*z, x*y^2*z^2, y^3*z^2, y^2*z^3);

o38 : MonomialIdeal of S

i39 : scan(time stdPairs I, P -> << endl << P << endl);
      -- used 0.66 seconds

{y, {x, z}}

{1, {x, z}}

  2 2
{y z , {}}

{z, {y}}

  2
{y z, {x}}

{1, {x, y}}
```

Our second algorithm is taken from section 3.2 of Saito, Sturmfels and
Takayama [22]. The proposition below provides the main ingredient for this
algorithm. If I is a monomial ideal and $F \subseteq \{x_1, \ldots, x_n\}$, we write I_F for the
monomial ideal in $\mathbb{Q}[x_i : x_i \notin F]$ obtained by replacing each $x_i \in F$ with 1
in every minimal generator of I.

Proposition 3.2. *For (x^α, F) to be a standard pair of I, it is necessary and
sufficient that $(x^\alpha, \emptyset)$ be a standard pair of I_F.*

Proof. Lemma 3.1 in Sturmfels, Trung and Vogel [27]. $\qquad\qquad$ □

The definition of a standard pair implies that $(x^\alpha, \emptyset)$ is a standard pair
of I_F if and only if x^α is one of the finitely many monomials contained in
$(I_F : P_F^\infty)$ but not contained in I_F, where $P_F = \langle x_i : x_i \notin F \rangle$. Since ideal
quotients and saturations are implemented in *Macaulay 2*, this reduces the
problem to finding a set D which contains F for every associated prime P_F
of I. One approach is to simply compute the associated primes of I from a
primary decomposition.

The method **standardPairs** uses this algorithm to determine the set of
standard pairs for a monomial ideal.

```
i40 : code(standardPairs, MonomialIdeal, List)
```

```
o40 = -- ../../../m2/monideal.m2:318-341
     standardPairs(MonomialIdeal, List) := (I,D) -> (
          R := ring I;
          X := gens R;
          S := {};
          k := coefficientRing R;
          scan(D, L -> (
                    Y := X;
                    m := vars R;
                    Lset := set L;
                    Y = select(Y, r -> not Lset#?r);
                    m = substitute(m, apply(L, r -> r => 1));
                    -- using monoid to create ring to avoid
                    -- changing global ring.
                    A := k (monoid [Y]);
                    phi := map(A, R, substitute(m, A));
                    J := ideal mingens ideal phi gens I;
                    Jsat := saturate(J, ideal vars A);
                    if Jsat != J then (
                         B := flatten entries super basis (
                              trim (Jsat / J));
                         psi := map(R, A, matrix{Y});
                         S = join(S, apply(B, b -> {psi(b), L}));
                    )));
          S)

i41 : time standardPairs I;
     -- used 0.83 seconds
```

As an example, we will compute the standard pairs of the permutahedron ideal. Let $S = \mathbb{Q}[x_1, \ldots, x_n]$ and let $\mathfrak{S}_n$ be the symmetric group of order n. We write ρ for the vector $(1, 2, \ldots, n)$ and $\sigma(\rho)$ for the vector obtained by applying $\sigma \in \mathfrak{S}_n$ to the coordinates of ρ. The n-th permutahedron ideal is $\langle x^{\sigma(\rho)} : \sigma \in \mathfrak{S}_n \rangle$. We compute the number of standard pairs for $2 \leq n \leq 5$.

```
i42 : permutohedronIdeal = n -> (
          S := QQ[X_1..X_n];
          monomialIdeal terms det matrix table(n ,gens S,
               (i,r) -> r^(i+1)));

i43 : L = apply({2,3,4,5}, j -> standardPairs(permutohedronIdeal(j)));

i44 : apply(L, i -> #i)

o44 = {3, 10, 53, 446}

o44 : List

i45 : erase symbol x; erase symbol z;
```

Example: Integer Programming Problems

As an application of standard pairs, we show how to solve integer linear programming problems. Let A be a $d \times n$ matrix of nonnegative integers, let $\omega \in \mathbb{R}^n$ and fix $\beta \in \mathbb{Z}^d$. We focus on the following optimization problem

$$\mathrm{IP}_{A,\omega}(\beta) : \quad \text{minimize } \omega \cdot \alpha \text{ subject to } A\alpha = \beta, \; \alpha \in \mathbb{N}^n.$$

We view this integer linear program as a family depending on the vector β. The algorithm we present for solving $\mathrm{IP}_{A,\omega}(\beta)$ depends on the proposition below.

The toric ideal $I_A \subseteq S = \mathbb{Q}[x_1, \ldots, x_n]$ associated to A is the binomial ideal generated by $x^\gamma - x^\delta$ where $\gamma, \delta \in \mathbb{N}^n$ and $A\gamma = A\delta$. We write $\mathrm{in}_\omega(I_A)$ for the initial ideal of I_A with respect to the following order:

$$x^\gamma \prec_\omega x^\delta \iff \begin{cases} \omega \cdot \gamma < \omega \cdot \delta & \text{or} \\ \omega \cdot \gamma = \omega \cdot \delta & \text{and } x^\alpha \prec_{\mathrm{rlex}} x^\gamma. \end{cases}$$

For more information on toric ideals and their initial ideals see Sturmfels [26].

Proposition 3.3. (1) *A monomial x^α is a standard monomial of $\mathrm{in}_\omega(I_A)$ if and only if α is the optimal solution to the integer program $\mathrm{IP}_{A,\omega}(A\alpha)$.*
(2) *If $(\bullet, F)$ is a standard pair of $\mathrm{in}_\omega(I_A)$, then the columns of A corresponding to F are linearly independent.*

Proof. See Proposition 2.1 in Hoşten and Thomas [14] for the proof of the first statement. The second statement follows from Corollary 2.9 of the same article. □

The first statement implies that the standard pairs of $\mathrm{in}_\omega(I_A)$ cover all optimal solutions to all integer programs in $\mathrm{IP}_{A,\omega}$. If α is the optimal solution to $\mathrm{IP}_{A,\omega}(\beta)$ covered by the standard pair (x^γ, F), then the second statement guarantees there exists a unique $\delta \in \mathbb{N}^n$ such that $A\delta = \beta - A\gamma$. Therefore, $\alpha = \delta + \gamma$. We point out that the complexity of this algorithm is dominated by determining the set of standard pairs of $\mathrm{in}_\omega(I_A)$ which depends only on A and ω. As a result, this method is particularly well suited to solving $\mathrm{IP}_{A,\omega}(\beta)$ as β varies.

To implement this algorithm in *Macaulay 2*, we need a function which returns the toric ideal I_A. Following Algorithm 12.3 in Sturmfels [26], we have

```
i47 : toBinomial = (b, S) -> (
          pos := 1_S;
          neg := 1_S;
          scan(#b, i -> if b_i > 0 then pos = pos*S_i^(b_i)
                        else if b_i < 0 then neg = neg*S_i^(-b_i));
          pos - neg);

i48 : toricIdeal = (A, omega) -> (
          n := rank source A;
          S = QQ[x_1..x_n, Weights => omega, MonomialSize => 16];
          B := transpose matrix syz A;
          J := ideal apply(entries B, b -> toBinomial(b, S));
          scan(gens S, r -> J = saturate(J, r));
          J);
```

Thus, we can solve $\mathrm{IP}_{A,\omega}(\beta)$ using the following function.

```
i49 : IP = (A, omega, beta) -> (
          std := standardPairs monomialIdeal toricIdeal(A, omega);
          n := rank source A;
          alpha := {};
          Q := first select(1, std, P -> (
              F := apply(last P, r -> index r);
              gamma := transpose matrix exponents first P;
              K := transpose syz (submatrix(A,F) | (A*gamma-beta));
              X := select(entries K, k -> abs last(k) === 1);
              scan(X, k -> if all(k, j -> j>=0) or all(k, j -> j<=0)
                  then alpha = apply(n, j -> if member(j, F)
                      then last(k)*k_(position(F, i -> i === j))
                      else 0));
              #alpha > 0));
          if #Q > 0 then (matrix {alpha})+(matrix exponents first Q)
          else 0);
```

We illustrate this with some examples.

```
i50 : A = matrix{{1,1,1,1,1},{1,2,4,5,6}}

o50 = | 1 1 1 1 1 |
      | 1 2 4 5 6 |

                2        5
o50 : Matrix ZZ  <--- ZZ

i51 : w1 = {1,1,1,1,1};

i52 : w2 = {2,3,5,7,11};

i53 : b1 = transpose matrix{{3,9}}

o53 = | 3 |
      | 9 |

                2        1
o53 : Matrix ZZ  <--- ZZ

i54 : b2 = transpose matrix{{5,16}}

o54 = | 5  |
      | 16 |

                2        1
o54 : Matrix ZZ  <--- ZZ

i55 : IP(A, w1, b1)

o55 = | 1 1 0 0 1 |

                1        5
o55 : Matrix ZZ  <--- ZZ

i56 : IP(A, w2, b1)

o56 = | 1 0 2 0 0 |

                1        5
o56 : Matrix ZZ  <--- ZZ

i57 : IP(A, w1, b2)

o57 = | 2 1 0 0 2 |
```

```
                        1       5
o57 : Matrix ZZ  <--- ZZ

i58 : IP(A, w2, b2)

o58 = | 2 0 1 2 0 |

                        1       5
o58 : Matrix ZZ  <--- ZZ
```

4   Generic Initial Ideals

Gröbner basis calculations and initial ideals depend heavily on the given co-ordinate system. By making a generic change of coordinates before taking the initial ideal, we may eliminate this dependence. This procedure also endows the resulting monomial ideal with a rich combinatorial structure.

To describe this structure, we introduce the following definitions and notation. Let $S = \mathbb{Q}[x_0, \ldots, x_n]$. If $g = [g_{i,j}] \in \mathrm{GL}_{n+1}(\mathbb{Q})$ and $f \in S$, then $g \cdot f$ denotes the standard action of the general linear group on S: $x_i \mapsto \sum_{j=0}^{n} g_{i,j} x_j$. For an ideal $I \subseteq S$, we define $g \cdot I = \{g \cdot f \,|\, f \in I\}$. Let B denote the Borel subgroup of $\mathrm{GL}_{n+1}(\mathbb{Q})$ consisting of upper triangular matrices. A monomial ideal I is called *Borel-fixed* if it satisfies any of the following equivalent conditions:

(a) $g \cdot I = I$ for every $g \in B$;
(b) if r is a generator of I divisible by x_j then $\frac{r x_i}{x_j} \in I$ for all $i < j$;
(c) $\mathrm{in}(g \cdot I) = I$ for every g is some open neighborhood of the identity in B.

For a proof that these conditions are equivalent, see Propositon 1.25 in Green [9].

In *Macaulay 2*, the function isBorel tests whether a monomial ideal is Borel-fixed.

```
i59 : S = QQ[a,b,c,d];

i60 : isBorel monomialIdeal(a^2, a*b, b^2)

o60 = true

i61 : isBorel monomialIdeal(a^2, b^2)

o61 = false
```

The function borel generates the smallest Borel-fixed ideal containing the given monomial ideal.

```
i62 : borel monomialIdeal(b*c)

                      2          2
o62 = monomialIdeal (a , a*b, b , a*c, b*c)

o62 : MonomialIdeal of S

i63 : borel monomialIdeal(a,c^3)

                      3   2       2   3
o63 = monomialIdeal (a, b , b c, b*c , c )

o63 : MonomialIdeal of S
```

The next theorem provides the main source of Borel-fixed ideals.

Theorem 4.1 (Galligo). *Fix a term order on $S = \mathbb{Q}[x_0, \ldots, x_n]$ such that $x_0 > \ldots > x_n$. If I is a homogeneous ideal in S, then there is a Zariski open subset $U \subseteq \mathrm{GL}_{n+1}(\mathbb{Q})$ such that*

(1) *there is a monomial ideal $J \subseteq S$ such that $J = \mathrm{in}(\mathrm{g} \cdot I)$ for all $\mathrm{g} \in U$;*
(2) *the ideal J is Borel-fixed.*

The ideal J is called the generic initial ideal of I.

Proof. See Theorem 1.27 in Green [9]. $\qquad\qquad\qquad\qquad\qquad\qquad$ □

The following method allows one to compute generic initial ideals.

```
i64 : gin = method();

i65 : gin Ideal := I -> (
            S := ring I;
            StoS := map(S, S, random(S^{0}, S^{numgens S:-1}));
            monomialIdeal StoS I);

i66 : gin MonomialIdeal := I -> gin ideal I;
```

This routine assumes that the random function generates a matrix in the Zariski open subset U. Since we are working over a field of characteristic zero this occurs with probability one. For example, we can determine the generic initial ideal of two generic homogeneous polynomials of degree p and q in $\mathbb{Q}[a, b, c, d]$.

```
i67 : genericForms = (p,q) -> ideal(random(p,S), random(q,S));

i68 : gin genericForms(2,2)

                 2       3
o68 = monomialIdeal (a , a*b, b )

o68 : MonomialIdeal of S

i69 : gin genericForms(2,3)

                 2     2   4
o69 = monomialIdeal (a , a*b , b )

o69 : MonomialIdeal of S
```

Although the generic initial ideal is Borel-fixed, some non-generic initial ideals may also be Borel-Fixed.

```
i70 : J = ideal(a^2, a*b+b^2, a*c)

            2           2
o70 = ideal (a , a*b + b , a*c)

o70 : Ideal of S

i71 : ginJ = gin J

                 2       2     2
o71 = monomialIdeal (a , a*b, b , a*c )

o71 : MonomialIdeal of S
```

```
i72 : inJ = monomialIdeal J

            2        3      2
o72 = monomialIdeal (a , a*b, b , a*c, b c)

o72 : MonomialIdeal of S

i73 : isBorel inJ and isBorel ginJ

o73 = true
```

Finally, we show that the generic initial ideal does depend on the term order by computing lexicographic generic initial ideal for two generic forms of degree p and q in $\mathbb{Q}[a, b, c, d]$

```
i74 : S = QQ[a,b,c,d, MonomialOrder => Lex];

i75 : gin genericForms(2,2)

            2        4      2
o75 = monomialIdeal (a , a*b, b , a*c )

o75 : MonomialIdeal of S

i76 : gin genericForms(2,3)

            2      2    6      2      6        2          4
o76 = monomialIdeal (a , a*b , b , a*b*c , a*c , a*b*c*d , a*b*d )

o76 : MonomialIdeal of S
```

A more comprehensive treatment of generic initial ideals can be found in Green [9]. The properties of Borel-fixed ideals in characteristic $p > 0$ are discussed in Eisenbud [7].

Example: Connectedness of the Hilbert Scheme

Generic initial ideals are a powerful tool for studying the structure of the Hilbert scheme. Intuitively, the Hilbert scheme $\mathrm{Hilb}^{p(t)}(\mathbb{P}^n)$ parameterizes subschemes $X \subseteq \mathbb{P}^n$ with Hilbert polynomial $p(t)$. For an introduction to Hilbert schemes see Harris and Morrison [11]. The construction of the Hilbert scheme $\mathrm{Hilb}^{p(t)}(\mathbb{P}^n)$ can be found in Grothendieck's original article [10] or Altman and Kleiman [1]. While much is known about specific Hilbert schemes, the general structure remain largely a mystery. In particular, the component structure — the number of irreducible components, their dimensions, how they intersect and what subschemes they parameterize — is not well understood.

Reeves [21] uses generic initial ideals to establish the most important theorem to date on the component structure. The incidence graph of $\mathrm{Hilb}^{p(t)}(\mathbb{P}^n)$ is defined as follows: to each irreducible component we assign a vertex and we connect two vertices if the corresponding components intersect. Reeves [21] proves that the distance (the number of edges in the shortest path) between any two vertices in the incidence graph of $\mathrm{Hilb}^{p(t)}(\mathbb{P}^n)$ is at most $2 \deg p(t) + 2$. Her proof can be divided into three major steps.

Step I: connect an arbitrary ideal to a Borel-fixed ideal. Passing to an initial ideal corresponds to taking the limit in a flat family, in other words a path on the Hilbert scheme; see Theorem 15.17 in Eisenbud [7]. Thus, Theorem 4.1 shows that generic initial ideals connect arbitrary ideals to Borel-fixed ideals.

Step II: connect Borel-fixed ideals by projection. For a homogeneous ideal $I \subseteq S = \mathbb{Q}[x_0, \ldots, x_n]$, let $\pi(I)$ denote the ideal obtained by setting $x_n = 1$ and $x_{n-1} = 1$ in I. With this notation, we have

Theorem 4.2. *If J is a Borel-fixed ideal, then the set of Borel-fixed ideals I, with Hilbert polynomial $p(t)$ and $\pi(I) = J$, consists of ideals defining subschemes of $\mathbb{P}^n$ which all lie on a single component of $\mathrm{Hilb}^{p(t)}(\mathbb{P}^n)$.*

Proof. See Theorem 6 in Reeves [21].   □

This gives an easy method for partitioning Borel-fixed ideals into classes, each of which must lie in a single component.

Step III: connect Borel-fixed ideals by distraction. Given a Borel-fixed ideal, we produce a new ideal via a two-step process called distraction. First, one polarizes the Borel-fixed ideal. The polarization of a monomial ideal $I \subset S$ is defined as:

$$\left\langle \prod_{i=0}^{n} \prod_{j=1}^{\alpha_i} z_{i,j} : \text{ where } x_0^{\alpha_0} \cdots x_n^{\alpha_n} \text{ is a minimal generator of } I \right\rangle .$$

One then pulls the result back to an ideal in the original variables by taking a linear section of the polarization. Theorem 4.10 in Hartshorne [12] shows that the distraction is connected to the original Borel-fixed ideal. Now, taking the lexicographic generic initial ideal of the distraction yields a second Borel-fixed ideal. Reeves [21] proves that repeating this process, at most $\deg p(t) + 1$ times, one arrives at a distinguished component of $\mathrm{Hilb}^{p(t)}(\mathbb{P}^n)$ called the lexicographic component. For more information on the lexicographic component see Reeves and Stillman [20].

We can implement these operations in *Macaulay 2* as follows:

```
i77 : projection = I -> (
         S := ring I;
         n := numgens S;
         X := gens S;
         monomialIdeal mingens substitute(ideal I,
             {X#(n-2) => 1, X#(n-1) => 1}));

i78 : polarization = I -> (
         n := numgens ring I;
         u := apply(numgens I, i -> first exponents I_i);
         I.lcm = max \ transpose u;
         Z := flatten apply(n, i -> apply(I.lcm#i, j -> z_{i,j}));
         R := QQ(monoid[Z]);
         Z = gens R;
         p := apply(n, i -> sum((I.lcm)_{0..i-1}));
         monomialIdeal apply(u, e -> product apply(n, i ->
                 product(toList(0..e#i-1), j -> Z#(p#i+j))))));
```

```
i79 : distraction = I -> (
         S := ring I;
         n := numgens S;
         X := gens S;
         J := polarization I;
         W := flatten apply(n, i -> flatten apply(I.lcm#i,
                 j -> X#i));
         section := map(S, ring J, apply(W, r -> r -
                 random(500)*X#(n-2) - random(500)*X#(n-1)));
         section ideal J);
```

For example, we have

```
i80 : S = QQ[x_0 .. x_4, MonomialOrder => GLex];

i81 : I = monomialIdeal(x_0^2, x_0*x_1^2*x_3, x_1^3*x_4)

                      2      2     3
o81 = monomialIdeal (x , x x x , x x )
                      0   0 1 3   1 4

o81 : MonomialIdeal of S

i82 : projection I

                      2    2   3
o82 = monomialIdeal (x , x x , x )
                      0   0 1   1

o82 : MonomialIdeal of S

i83 : polarization I

o83 = monomialIdeal (z       z       , z     z       z       z      , z   ···
                      {0, 0} {0, 1}    {0, 0} {1, 0} {1, 1} {3, 0}   {1 ···

o83 : MonomialIdeal of QQ [z       , z      , z      , z      , z      ···
                           {0, 0}    {0, 1}   {1, 0}   {1, 1}   {1, 2} ···

i84 : distraction I

             2                        2                       2    ···
o84 = ideal (x  - 398x x  - 584x x  + 36001x  + 92816x x  + 47239x , - ···
             0        0 3       0 4         3         3 4         4  ···

o84 : Ideal of S
```

To illustrate Reeves' method, we show that the incidence graph of the Hilbert scheme $\mathrm{Hilb}^{4t+1}(\mathbb{P}^4)$ has diameter at most 2. Note that the rational quartic curve in $\mathbb{P}^4$ has Hilbert polynomial $4t + 1$.

```
i85 : m =  matrix table({0,1,2}, {0,1,2}, (i,j) -> (gens S)#(i+j))

o85 = | x_0 x_1 x_2 |
      | x_1 x_2 x_3 |
      | x_2 x_3 x_4 |

              3       3
o85 : Matrix S <--- S

i86 : rationalQuartic = minors(2, m);

o86 : Ideal of S

i87 : H = hilbertPolynomial(S/rationalQuartic);
```

```
i88 : hilbertPolynomial(S/rationalQuartic, Projective => false)

o88 = 4$i + 1

o88 : QQ [$i]
```

There are 12 Borel-fixed ideals with Hilbert polynomial $4t+1$; see Example 1 in Reeves [21].

```
i89 : L = {monomialIdeal(x_0^2, x_0*x_1, x_0*x_2, x_1^2, x_1*x_2, x_2^ ...

i90 : scan(#L, i -> << endl << i+1 << " : " << L#i << endl);

                2          2              2
1 : monomialIdeal (x , x x , x , x x , x x , x )
                0     0 1   1    0 2   1 2   2

                2          2              3
2 : monomialIdeal (x , x x , x , x x , x x , x , x x )
                0     0 1   1    0 2   1 2   2    0 3

                   2     2   3
3 : monomialIdeal (x , x , x x , x , x x x )
                0   1   1 2   2   1 2 3

                   2       4   3
4 : monomialIdeal (x , x , x x , x , x x )
                0   1   1 2   2   2 3

                      5   4 3
5 : monomialIdeal (x , x , x , x x )
                0   1   2   2 3

                   2       5       4 2
6 : monomialIdeal (x , x , x x , x , x x , x x )
                0   1   1 2   2   1 3   2 3

                2          2              5          4
7 : monomialIdeal (x , x x , x , x x , x x , x , x x , x x , x x )
                0     0 1   1    0 2   1 2   2    0 3   1 3   2 3

                   2       5 4       2
8 : monomialIdeal (x , x , x x , x , x x , x x )
                0   1   1 2   2   2 3   1 3

                2          2          4           2
9 : monomialIdeal (x , x x , x , x x , x x , x , x x , x x )
                0     0 1   1    0 2   1 2   2    0 3   1 3

                    2     2   4           2
10 : monomialIdeal (x , x , x x , x , x x x , x x )
                 0   1   1 2   2   1 2 3   1 3

                    2       4   3
11 : monomialIdeal (x , x , x x , x , x x )
                 0   1   1 2   2   1 3

                    6   5       4 2
12 : monomialIdeal (x , x , x , x x , x x )
                 0   1   2   2 3   2 3

i91 : all(L, I -> isBorel I and hilbertPolynomial(S/I) == H)

o91 = true
```

The projection operation partitions the list L into 3 classes:

```
i92 : class1 = projection L#0

                2           2                    2
o92 = monomialIdeal (x , x x , x , x x , x x , x )
                      0     0 1   1     0 2   1 2   2

o92 : MonomialIdeal of S

i93 : class2 = projection L#1

                2             3
o93 = monomialIdeal (x , x , x x , x )
                      0   1   1 2   2

o93 : MonomialIdeal of S

i94 : class3 = projection L#4

                        4
o94 = monomialIdeal (x , x , x )
                      0   1   2

o94 : MonomialIdeal of S

i95 : all(1..3, i -> projection L#i == class2)

o95 = true

i96 : all(4..11, i -> projection L#i == class3)

o96 = true
```

Finally, we use the distraction to connect the classes.

```
i97 : all(L, I -> I == monomialIdeal distraction I)

o97 = true

i98 : all(0..3, i -> projection gin distraction L#i == class3)

o98 = true
```

Therefore, the components corresponding to class1 and class2 intersect the one corresponding to class3. Note that class3 corresponds to the lexicographic component.

5   The Chain Property

Hoşten and Thomas [14] recently established that the initial ideals of a toric ideal have an interesting combinatorial structure called the chain property. This structure is on the poset of associated primes where the partial order is given by inclusion. Since a monomial ideal $I \subset S = \mathbb{Q}[x_1, \ldots, x_n]$ is prime if and only if it is generated by a subset of the variables $\{x_1, \ldots, x_n\}$, the poset of associated primes of I is contained in the power set of the variables. We say that a monomial ideal I has the chain property if the following condition holds:

For any embedded prime $P_F = \langle x_i : x_i \notin F \rangle$ of I, there exists an associated prime $P_G \subset P_F$ such that $|G| = |F| - 1$.

In other words, there is a saturated chain from every embedded prime to some minimal prime. Experimental evidence suggests that, in fact, most initial ideals of prime ideals satisfy this saturated chain condition. Because of ubiquity and simplicity of this condition, we are interested in understanding which classes of initial ideals (or more generally monomial ideals) have the chain property.

More recently, Miller, Sturmfels and Yanagawa [18] provided a large class of monomial ideals with the chain property. A monomial ideal I is called generic when the following condition holds: if two distinct minimal generators r_1 and r_2 of I have the same positive degree in some variable x_i, there is a third generator r_3 which strictly divides the least common multiple of r_1 and r_2. In particular, if no two distinct minimal generators have the same positive degree in any variable, then the monomial ideal is generic. Theorem 2.2 in Miller, Sturmfels and Yanagawa [18] shows that generic monomial ideals have the chain property.

Examples and Counterexamples

In this final section, we illustrate how to use *Macaulay 2* for further experimentation and investigation of the chain property. The following function determines whether a monomial ideal has the chain property:

```
i99 : hasChainProperty = I -> (
          L := ass I;
          radI := radical I;
          all(L, P -> radI : (radI : P) == P or (
                  gensP := first entries gens P;
                  all(gensP, r -> (
                          Q := monomialIdeal delete(r, gensP);
                          I : (I : Q) == Q)))));
```

Using hasChainProperty, we examine the initial ideals of four interesting classes of ideals related to toric ideals.

An Initial Ideal of a Toric Ideal.

As mentioned above, Hoşten and Thomas proved that any initial ideal of a toric ideal satisfies the saturated chain condition. The following example demonstrates this phenomenon. Consider the matrix A:

```
i100 : A = matrix{{1,1,1,1,1,1,1}, {2,0,0,0,1,0,0}, {0,2,0,0,0,1,0}, { · · ·

o100 = | 1 1 1 1 1 1 1 |
       | 2 0 0 0 1 0 0 |
       | 0 2 0 0 0 1 0 |
       | 2 2 0 2 1 1 1 |

                 4          7
o100 : Matrix ZZ  <--- ZZ

i101 : IA = toricIdeal(A, {1,1,1,1,1,1,1})
```

```
                2              2              2
o101 = ideal (x x   - x , x x   - x , x x  - x )
              3 4     7   2 3     6   1 3    5

o101 : Ideal of S

i102 : inIA = monomialIdeal IA

                                      2     2     2
o102 = monomialIdeal (x x , x x , x x , x x , x x , x x )
                       1 3   2 3   3 4   2 5   4 5   4 6

o102 : MonomialIdeal of S

i103 : hasChainProperty inIA

o103 = true
```

An Initial Ideal of a Prime Ideal.

Since toric ideals are prime, one naturally asks if the initial ideal of any prime ideal has the chain property. By modifying the previous example, we can show that this is not the case. In particular, making the linear change of coordinates by $x_4 \mapsto x_3 - x_4$, we obtain a new prime ideal J.

```
i104 : StoS = map(S, S, {x_1, x_2, x_3, x_3 - x_4, x_5, x_6, x_7});

o104 : RingMap S <--- S

i105 : J = StoS IA

            2            2            2            2
o105 = ideal (x   - x x   - x , x x   - x , x x  - x )
              3     3 4    7   2 3     6   1 3    5

o105 : Ideal of S
```

Taking the initial ideal with respect to the reverse lexicographic term order (the default order), we have

```
i106 : inJ = monomialIdeal J

                      2     2     2          2     2          2 ···
o106 = monomialIdeal (x x , x x , x , x x , x x , x x x , x x , x x x ···
                       1 3   2 3   3   2 5   3 5   1 4 5   3 6   1 4 6 ···

o106 : MonomialIdeal of S

i107 : hasChainProperty inJ

o107 = false
```

An A-graded Monomial Ideal.

Let A be a $d \times n$ matrix of nonnegative integers and let $\mathbf{a}_i$ denote the i-th column of A. Consider the polynomial ring $S = \mathbb{Q}[x_1, \ldots, x_n]$ with the $\mathbb{Z}^d$-grading defined by $\deg x_i = \mathbf{a}_i$. An ideal $I \subset \mathbb{Q}[x_1, \ldots, x_n]$ is called A-graded provided it is homogeneous with respect to the A-grading and

$$\dim_{\mathbb{Q}} \left(\frac{S}{I} \right)_{\mathbf{b}} = \begin{cases} 1 & \text{if } \mathbf{b} \in \mathbb{N}A \\ 0 & \text{otherwise} \end{cases}$$

for all $\mathbf{b} \in \mathbb{N}^d$. Remark 10.1 in Sturmfels [26] shows that the initial ideal of the toric ideal I_A is A-graded. Altmann [2] shows that when A has rank 2 every A-graded monomial ideal has the chain property. However, Altmann [2] also provides a counterexample when A has rank 3. We can verify his example in *Macaulay 2* as follows:

```
i108 : A = matrix{{2,0,0,1,0,0,2,1,1,3,2,2,2,3,3,3},
                  {0,2,0,0,1,0,1,2,1,2,3,2,3,2,3,3},
                  {0,0,2,0,0,1,1,1,2,2,2,3,3,3,2,3}};

              3          16
o108 : Matrix ZZ  <--- ZZ
```

In *Macaulay 2*, the first entry in degree vector of each variable must be positive. Hence, we append to A the sum of its rows to get a matrix whose columns will serve as the degrees of the variables.

```
i109 : D = A^{0}+A^{1}+A^{2} || A

o109 = | 2 2 2 1 1 1 4 4 4 7 7 7 8 8 8 9 |
       | 2 0 0 1 0 0 2 1 1 3 2 2 2 3 3 3 |
       | 0 2 0 0 1 0 1 2 1 2 3 2 3 2 3 3 |
       | 0 0 2 0 0 1 1 1 2 2 2 3 3 3 2 3 |

              4          16
o109 : Matrix ZZ  <--- ZZ

i110 : D = entries transpose D;

i111 : S = QQ[vars(0..15), Degrees => D, MonomialSize => 16];

i112 : I = monomialIdeal(d*j, d*k, d*l, d*m, d*n, d*o, d*p, e*j, e*k,
           e*l, e*m, e*n, e*o, e*p, f*j, f*k, f*l, f*m, f*n, f*o, f*p,
           g*j, g*k, g*l, g*m, g*n, g*o, g*p, h*j, h*k, h*l, h*m, h*n,
           h*o, h*p, i*j, i*k, i*l, i*m, i*n, i*o, i*p, g^2, g*h, g*i,
           h^2, h*i, i^2, j^2, j*k, j*l, j*m, j*n, j*o, j*p, k^2, k*l,
           k*m, k*n, k*o, k*p, l^2, l*m, l*n, l*o, l*p, m^2, m*n, m*o,
           m*p, n^2, n*o, n*p, o^2, o*p, p^2, d^2, e^2, f^2, d*h, e*i,
           f*g, f*d*i, d*e*g, e*f*h, c*d*g, a*e*h, b*f*i, c*e*g,
           a*f*h, b*d*i, c*d*e, a*e*f, b*f*d, c*b*d, a*c*e, b*a*f,
           c*b*g, a*c*h, b*a*i);

o112 : MonomialIdeal of S
```

To help convince you that I is an A-graded ideal, we compute the $\dim_{\mathbb{Q}} \left(\frac{S}{I} \right)_{\mathbf{a}_i}$ for $1 \leq i \leq 16$.

```
i113 : apply(D, d -> rank source basis(d, (S^1)/ ideal I))

o113 = {1, 1, 1, 1, 1, 1, 1, 1, 1, 1, 1, 1, 1, 1, 1, 1}

o113 : List
```

Finally, we check the chain property.

```
i114 : hasChainProperty I

o114 = false
```

The Vertex Ideal. Lastly, we consider a different family of monomials ideals arising from toric ideals. The vertex ideal V_A is defined as intersection all the monomial initial ideals of the toric ideal I_A. Although there are (in general) infinitely many distinct term orders on a polynomial ring, an ideal has only finitely many initial ideals; see Theorem 1.2 in Sturmfels [26]. In particular, the above intersection is finite. Vertex ideals were introduced and studied by Hoşten and Maclagan [13]. However, the question "Does the vertex ideal V_A have the chain property?" remains open.

References

1. Allen B. Altman and Steven L. Kleiman: Compactifying the Picard scheme. *Adv. in Math.*, 35(1):50–112, 1980.
2. Klaus Altmann: The chain property for the associated primes of A-graded ideals. arXiv:math.AG/0004142.
3. Dave Bayer and Mike Stillman: Computation of Hilbert functions. *J. Symbolic Comput.*, 14(1):31–50, 1992.
4. Anders Björner: Nonpure shellability, f-vectors, subspace arrangements and complexity. In *Formal power series and algebraic combinatorics (New Brunswick, NJ, 1994)*, pages 25–53. Amer. Math. Soc., Providence, RI, 1996.
5. Anders Björner and Gil Kalai: An extended Euler-Poincaré theorem. *Acta Math.*, 161(3-4):279–303, 1988.
6. Winfried Bruns and Jürgen Herzog: *Cohen-Macaulay rings*. Cambridge University Press, Cambridge, 1993.
7. David Eisenbud: *Commutative algebra with a view toward algebraic geometry*. Springer-Verlag, New York, 1995.
8. Michael R. Garey and David S. Johnson: *Computers and intractability*. W. H. Freeman and Co., San Francisco, Calif., 1979. A guide to the theory of NP-completeness, A Series of Books in the Mathematical Sciences.
9. Mark L. Green: Generic initial ideals. In *Six lectures on commutative algebra (Bellaterra, 1996)*, pages 119–186. Birkhäuser, Basel, 1998.
10. Alexander Grothendieck: *Fondements de la géométrie algébrique. [Extraits du Séminaire Bourbaki, 1957-1962.]*. Secrétariat mathématique, Paris, 1962.
11. Joe Harris and Ian Morrison: *Moduli of curves*. Springer-Verlag, New York, 1998.
12. Robin Hartshorne: Connectedness of the Hilbert scheme. *Inst. Hautes Études Sci. Publ. Math.*, 29:5–48, 1966.
13. Serkan Hoşten and Diane Maclagan: The vertex ideal of a lattice. 20 pages, (2000), preprint.
14. Serkan Hoşten and Rekha R. Thomas: The associated primes of initial ideals of lattice ideals. *Math. Res. Lett.*, 6(1):83–97, 1999.
15. Serkan Hoşten and Rekha R. Thomas: Standard pairs and group relaxations in integer programming. *J. Pure Appl. Algebra*, 139(1-3):133–157, 1999. Effective methods in algebraic geometry (Saint-Malo, 1998).
16. Ezra Miller: Alexander Duality for Monomial Ideals and Their Resolutions. arXiv:math.AG/9812095.
17. Ezra Miller: The Alexander duality functors and local duality with monomial support. *J. Algebra*, 231(1):180–234, 2000.
18. Ezra Miller, Bernd Sturmfels, and Kohji Yanagawa: Generic and cogeneric monomial ideals. *J. Symbolic Comput.*, 29(4-5):691–708, 2000. Symbolic computation in algebra, analysis, and geometry (Berkeley, CA, 1998).

19. Hidefumi Ohsugi and Takayuki Hibi: Normal polytopes arising from finite graphs. *J. Algebra*, 207(2):409–426, 1998.
20. Alyson Reeves and Mike Stillman: Smoothness of the lexicographic point. *J. Algebraic Geom.*, 6(2):235–246, 1997.
21. Alyson A. Reeves: The radius of the Hilbert scheme. *J. Algebraic Geom.*, 4(4):639–657, 1995.
22. Mutsumi Saito, Bernd Sturmfels, and Nobuki Takayama: *Gröbner deformations of hypergeometric differential equations*. Springer-Verlag, Berlin, 2000.
23. Aron Simis, Wolmer V. Vasconcelos, and Rafael H. Villarreal: The integral closure of subrings associated to graphs. *J. Algebra*, 199(1):281–289, 1998.
24. Richard P. Stanley: *Combinatorics and commutative algebra*. Birkhäuser Boston Inc., Boston, MA, second edition, 1996.
25. Bernd Sturmfels: Gröbner bases and Stanley decompositions of determinantal rings. *Math. Z.*, 205(1):137–144, 1990.
26. Bernd Sturmfels: *Gröbner bases and convex polytopes*. American Mathematical Society, Providence, RI, 1996.
27. Bernd Sturmfels, Ngô Viêt Trung, and Wolfgang Vogel: Bounds on degrees of projective schemes. *Math. Ann.*, 302(3):417–432, 1995.
28. Rafael H. Villarreal: Cohen-Macaulay graphs. *Manuscripta Math.*, 66(3):277–293, 1990.
29. Günter M. Ziegler: *Lectures on polytopes*. Springer-Verlag, New York, 1995.

From Enumerative Geometry to Solving Systems of Polynomial Equations

Frank Sottile[*]

Solving a system of polynomial equations is a ubiquitous problem in the applications of mathematics. Until recently, it has been hopeless to find explicit solutions to such systems, and mathematics has instead developed deep and powerful theories about the solutions to polynomial equations. Enumerative Geometry is concerned with counting the number of solutions when the polynomials come from a geometric situation and Intersection Theory gives methods to accomplish the enumeration.

We use *Macaulay 2* to investigate some problems from enumerative geometry, illustrating some applications of symbolic computation to this important problem of solving systems of polynomial equations. Besides enumerating solutions to the resulting polynomial systems, which include overdetermined, deficient, and improper systems, we address the important question of real solutions to these geometric problems.

1 Introduction

A basic question to ask about a system of polynomial equations is its number of solutions. For this, the fundamental result is the following Bézout Theorem.

Theorem 1.1. *The number of isolated solutions to a system of polynomial equations*

$$f_1(x_1, \ldots, x_n) = f_2(x_1, \ldots, x_n) = \cdots = f_n(x_1, \ldots, x_n) = 0$$

is bounded by $d_1 d_2 \cdots d_n$, where $d_i := \deg f_i$. If the polynomials are generic, then this bound is attained for solutions in an algebraically closed field.

Here, isolated is taken with respect to the algebraic closure. This Bézout Theorem is a consequence of the refined Bézout Theorem of Fulton and MacPherson [12, §1.23].

A system of polynomial equations with fewer than this degree bound or Bézout number of solutions is called *deficient*, and there are well-defined classes of deficient systems that satisfy other bounds. For example, fewer monomials lead to fewer solutions, for which polyhedral bounds [4] on the number of solutions are often tighter (and no weaker than) the Bézout number, which applies when all monomials are present. When the polynomials

---

[*] Supported in part by NSF grant DMS-0070494.

come from geometry, determining the number of solutions is the central problem in enumerative geometry.

Symbolic computation can help compute the solutions to a system of equations that has only isolated solutions. In this case, the polynomials generate a zero-dimensional ideal I. The *degree* of I is $\dim_k k[X]/I$, the dimension of the k-vector space $k[X]/I$, which is also the number of standard monomials in any term order. This degree gives an upper bound on the number of solutions, which is attained when I is radical.

Example 1.2. We illustrate this discussion with an example. Let f_1, f_2, f_3, and f_4 be random quadratic polynomials in the ring $\mathbb{F}_{101}[y_{11}, y_{12}, y_{21}, y_{22}]$.

```
i1 : R = ZZ/101[y11, y12, y21, y22];
```

```
i2 : PolynomialSystem = apply(1..4, i ->
                 random(0, R) + random(1, R) + random(2, R));
```

The ideal they generate has dimension 0 and degree $16 = 2^4$, which is the Bézout number.

```
i3 : I = ideal PolynomialSystem;

o3 : Ideal of R

i4 : dim I, degree I

o4 = (0, 16)

o4 : Sequence
```

If we restrict the monomials which appear in the f_i to be among

$$1, \quad y_{11}, \quad y_{12}, \quad y_{21}, \quad y_{22}, \quad y_{11}y_{22}, \quad \text{and} \quad y_{12}y_{21},$$

then the ideal they generate again has dimension 0, but its degree is now 4.

```
i5 : J = ideal (random(R^4, R^7) * transpose(
              matrix{{1, y11, y12, y21, y22, y11*y22, y12*y21}}));

o5 : Ideal of R

i6 : dim J, degree J

o6 = (0, 4)

o6 : Sequence
```

If we further require that the coefficients of the quadratic terms sum to zero, then the ideal they generate now has degree 2.

```
i7 : K = ideal (random(R^4, R^6) * transpose(
              matrix{{1, y11, y12, y21, y22, y11*y22 - y12*y21}}));

o7 : Ideal of R

i8 : dim K, degree K

o8 = (0, 2)

o8 : Sequence
```

In Example 4.2, we shall see how this last specialization is geometrically meaningful.

For us, enumerative geometry is concerned with *enumerating geometric figures of some kind having specified positions with respect to general fixed figures*. That is, counting the solutions to a geometrically meaningful system of polynomial equations. We use *Macaulay 2* to investigate some enumerative geometric problems from this point of view. The problem of enumeration will be solved by computing the degree of the (0-dimensional) ideal generated by the polynomials.

2   Solving Systems of Polynomials

We briefly discuss some aspects of solving systems of polynomial equations. For a more complete survey, see the relevant chapters in [6,7].

Given an ideal I in a polynomial ring $k[X]$, set $\mathcal{V}(I) := \operatorname{Spec} k[X]/I$. When I is generated by the polynomials $f_1, \ldots, f_N$, $\mathcal{V}(I)$ gives the set of solutions in affine space to the system

$$f_1(X) = \cdots = f_N(X) = 0 \qquad (1)$$

a geometric structure. These solutions are the *roots* of the ideal I. The degree of a zero-dimensional ideal I provides an algebraic count of its roots. The degree of its radical counts roots in the algebraic closure, ignoring multiplicities.

2.1   Excess Intersection

Sometimes, only a proper (open) subset of affine space is geometrically meaningful, and we want to count only the meaningful roots of I. Often the roots $\mathcal{V}(I)$ has positive dimensional components that lie in the complement of the meaningful subset. One way to treat this situation of excess or improper intersection is to saturate I by a polynomial f vanishing on the extraneous roots. This has the effect of working in $k[X][f^{-1}]$, the coordinate ring of the complement of $\mathcal{V}(f)$ [9, Exer. 2.3].

Example 2.1. We illustrate this with an example. Consider the following ideal in $\mathbb{F}_7[x,y]$.

```
i9 : R = ZZ/7[y, x, MonomialOrder=>Lex];

i10 : I = ideal (y^3*x^2 + 2*y^2*x + 3*x*y,  3*y^2 + x*y - 3*y);

o10 : Ideal of R
```

Since the generators have greatest common factor y, I defines finitely many points together with the line $y = 0$. Saturate I by the variable y to obtain the ideal J of isolated roots.

```
i11 : J = saturate(I, ideal(y))
              4    3    2
o11 = ideal (x  + x  + 3x  + 3x, y - 2x - 1)

o11 : Ideal of R
```

The first polynomial factors completely in $\mathbb{F}_7[x]$,

```
i12 : factor(J_0)

o12 = (x)(x - 2)(x + 2)(x + 1)

o12 : Product
```

and so the isolated roots of I are $(2,5), (-1,-1), (0,1)$, and $(-2,-3)$.

Here, the extraneous roots came from a common factor in both equations. A less trivial example of this phenomenon will be seen in Section 5.2.

2.2   Elimination, Rationality, and Solving

Elimination theory can be used to study the roots of a zero-dimensional ideal $I \subset k[X]$. A polynomial $h \in k[X]$ defines a map $k[y] \to k[X]$ (by $y \mapsto h$) and a corresponding projection $h \colon \operatorname{Spec} k[X] \twoheadrightarrow \mathbb{A}^1$. The generator $g(y) \in k[y]$ of the kernel of the map $k[y] \to k[X]/I$ is called an *eliminant* and it has the property that $\mathcal{V}(g) = h(\mathcal{V}(I))$. When h is a coordinate function x_i, we may consider the eliminant to be in the polynomial ring $k[x_i]$, and we have $\langle g(x_i) \rangle = I \cap k[x_i]$. The most important result concerning eliminants is the Shape Lemma [2].

Shape Lemma. *Suppose h is a linear polynomial and g is the corresponding eliminant of a zero-dimensional ideal $I \subset k[X]$ with $\deg(I) = \deg(g)$. Then the roots of I are defined in the splitting field of g and I is radical if and only if g is square-free.*

Suppose further that $h = x_1$ so that $g = g(x_1)$. Then, in the lexicographic term order with $x_1 < x_2 < \cdots < x_n$, I has a Gröbner basis of the form:

$$g(x_1), \quad x_2 - g_2(x_1), \quad \ldots, \quad x_n - g_n(x_1), \tag{2}$$

where $\deg(g) > \deg(g_i)$ for $i = 2, \ldots, n$.

When k is infinite and I is radical, an eliminant g given by a generic linear polynomial h will satisfy $\deg(g) = \deg(I)$. Enumerative geometry counts solutions when the fixed figures are generic. We are similarly concerned with the generic situation of $\deg(g) = \deg(I)$. In this case, eliminants provide a useful computational device to study further questions about the roots of I. For instance, the Shape Lemma holds for the saturated ideal of Example 2.1. Its eliminant, which is the polynomial J_0, factors completely over the ground field $\mathbb{F}_7$, so all four solutions are defined in $\mathbb{F}_7$. In Section 4.3, we will use eliminants in another way, to show that an ideal is radical.

Given a polynomial h in a zero-dimensional ring $k[X]/I$, the procedure eliminant(h, k[y]) finds a linear relation modulo I among the powers $1, h, h^2, \ldots, h^d$ of h with d minimal and returns this as a polynomial in $k[y]$. This procedure is included in the Macaulay 2 package realroots.m2.

```
i13 : load "realroots.m2"

i14 : code eliminant

o14 = -- realroots.m2:65-80
      eliminant = (h, C) -> (
          Z := C_0;
          A := ring h;
          assert( dim A == 0 );
          F := coefficientRing A;
          assert( isField F );
          assert( F == coefficientRing C );
          B := basis A;
          d := numgens source B;
          M := fold((M, i) -> M ||
                  substitute(contract(B, h^(i+1)), F),
                  substitute(contract(B, 1_A), F),
                  flatten subsets(d, d));
          N := ((ker transpose M)).generators;
          P := matrix {toList apply(0..d, i -> Z^i)} * N;
          (flatten entries(P))_0)
```

Here, M is a matrix whose rows are the normal forms of the powers 1, h, h^2, ..., h^d of h, for d the degree of the ideal. The columns of the kernel N of transpose M are a basis of the linear relations among these powers. The matrix P converts these relations into polynomials. Since N is in column echelon form, the initial entry of P is the relation of minimal degree. (This method is often faster than naïvely computing the kernel of the map $k[Z] \to A$ given by $Z \mapsto h$, which is implemented by eliminantNaive(h, Z).)

Suppose we have an eliminant $g(x_1)$ of a zero-dimensional ideal $I \subset k[X]$ with $\deg(g) = \deg(I)$, and we have computed the lexicographic Gröbner basis (2). Then the roots of I are

$$\{(\xi_1, g_2(\xi_1), \ldots, g_n(\xi_1)) \mid g(\xi_1) = 0\}. \tag{3}$$

Suppose now that $k = \mathbb{Q}$ and we seek floating point approximations for the (complex) roots of I. Following this method, we first compute floating point solutions to $g(\xi) = 0$, which give all the x_1-coordinates of the roots of I, and then use (3) to find the other coordinates. The difficulty here is that enough precision may be lost in evaluating $g_i(\xi_1)$ so that the result is a poor approximation for the other components ξ_i.

2.3 Solving with Linear Algebra

We describe another method based upon numerical linear algebra. When $I \subset k[X]$ is zero-dimensional, $A = k[X]/I$ is a finite-dimensional k-vector space, and any Gröbner basis for I gives an efficient algorithm to compute ring operations using linear algebra. In particular, multiplication by $h \in A$ is a linear transformation $m_h : A \to A$ and the command `regularRep(h)` from `realroots.m2` gives the matrix of m_h in terms of the standard basis of A.

```
i15 : code regularRep

o15 = -- realroots.m2:96-100
      regularRep = f -> (
          assert( dim ring f == 0 );
          b := basis ring f;
          k := coefficientRing ring f;
          substitute(contract(transpose b, f*b), k))
```

Since the action of A on itself is faithful, the minimal polynomial of m_h is the eliminant corresponding to h. The procedure `charPoly(h, Z)` in `realroots.m2` computes the characteristic polynomial $\det(Z \cdot Id - m_h)$ of h.

```
i16 : code charPoly

o16 = -- realroots.m2:106-113
      charPoly = (h, Z) -> (
          A := ring h;
          F := coefficientRing A;
          S := F[Z];
          Z = value Z;
          mh := regularRep(h) ** S;
          Idz := S_0 * id_(S^(numgens source mh));
          det(Idz - mh))
```

When this is the minimal polynomial (the situation of the Shape Lemma), this procedure often computes the eliminant faster than does `eliminant`, and for systems of moderate degree, much faster than naïvely computing the kernel of the map $k[Z] \to A$ given by $Z \mapsto h$.

The eigenvalues and eigenvectors of m_h give another algorithm for finding the roots of I. The engine for this is the following result.

Stickelberger's Theorem. *Let $h \in A$ and m_h be as above. Then there is a one-to-one correspondence between eigenvectors $\mathbf{v}_\xi$ of m_h and roots ξ of I, the eigenvalue of m_h on $\mathbf{v}_\xi$ is the value $h(\xi)$ of h at ξ, and the multiplicity of this eigenvalue (on the eigenvector $\mathbf{v}_\xi$) is the multiplicity of the root ξ.*

Since the linear transformations m_h for $h \in A$ commute, the eigenvectors $\mathbf{v}_\xi$ are common to all m_h. Thus we may compute the roots of a zero-dimensional ideal $I \subset k[X]$ by first computing floating-point approximations to the eigenvectors $\mathbf{v}_\xi$ of m_{x_1}. Then the root $\xi = (\xi_1, \ldots, \xi_n)$ of I corresponding to the eigenvector $\mathbf{v}_\xi$ has ith coordinate satisfying

$$m_{x_i} \cdot \mathbf{v}_\xi = \xi_i \cdot \mathbf{v}_\xi. \tag{4}$$

An advantage of this method is that we may use structured numerical linear algebra after the matrices m_{x_i} are precomputed using exact arithmetic. (These matrices are typically sparse and have additional structures which may be exploited.) Also, the coordinates ξ_i are *linear* functions of the floating point entries of $\mathbf{v}_\xi$, which affords greater precision than the non-linear evaluations $g_i(\xi_1)$ in the method based upon elimination. While in principle only one of the $\deg(I)$ components of the vectors in (4) need be computed, averaging the results from all components can improve precision.

2.4   Real Roots

Determining the real roots of a polynomial system is a challenging problem with real world applications. When the polynomials come from geometry, this is the main problem of real enumerative geometry. Suppose $k \subset \mathbb{R}$ and $I \subset k[X]$ is zero-dimensional. If g is an eliminant of $k[X]/I$ with $\deg(g) = \deg(I)$, then the real roots of g are in 1-1 correspondence with the real roots of I. Since there are effective methods for counting the real roots of a univariate polynomial, eliminants give a naïve, but useful method for determining the number of real roots to a polynomial system. (For some applications of this technique in mathematics, see [20,23,25].)

The classical symbolic method of Sturm, based upon Sturm sequences, counts the number of real roots of a univariate polynomial in an interval. When applied to an eliminant satisfying the Shape Lemma, this method counts the number of real roots of the ideal. This is implemented in *Macaulay 2* via the command SturmSequence(f) of realroots.m2

```
i17 : code SturmSequence

o17 = -- realroots.m2:117-131
      SturmSequence = f -> (
          assert( isPolynomialRing ring f );
          assert( numgens ring f === 1 );
          R := ring f;
          assert( char R == 0 );
          x := R_0;
          n := first degree f;
          c := new MutableList from toList (0 .. n);
          if n >= 0 then (
              c#0 = f;
              if n >= 1 then (
                  c#1 = diff(x,f);
                  scan(2 .. n, i -> c#i = - c#(i-2) % c#(i-1));
                  ));
          toList c)
```

The last few lines of SturmSequence construct the Sturm sequence of the univariate argument f: This is $(f_0, f_1, f_2, \ldots)$ where $f_0 = f$, $f_1 = f'$, and for $i > 1$, f_i is the normal form reduction of $-f_{i-2}$ modulo f_{i-1}. Given any real number x, the *variation* of f at x is the number of changes in sign of the sequence $(f_0(x), f_1(x), f_2(x), \ldots)$ obtained by evaluating the Sturm sequence of f at x. Then the number of real roots of f over an interval $[x, y]$ is the difference of the variation of f at x and at y.

The *Macaulay 2* commands numRealSturm and numPosRoots (and also numNegRoots) use this method to respectively compute the total number of real roots and the number of positive roots of a univariate polynomial.

```
i18 : code numRealSturm

o18 = -- realroots.m2:160-163
       numRealSturm = f -> (
           c := SturmSequence f;
           variations (signAtMinusInfinity \ c)
               - variations (signAtInfinity \ c))

i19 : code numPosRoots

o19 = -- realroots.m2:168-171
       numPosRoots = f -> (
           c := SturmSequence f;
           variations (signAtZero \ c)
               - variations (signAtInfinity \ c))
```

These use the commands signAt*(f), which give the sign of f at *. (Here, *
is one of Infinity, zero, or MinusInfinity.) Also variations(c) computes
the number of sign changes in the sequence c.

```
i20 : code variations

o20 = -- realroots.m2:183-191
       variations = c -> (
           n := 0;
           last := 0;
           scan(c, x -> if x != 0 then (
                       if last < 0 and x > 0 or last > 0
                           and x < 0 then n = n+1;
                       last = x;
                       ));
           n)
```

A more sophisticated method to compute the number of real roots which
can also give information about their location uses the rank and signature
of the symmetric trace form. Suppose $I \subset k[X]$ is a zero-dimensional ideal
and set $A := k[X]/I$. For $h \in k[X]$, set $S_h(f, g) := \text{trace}(m_{hfg})$. It is an
easy exercise that S_h is a symmetric bilinear form on A. The procedure
traceForm(h) in realroots.m2 computes this trace form S_h.

```
i21 : code traceForm

o21 = -- realroots.m2:196-203
       traceForm = h -> (
           assert( dim ring h == 0 );
           b   := basis ring h;
           k   := coefficientRing ring h;
           mm  := substitute(contract(transpose b, h * b ** b), k);
           tr  := matrix {apply(first entries b, x ->
                       trace regularRep x)};
           adjoint(tr * mm, source tr, source tr))
```

The value of this construction is the following theorem.

Theorem 2.2 ([3,19]). *Suppose $k \subset \mathbb{R}$ and I is a zero-dimensional ideal
in $k[x_1, \ldots, x_n]$ and consider $\mathcal{V}(I) \subset \mathbb{C}^n$. Then, for $h \in k[x_1, \ldots, x_n]$, the
signature $\sigma(S_h)$ and rank $\rho(S_h)$ of the bilinear form S_h satisfy*

$$\sigma(S_h) = \#\{a \in \mathcal{V}(I) \cap \mathbb{R}^n : h(a) > 0\} - \#\{a \in \mathcal{V}(I) \cap \mathbb{R}^n : h(a) < 0\}$$
$$\rho(S_h) = \#\{a \in \mathcal{V}(I) : h(a) \neq 0\}.$$

That is, the rank of S_h counts roots in $\mathbb{C}^n - \mathcal{V}(h)$, and its signature counts the real roots weighted by the sign of h (which is -1, 0, or 1) at each root. The command `traceFormSignature(h)` in `realroots.m2` returns the rank and signature of the trace form S_h.

```
i22 : code traceFormSignature

o22 = -- realroots.m2:208-218
      traceFormSignature = h -> (
          A := ring h;
          assert( dim A == 0 );
          assert( char A == 0 );
          S := QQ[Z];
          TrF := traceForm(h) ** S;
          IdZ := Z * id_(S^(numgens source TrF));
          f := det(TrF - IdZ);
          << "The trace form S_h with h = " << h <<
             " has rank " << rank(TrF) << " and signature " <<
             numPosRoots(f) - numNegRoots(f) << endl; )
```

The *Macaulay 2* command `numRealTrace(A)` simply returns the number of real roots of I, given $A = k[X]/I$.

```
i23 : code numRealTrace

o23 = -- realroots.m2:223-230
      numRealTrace = A -> (
          assert( dim A == 0 );
          assert( char A == 0 );
          S := QQ[Z];
          TrF := traceForm(1_A) ** S;
          IdZ := Z * id_(S^(numgens source TrF));
          f := det(TrF - IdZ);
          numPosRoots(f)-numNegRoots(f))
```

Example 2.3. We illustrate these methods on the following polynomial system.

```
i24 : R = QQ[x, y];

i25 : I = ideal (1 - x^2*y + 2*x*y^2,  y - 2*x - x*y + x^2);

o25 : Ideal of R
```

The ideal I has dimension zero and degree 5.

```
i26 : dim I, degree I

o26 = (0, 5)

o26 : Sequence
```

We compare the two methods to compute the eliminant of x in the ring R/I.

```
i27 : A = R/I;

i28 : time g = eliminant(x, QQ[Z])
      -- used 0.09 seconds

            5      4      3     2
o28 = Z    - 5Z   + 6Z   + Z   - 2Z + 1

o28 : QQ [Z]
```

```
i29 : time g = charPoly(x, Z)
        -- used 0.02 seconds

        5      4      3    2
o29 = Z   - 5Z   + 6Z   + Z   - 2Z + 1

o29 : QQ [Z]
```

The eliminant has 3 real roots, which we test in two different ways.

```
i30 : numRealSturm(g), numRealTrace(A)

o30 = (3, 3)

o30 : Sequence
```

We use Theorem 2.2 to isolate these roots in the x, y-plane.

```
i31 : traceFormSignature(x*y);
The trace form S_h with h = x*y has rank 5 and signature 3
```

Thus all 3 real roots lie in the first and third quadrants (where $xy > 0$). We isolate these further.

```
i32 : traceFormSignature(x - 2);
The trace form S_h with h = x - 2 has rank 5 and signature 1
```

This shows that two roots lie in the first quadrant with $x > 2$ and one lies in the third. Finally, one of the roots lies in the triangle $y > 0$, $x > 2$, and $x + y < 3$.

```
i33 : traceFormSignature(x + y - 3);
The trace form S_h with h = x + y - 3 has rank 5 and signature -1
```

Figure 1 shows these three roots (dots), as well as the lines $x + y = 3$ and $x = 2$.

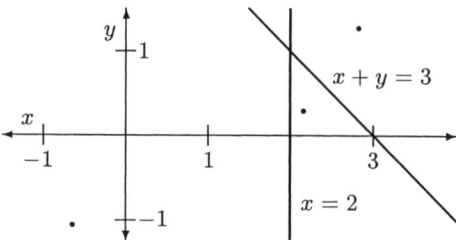

Fig. 1. Location of roots

2.5   Homotopy Methods

We describe symbolic-numeric *homotopy continuation methods* for finding approximate complex solutions to a system of equations. These exploit the traditional principles of conservation of number and specialization from enumerative geometry.

Suppose we seek the isolated solutions of a system $F(X) = 0$ where $F = (f_1, \ldots, f_n)$ are polynomials in the variables $X = (x_1, \ldots, x_N)$. First, a *homotopy* $H(X, t)$ is found with the following properties:

1. $H(X, 1) = F(X)$.
2. The isolated solutions of the *start system* $H(X, 0) = 0$ are known.
3. The system $H(X, t) = 0$ defines finitely many (complex) curves, and each isolated solution of the original system $F(X) = 0$ is connected to an isolated solution $\sigma_i(0)$ of $H(X, 0) = 0$ along one of these curves.

Next, choose a generic smooth path $\gamma(t)$ from 0 to 1 in the complex plane. Lifting γ to the curves $H(X, t) = 0$ gives smooth paths $\sigma_i(t)$ connecting each solution $\sigma_i(0)$ of the start system to a solution of the original system. The path γ must avoid the finitely many points in $\mathbb{C}$ over which the curves are singular or meet other components of the solution set $H(X, t) = 0$.

Numerical path continuation is used to trace each path $\sigma_i(t)$ from $t = 0$ to $t = 1$. When there are fewer solutions to $F(X) = 0$ than to $H(X, 0) = 0$, some paths will diverge or become singular as $t \to 1$, and it is expensive to trace such a path. The homotopy is *optimal* when this does not occur.

When $N = n$ and the f_i are generic, set $G(X) := (g_1, \ldots, g_n)$ with $g_i = (x_i - 1)(x_i - 2) \cdots (x_i - d_i)$ where $d_i := \deg(f_i)$. Then the *Bézout homotopy*

$$H(X, t) \quad := \quad tF(X) + (1 - t)G(X)$$

is optimal. This homotopy furnishes an effective demonstration of the bound in Bézout's Theorem for the number of solutions to $F(X) = 0$.

When the polynomial system is deficient, the Bézout homotopy is not optimal. When $n > N$ (often the case in geometric examples), the Bézout homotopy does not apply. In either case, a different strategy is needed. Present optimal homotopies for such systems all exploit some structure of the systems they are designed to solve. The current state-of-the-art is described in [29].

Example 2.4. The Gröbner homotopy [14] is an optimal homotopy that exploits a square-free initial ideal. Suppose our system has the form

$$F := g_1(X), \ldots, g_m(X),\ \Lambda_1(X), \ldots, \Lambda_d(X)$$

where $g_1(X), \ldots, g_m(X)$ form a Gröbner basis for an ideal I with respect to a given term order $\prec$, $\Lambda_1, \ldots, \Lambda_d$ are linear forms with $d = \dim(\mathcal{V}(I))$, *and* we assume that the initial ideal $\mathrm{in}_{\prec} I$ is square-free. This last, restrictive, hypothesis occurs for certain determinantal varieties.

As in [9, Chapter 15], there exist polynomials $g_i(X, t)$ interpolating between $g_i(X)$ and their initial terms $\mathrm{in}_{\prec} g_i(X)$

$$g_i(X; 1) = g_i(X) \quad \text{and} \quad g_i(X; 0) = \mathrm{in}_{\prec} g_i(X)$$

so that $\langle g_1(X, t), \ldots, g_m(X, t) \rangle$ is a flat family with generic fibre isomorphic to I and special fibre $\mathrm{in}_{\prec} I$. The *Gröbner homotopy* is

$$H(X, t) := g_1(X, t), \ldots, g_m(X, t),\ \Lambda_1(X), \ldots, \Lambda_d(X).$$

Since in$_\prec I$ is square-free, $\mathcal{V}(\text{in}_\prec I)$ is a union of $\deg(I)$-many coordinate d-planes. We solve the start system by linear algebra. This conceptually simple homotopy is in general not efficient as it is typically overdetermined.

3   Some Enumerative Geometry

We use the tools we have developed to explore the enumerative geometric problems of cylinders meeting 5 general points and lines tangent to 4 spheres.

3.1   Cylinders Meeting 5 Points

A *cylinder* is the locus of points equidistant from a fixed line in $\mathbb{R}^3$. The Grassmannian of lines in 3-space is 4-dimensional, which implies that the space of cylinders is 5-dimensional, and so we expect that 5 points in $\mathbb{R}^3$ will determine finitely many cylinders. That is, there should be finitely many lines equidistant from 5 general points. The question is: How many cylinders/lines, and how many of them can be real?

Bottema and Veldkamp [5] show there are 6 *complex* cylinders and Lichtblau [17] observes that if the 5 points are the vertices of a bipyramid consisting of 2 regular tetrahedra sharing a common face, then all 6 will be real. We check this reality on a configuration with less symmetry (so the Shape Lemma holds).

If the axial line has direction $\mathbf{V}$ and contains the point $\mathbf{P}$ (and hence has parameterization $\mathbf{P} + t\mathbf{V}$), and if r is the squared radius, then the cylinder is the set of points $\mathbf{X}$ satisfying

$$0 = r - \left\| \mathbf{X} - \mathbf{P} - \frac{\mathbf{V} \cdot (\mathbf{X} - \mathbf{P})}{\|\mathbf{V}\|^2} \mathbf{V} \right\|^2 .$$

Expanding and clearing the denominator of $\|\mathbf{V}\|^2$ yields

$$0 = r\|\mathbf{V}\|^2 + [\mathbf{V} \cdot (\mathbf{X} - \mathbf{P})]^2 - \|\mathbf{X} - \mathbf{P}\|^2 \|\mathbf{V}\|^2 . \tag{5}$$

We consider cylinders containing the following 5 points, which form an asymmetric bipyramid.

```
i34 : Points = {{2, 2,  0 }, {1, -2,  0}, {-3, 0, 0},
                {0, 0, 5/2}, {0,  0, -3}};
```

Suppose that $\mathbf{P} = (0, y_{11}, y_{12})$ and $\mathbf{V} = (1, y_{21}, y_{22})$.

```
i35 : R = QQ[r, y11, y12, y21, y22];

i36 : P = matrix{{0, y11, y12}};

              1       3
o36 : Matrix R  <--- R

i37 : V = matrix{{1, y21, y22}};

              1       3
o37 : Matrix R  <--- R
```

We construct the ideal given by evaluating the polynomial (5) at each of the five points.

```
i38 : Points = matrix Points ** R;

           5       3
o38 : Matrix R  <--- R

i39 : I = ideal apply(0..4, i -> (
            X := Points^{i};
            r * (V * transpose V)  +
            ((X - P) * transpose V)^2) -
            ((X - P) * transpose(X - P)) * (V * transpose V)
          );

o39 : Ideal of R
```

This ideal has dimension 0 and degree 6.

```
i40 : dim I, degree I

o40 = (0, 6)

o40 : Sequence
```

There are 6 real roots, and they correspond to real cylinders (with $r > 0$).

```
i41 : A = R/I; numPosRoots(charPoly(r, Z))

o42 = 6
```

3.2   Lines Tangent to 4 Spheres

We now ask for the lines having a fixed distance from 4 general points. Equivalently, these are the lines mutually tangent to 4 spheres of equal radius. Since the Grassmannian of lines is four-dimensional, we expect there to be only finitely many such lines. Macdonald, Pach, and Theobald [18] show that there are indeed 12 lines, and that all 12 may be real. This problem makes geometric sense over any field k not of characteristic 2, and the derivation of the number 12 is also valid for algebraically closed fields not of characteristic 2.

A sphere in k^3 is given by $\mathcal{V}(q(1, \mathbf{x}))$, where q is some quadratic form on k^4. Here $\mathbf{x} \in k^3$ and we note that not all quadratic forms give spheres. If our field does not have characteristic 2, then there is a symmetric 4×4 matrix M such that $q(\mathbf{u}) = \mathbf{u}M\mathbf{u}^t$.

A line ℓ having direction $\mathbf{V}$ and containing the point $\mathbf{P}$ is tangent to the sphere defined by q when the univariate polynomial in s

$$q((1, \mathbf{P}) + s(0, \mathbf{V})) \ = \ q(1, \mathbf{P}) + 2s(1, \mathbf{P})M(0, \mathbf{V})^t + s^2 q(0, \mathbf{V}),$$

has a double root. Thus its discriminant vanishes, giving the equation

$$\left((1, \mathbf{P})M(0, \mathbf{V})^t\right)^2 \ - \ (1, \mathbf{P})M(1, \mathbf{P})^t \cdot (0, \mathbf{V})M(0, \mathbf{V})^t \ = \ 0. \qquad (6)$$

The matrix M of the quadratic form q of the sphere with center (a, b, c) and squared radius r is constructed by Sphere(a,b,c,r).

```
i43 : Sphere = (a, b, c, r) -> (
            matrix{{a^2 + b^2 + c^2 - r ,-a ,-b ,-c },
                   {            -a        , 1 , 0 , 0 },
                   {            -b        , 0 , 1 , 0 },
                   {            -c        , 0 , 0 , 1 }}
        );
```

If a line ℓ contains the point $\mathbf{P} = (0, y_{11}, y_{12})$ and ℓ has direction $\mathbf{V} = (1, y_{21}, y_{22})$, then `tangentTo(M)` is the equation for ℓ to be tangent to the quadric $uMu^T = 0$ determined by the matrix M.

```
i44 : R = QQ[y11, y12, y21, y22];
```

```
i45 : tangentTo = (M) -> (
            P := matrix{{1, 0, y11, y12}};
            V := matrix{{0, 1, y21, y22}};
            (P * M * transpose V)^2 -
            (P * M * transpose P) * (V * M * transpose V)
        );
```

The ideal of lines having distance $\sqrt{5}$ from the four points $(0, 0, 0)$, $(4, 1, 1)$, $(1, 4, 1)$, and $(1, 1, 4)$ has dimension zero and degree 12.

```
i46 : I = ideal (tangentTo(Sphere(0,0,0,5)),
                 tangentTo(Sphere(4,1,1,5)),
                 tangentTo(Sphere(1,4,1,5)),
                 tangentTo(Sphere(1,1,4,5)));
```

```
o46 : Ideal of R
```

```
i47 : dim I, degree I
```

```
o47 = (0, 12)
```

```
o47 : Sequence
```

Thus there are 12 lines whose distance from those 4 points is $\sqrt{5}$. We check that all 12 are real.

```
i48 : A = R/I;
```

```
i49 : numRealSturm(eliminant(y11 - y12 + y21 + y22, QQ[Z]))
```

```
o49 = 12
```

Since no eliminant given by a coordinate function satisfies the hypotheses of the Shape Lemma, we took the eliminant with respect to the linear form $y_{11} - y_{12} + y_{21} + y_{22}$.

This example is an instance of Lemma 3 of [18]. These four points define a regular tetrahedron with volume $V = 9$ where each face has area $A = \sqrt{3^5}/2$ and each edge has length $e = \sqrt{18}$. That result guarantees that all 12 lines will be real when $e/2 < r < A^2/3V$, which is the case above.

4   Schubert Calculus

The classical Schubert calculus of enumerative geometry concerns linear subspaces having specified positions with respect to other, fixed subspaces. For instance, how many lines in $\mathbb{P}^3$ meet four given lines? (See Example 4.2.)

More generally, let $1 < r < n$ and suppose that we are given general linear subspaces $L_1, \ldots, L_m$ of k^n with $\dim L_i = n - r + 1 - l_i$. When $l_1 + \cdots + l_m = r(n - r)$, there will be a finite number $d(r, n; l_1, \ldots, l_m)$ of r-planes in k^n which meet each L_i non-trivially. This number may be computed using classical algorithms of Schubert and Pieri (see [16]).

The condition on r-planes to meet a fixed $(n-r+1-l)$-plane non-trivially is called a *(special) Schubert condition*, and we call the data $(r, n; l_1, \ldots, l_m)$ *(special) Schubert data*. The *(special) Schubert calculus* concerns this class of enumerative problems. We give two polynomial formulations of this special Schubert calculus, consider their solutions over $\mathbb{R}$, and end with a question for fields of arbitrary characteristic.

4.1   Equations for the Grassmannian

The ambient space for the Schubert calculus is the Grassmannian of r-planes in k^n, denoted $\mathbf{G}_{r,n}$. For $H \in \mathbf{G}_{r,n}$, the rth exterior product of the embedding $H \to k^n$ gives a line

$$k \simeq \wedge^r H \longrightarrow \wedge^r k^n \simeq k^{\binom{n}{r}}.$$

This induces the Plücker embedding $\mathbf{G}_{r,n} \hookrightarrow \mathbb{P}^{\binom{n}{r}-1}$. If H is the row space of an r by n matrix, also written H, then the Plücker embedding sends H to its vector of $\binom{n}{r}$ maximal minors. Thus the r-subsets of $\{0, \ldots, n-1\}$, $\mathbb{Y}_{r,n} := \mathsf{subsets}(\mathsf{n}, \mathsf{r})$, index Plücker coordinates of $\mathbf{G}_{r,n}$. The Plücker ideal of $\mathbf{G}_{r,n}$ is therefore the ideal of algebraic relations among the maximal minors of a generic r by n matrix.

We create the coordinate ring $k[p_\alpha \mid \alpha \in \mathbb{Y}_{2,5}]$ of $\mathbb{P}^9$ and the Plücker ideal of $\mathbf{G}_{2,5}$. The Grassmannian $\mathbf{G}_{r,n}$ of r-dimensional subspaces of k^n is also the Grassmannian of $r-1$-dimensional affine subspaces of $\mathbb{P}^{n-1}$. *Macaulay 2* uses this alternative indexing scheme.

```
i50 : R = ZZ/101[apply(subsets(5,2), i -> p_i )];

i51 : I = Grassmannian(1, 4, R)

o51 = ideal (p        p       - p       p       + p       p       , p       ...
             {2, 3} {1, 4}     {1, 3} {2, 4}     {1, 2} {3, 4}    {2, 3} ...

o51 : Ideal of R
```

This projective variety has dimension 6 and degree 5

```
i52 : dim(Proj(R/I)), degree(I)

o52 = (6, 5)

o52 : Sequence
```

This ideal has an important combinatorial structure [28, Example 11.9]. We write each $\alpha \in \mathbb{Y}_{r,n}$ as an increasing sequence $\alpha \colon \alpha_1 < \cdots < \alpha_r$. Given $\alpha, \beta \in \mathbb{Y}_{r,n}$, consider the two-rowed array with α written above β. We say $\alpha \leq \beta$ if each column weakly increases. If we sort the columns of an array with rows α and β, then the first row is the *meet* $\alpha \wedge \beta$ (greatest lower bound) and the second row the *join* $\alpha \vee \beta$ (least upper bound) of α and β. These definitions endow $\mathbb{Y}_{r,n}$ with the structure of a distributive lattice. Figure 2 shows $\mathbb{Y}_{2,5}$.

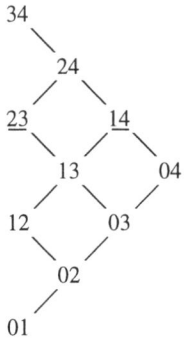

Fig. 2. $\mathbb{Y}_{2,5}$

We give $k[p_\alpha]$ the degree reverse lexicographic order, where we first order the variables p_α by lexicographic order on their indices α.

Theorem 4.1. *The reduced Gröbner basis of the Plücker ideal with respect to this degree reverse lexicographic term order consists of quadratic polynomials*

$$g(\alpha, \beta) \;=\; p_\alpha \cdot p_\beta \;-\; p_{\alpha \vee \beta} \cdot p_{\alpha \wedge \beta} \;+\; \text{lower terms in} \prec,$$

for each incomparable pair α, β in $\mathbb{Y}_{r,n}$, and all lower terms $\lambda p_\gamma \cdot p_\delta$ in $g(\alpha, \beta)$ satisfy $\gamma \leq \alpha \wedge \beta$ and $\alpha \vee \beta \leq \delta$.

The form of this Gröbner basis implies that the standard monomials are the sortable monomials, those $p_\alpha p_\beta \cdots p_\gamma$ with $\alpha \leq \beta \leq \cdots \leq \gamma$. Thus the Hilbert function of $\mathbf{G}_{r,n}$ may be expressed in terms of the combinatorics of $\mathbb{Y}_{r,n}$. For instance, the dimension of $\mathbf{G}_{r,n}$ is the rank of $\mathbb{Y}_{r,n}$, and its degree is the number of maximal chains. From Figure 2, these are 6 and 5 for $\mathbb{Y}_{2,5}$, confirming our previous calculations.

Since the generators $g(\alpha, \beta)$ are linearly independent, this Gröbner basis is also a minimal generating set for the ideal. The displayed generator in o51,

$$p_{\{2,3\}} p_{\{1,4\}} \;-\; p_{\{1,3\}} p_{\{2,4\}} \;-\; p_{\{1,2\}} p_{\{3,4\}} \;,$$

is $g(23, 14)$, and corresponds to the underlined incomparable pair in Figure 2. Since there are 5 such incomparable pairs, the Gröbner basis has 5 generators. As $\mathbf{G}_{2,5}$ has codimension 3, it is not a complete intersection. This shows how the general enumerative problem from the Schubert calculus gives rise to an overdetermined system of equations in this global formulation.

The Grassmannian has a useful system of local coordinates given by $\text{Mat}_{r,n-r}$ as follows

$$Y \in \text{Mat}_{r,n-r} \longmapsto \text{rowspace } [I_r : Y] \in \mathbf{G}_{r,n}. \tag{7}$$

Let L be a $(n-r+1-l)$-plane in k^n which is the row space of a $n-r+1-l$ by n matrix, also written L. Then L meets $X \in \mathbf{G}_{r,n}$ non-trivially if

$$\text{maximal minors of } \begin{bmatrix} L \\ X \end{bmatrix} = 0.$$

Laplace expansion of each minor along the rows of X gives a linear equation in the Plücker coordinates. In the local coordinates (substituting $[I_r : Y]$ for X), we obtain multilinear equations of degree $\min\{r, n-r\}$. These equations generate a prime ideal of codimension l.

Suppose each $l_i = 1$ in our enumerative problem. Then in the Plücker coordinates, we have the Plücker ideal of $\mathbf{G}_{r,n}$ together with $r(n-r)$ linear equations, one for each $(n-r)$-plane L_i. By Theorem 4.1, the Plücker ideal has a square-free initial ideal, and so the Gröbner homotopy of Example 2.4 may be used to solve this enumerative problem.

Example 4.2. $\mathbf{G}_{2,4} \subset \mathbb{P}^5$ has equation

$$p_{\{1,2\}}p_{\{0,3\}} - p_{\{1,3\}}p_{\{0,2\}} + p_{\{2,3\}}p_{\{0,1\}} = 0. \tag{8}$$

The condition for $H \in \mathbf{G}_{2,4}$ to meet a 2-plane L is the vanishing of

$$p_{\{1,2\}}L_{34} - p_{\{1,3\}}L_{24} + p_{\{2,3\}}L_{14} + p_{\{1,4\}}L_{23} - p_{\{2,4\}}L_{13} + p_{\{3,4\}}L_{12}, \tag{9}$$

where L_{ij} is the (i,j)th maximal minor of L.

If $l_1 = \cdots = l_4 = 1$, we have 5 equations in $\mathbb{P}^5$, one quadratic and 4 linear, and so by Bézout's Theorem there are two 2-planes in k^4 that meet 4 general 2-planes non-trivially. This means that there are 2 lines in $\mathbb{P}^3$ meeting 4 general lines. In local coordinates, (9) becomes

$$L_{34} - L_{14}y_{11} + L_{13}y_{12} - L_{24}y_{21} + L_{23}y_{22} + L_{12}(y_{11}y_{22} - y_{12}y_{21}).$$

This polynomial has the form of the last specialization in Example 1.2.

4.2   Reality in the Schubert Calculus

Like the other enumerative problems we have discussed, enumerative problems in the special Schubert calculus are fully real in that all solutions can be real [22]. That is, given any Schubert data $(r, n; l_1, \ldots, l_m)$, there exist subspaces $L_1, \ldots, L_m \subset \mathbb{R}^n$ such that each of the $d(r, n; l_1, \ldots, l_m)$ r-planes that meet each L_i are themselves real.

This result gives some idea of which choices of the L_i give all r-planes real. Let γ be a fixed rational normal curve in $\mathbb{R}^n$. Then the L_i are linear subspaces osculating γ. More concretely, suppose that γ is the standard rational normal curve, $\gamma(s) = (1, s, s^2, \ldots, s^{n-1})$. Then the i-plane $L_i(s) := \langle \gamma(s), \gamma'(s), \ldots, \gamma^{(i-1)}(s) \rangle$ osculating γ at $\gamma(s)$ is the row space of the matrix given by oscPlane(i, n, s).

```
i53 : oscPlane = (i, n, s) -> (
            gamma := matrix {toList apply(1..n, i -> s^(i-1))};
            L := gamma;
            j := 0;
            while j < i-1 do (gamma = diff(s, gamma);
                    L = L || gamma;
                    j = j+1);
            L);

i54 : QQ[s]; oscPlane(3, 6, s)

o55 = | 1 s s2 s3  s4   s5  |
      | 0 1 2s 3s2 4s3  5s4 |
      | 0 0 2  6s  12s2 20s3 |

                   3        6
o55 : Matrix QQ [s]  <--- QQ [s]
```

(In o55, the exponents of s are displayed in line: s^2 is written s2. *Macaulay 2* uses this notational convention to display matrices efficiently.)

Theorem 4.3 ([22]). *For any Schubert data* $(r, n; l_1, \ldots, l_m)$, **there exist** *real numbers* $s_1, s_2, \ldots, s_m$ *such that there are* $d(r, n; l_1, \ldots, l_m)$ *r-planes that meet each osculating plane* $L_i(s_i)$, *and all are real.*

The inspiration for looking at subspaces osculating the rational normal curve to study real enumerative geometry for the Schubert calculus is the following very interesting conjecture of Boris Shapiro and Michael Shapiro, or more accurately, extensive computer experimentation based upon their conjecture [20,23,25,30].

Shapiros's Conjecture. *For any Schubert data* $(r, n; l_1, \ldots, l_m)$ *and* **for all** *real numbers* $s_1, s_2, \ldots, s_m$ *there are* $d(r, n; l_1, \ldots, l_m)$ *r-planes that meet each osculating plane* $L_i(s_i)$, *and all are real.*

In addition to Theorem 4.3, (which replaces the quantifier *for all* by *there exist*), the strongest evidence for this Conjecture is the following result of Eremenko and Gabrielov [10].

Theorem 4.4. *Shapiros's Conjecture is true when either r or $n - r$ is 2.*

We test an example of this conjecture for the Schubert data $(3, 6; 1^3, 2^3)$, (where a^b is a repeated b times). The algorithms of the Schubert calculus predict that $d(3, 6; 1^3, 2^3) = 6$. The function spSchub(r, L, P) computes the ideal of r-planes meeting the row space of L in the Plücker coordinates P_α.

```
i56 : spSchub = (r, L, P) -> (
          I := ideal apply(subsets(numgens source L,
                            r + numgens target L), S ->
                fold((sum, U) -> sum +
                  fold((term,i) -> term*(-1)^i, P_(S_U) * det(
                  submatrix(L, sort toList(set(S) - set(S_U)))), U),
                  0, subsets(#S, r))));
```

We are working in the Grassmannian of 3-planes in $\mathbb{C}^6$.

```
i57 : R = QQ[apply(subsets(6,3), i -> p_i )];
```

The ideal I consists of the special Schubert conditions for the 3-planes to meet the 3-planes osculating the rational normal curve at the points 1, 2, and 3, and to also meet the 2-planes osculating at 4, 5, and 6, together with the Plücker ideal Grassmannian(2, 5, R). Since this is a 1-dimensional homogeneous ideal, we add the linear form p_{0,1,5} - 1 to make the ideal zero-dimensional. As before, Grassmannian(2, 5, R) creates the Plücker ideal of $\mathbf{G}_{3,6}$.

```
i58 : I = fold((J, i) -> J +
            spSchub(3, substitute(oscPlane(3, 6, s), {s=> 1+i}), p) +
            spSchub(3, substitute(oscPlane(2, 6, s), {s=> 4+i}), p),
            Grassmannian(2, 5, R), {0,1,2}) +
            ideal (p_{0,1,5} - 1);

o58 : Ideal of R
```

This has dimension 0 and degree 6, in agreement with the Schubert calculus.

```
i59 : dim I, degree I

o59 = (0, 6)

o59 : Sequence
```

As expected, all roots are real.

```
i60 : A = R/I; numRealSturm(eliminant(p_{2,3,4}, QQ[Z]))

o61 = 6
```

There have been many checked instances of this conjecture [23,25,30], and it has some geometrically interesting generalizations [26].

The question remains for which numbers $0 \leq d \leq d(r, n; l_1, \ldots, l_m)$ do there exist real planes L_i with $d(r, n; l_1, \ldots, l_m)$ r-planes meeting each L_i, and exactly d of them are real. Besides Theorem 4.3 and the obvious parity condition, nothing is known in general. In every known case, every possibility occurs—which is not the case in all enumerative problems, even those that are fully real[1]. Settling this (for $d = 0$) has implications for linear systems theory [20].[2]

---

[1] For example, of the 12 rational plane cubics containing 8 real points in $\mathbb{P}^2$, either 8, 10 or 12 can be real, and there are 8 points with all 12 real [8, Proposition 4.7.3].

[2] After this was written, Eremenko and Gabrielov [11] showed that d can be zero for the enumerative problems given by data $(2, 2n, 1^{4n-4})$ and $(2n-2, 2n, 1^{4n-4})$.

4.3   Transversality in the Schubert Calculus

A basic principle of the classical Schubert calculus is that the intersection number $d(r, n; l_1, \ldots, l_m)$ has enumerative significance—that is, for general linear subspaces L_i, all solutions appear with multiplicity 1. This basic principle is not known to hold in general. For fields of characteristic zero, Kleiman's Transversality Theorem [15] establishes this principle. When r or $n-r$ is 2, then Theorem E of [21] establishes this principle in arbitrary characteristic. We conjecture that this principle holds in general; that is, for arbitrary infinite fields and any Schubert data, if the planes L_i are in general position, then the resulting zero-dimensional ideal is radical.

We test this conjecture on the enumerative problem of Section 4.2, which is not covered by Theorem E of [21]. The function testTransverse(F) tests transversality for this enumerative problem, for a given field F. It does this by first computing the ideal of the enumerative problem using random planes L_i.

```
i62 : randL = (R, n, r, l) ->
              matrix table(n-r+1-1, n, (i, j) -> random(0, R));
```

and the Plücker ideal of the Grassmannian $G_{3,6}$ Grassmannian(2, 5, R).) Then it adds a random (inhomogeneous) linear relation 1 + random(1, R) to make the ideal zero-dimensional for generic L_i. When this ideal is zero dimensional and has degree 6 (the expected degree), it computes the characteristic polynomial g of a generic linear form. If g has no multiple roots, 1 == gcd(g, diff(Z, g)), then the Shape Lemma guarantees that the ideal was radical. testTransverse exits either when it computes a radical ideal, or after limit iterations (which is set to 5 for these examples), and prints the return status.

```
i63 : testTransverse = F -> (
           R := F[apply(subsets(6, 3), i -> q_i )];
           continue := true;
           j := 0;
           limit := 5;
           while continue and (j < limit) do (
                j = j + 1;
                I := fold((J, i) -> J +
                          spSchub(3, randL(R, 6, 3, 1), q) +
                          spSchub(3, randL(R, 6, 3, 2), q),
                          Grassmannian(2, 5, R) +
                          ideal (1 + random(1, R)),
                          {0, 1, 2});
                if (dim I == 0) and (degree I == 6) then (
                lin := promote(random(1, R), (R/I));
                g := charPoly(lin, Z);
                continue = not(1 == gcd(g, diff(Z, g)));
                ));
           if continue then << "Failed for the prime " << char F <<
                " with " << j << " iterations" << endl;
           if not continue then << "Succeeded for the prime " <<
                char F << " in " << j << " iteration(s)" << endl;
           );
```

Since 5 iterations do not show transversality for $\mathbb{F}_2$,

```
i64 : testTransverse(ZZ/2);
Failed for the prime 2 with 5 iterations
```

we can test transversality in characteristic 2 using the field with four elements, $\mathbb{F}_4 = \text{GF } 4$.

```
i65 : testTransverse(GF 4);
Succeeded for the prime 2 in 3 iteration(s)
```

We do find transversality for $\mathbb{F}_7$.

```
i66 : testTransverse(ZZ/7);
Succeeded for the prime 7 in 2 iteration(s)
```

We have tested transversality for all primes less than 100 in every enumerative problem involving Schubert conditions on 3-planes in k^6. These include the problem above as well as the problem of 42 3-planes meeting 9 general 3-planes.[3]

5   The 12 Lines: Reprise

The enumerative problems of Section 3 were formulated in local coordinates (7) for the Grassmannian of lines in $\mathbb{P}^3$ (Grassmannian of 2-dimensional subspaces in k^4). When we formulate the problem of Section 3.2 in the global Plücker coordinates of Section 4.1, we find some interesting phenomena. We also consider some related enumerative problems.

5.1   Global Formulation

A quadratic form q on a vector space V over a field k not of characteristic 2 is given by $q(\mathbf{u}) = (\varphi(\mathbf{u}), \mathbf{u})$, where $\varphi \colon V \to V^*$ is a *symmetric* linear map, that is $(\varphi(\mathbf{u}), \mathbf{v}) = (\varphi(\mathbf{v}), \mathbf{u})$. Here, V^* is the linear dual of V and $(\cdot, \cdot)$ is the pairing $V \otimes V^* \to k$. The map φ induces a quadratic form $\wedge^r q$ on the rth exterior power $\wedge^r V$ of V through the symmetric map $\wedge^r \varphi \colon \wedge^r V \to \wedge^r V^* = (\wedge^r V)^*$. The action of $\wedge^r V^*$ on $\wedge^r V$ is given by

$$(\mathbf{x}_1 \wedge \mathbf{x}_2 \wedge \cdots \wedge \mathbf{x}_r, \ \mathbf{y}_1 \wedge \mathbf{y}_2 \wedge \cdots \wedge \mathbf{y}_r) \ = \ \det |(\mathbf{x}_i, \mathbf{y}_j)|, \qquad (10)$$

where $\mathbf{x}_i \in V^*$ and $\mathbf{y}_j \in V$.

When we fix isomorphisms $V \simeq k^n \simeq V^*$, the map φ is given by a symmetric $n \times n$ matrix M as in Section 3.2. Suppose $r = 2$. Then for $\mathbf{u}, \mathbf{v} \in k^n$,

$$\wedge^2 q(\mathbf{u} \wedge \mathbf{v}) \ = \ \det \begin{bmatrix} \mathbf{u}M\mathbf{u}^t & \mathbf{u}M\mathbf{v}^t \\ \mathbf{v}M\mathbf{u}^t & \mathbf{v}M\mathbf{v}^t \end{bmatrix},$$

which is Equation (6) of Section 3.2.

---

[3] After this was written, we discovered an elementary proof of transversality for the enumerative problems given by data $(r, n; 1^{r(n-r)})$, where the conditions are all codimension 1 [24].

Proposition 5.1. *A line ℓ is tangent to a quadric $\mathcal{V}(q)$ in $\mathbb{P}^{n-1}$ if and only if its Plücker coordinate $\wedge^2\ell \in \mathbb{P}^{\binom{n}{2}-1}$ lies on the quadric $\mathcal{V}(\wedge^2 q)$.*

Thus the Plücker coordinates for the set of lines tangent to 4 general quadrics in $\mathbb{P}^3$ satisfy 5 quadratic equations: The single Plücker relation (8) together with one quadratic equation for each quadric. Thus we expect the Bézout number of $2^5 = 32$ such lines. We check this.

The procedure `randomSymmetricMatrix(R, n)` generates a random symmetric $n \times n$ matrix with entries in the base ring of R.

```
i67 : randomSymmetricMatrix = (R, n) -> (
          entries := new MutableHashTable;
          scan(0..n-1, i -> scan(i..n-1, j ->
                       entries#(i, j) = random(0, R)));
          matrix table(n, n, (i, j) -> if i > j then
                       entries#(j, i) else entries#(i, j))
      );
```

The procedure `tangentEquation(r, R, M)` gives the equation in Plücker coordinates for a point in $\mathbb{P}^{\binom{n}{r}-1}$ to be isotropic with respect to the bilinear form $\wedge^r M$ (R is assumed to be the coordinate ring of $\mathbb{P}^{\binom{n}{r}-1}$). This is the equation for an r-plane to be tangent to the quadric associated to M.

```
i68 : tangentEquation = (r, R, M) -> (
          g := matrix {gens(R)};
          (entries(g * exteriorPower(r, M) * transpose g))_0_0
      );
```

We construct the ideal of lines tangent to 4 general quadrics in $\mathbb{P}^3$.

```
i69 : R = QQ[apply(subsets(4, 2), i -> p_i )];
```

```
i70 : I = Grassmannian(1, 3, R) + ideal apply(0..3, i ->
              tangentEquation(2, R, randomSymmetricMatrix(R, 4)));
```

```
o70 : Ideal of R
```

As expected, this ideal has dimension 0 and degree 32.

```
i71 : dim Proj(R/I), degree I
```

```
o71 = (0, 32)
```

```
o71 : Sequence
```

5.2   Lines Tangent to 4 Spheres

That calculation raises the following question: In Section 3.2, why did we obtain only 12 lines tangent to 4 spheres? To investigate this, we generate the global ideal of lines tangent to the spheres of Section 3.2.

```
i72 : I = Grassmannian(1, 3, R) +
              ideal (tangentEquation(2, R, Sphere(0,0,0,5)),
                     tangentEquation(2, R, Sphere(4,1,1,5)),
                     tangentEquation(2, R, Sphere(1,4,1,5)),
                     tangentEquation(2, R, Sphere(1,1,4,5)));
```

```
o72 : Ideal of R
```

We compute the dimension and degree of $\mathcal{V}(I)$.

```
i73 : dim Proj(R/I), degree I
```

```
o73 = (1, 4)
```

```
o73 : Sequence
```

The ideal is not zero dimensional; there is an extraneous one-dimensional component of zeroes with degree 4. Since we found 12 lines in Section 3.2 using the local coordinates (7), the extraneous component must lie in the complement of that coordinate patch, which is defined by the vanishing of the first Plücker coordinate, $p_{\{0,1\}}$. We saturate I by $p_{\{0,1\}}$ to obtain the desired lines.

```
i74 : Lines = saturate(I, ideal (p_{0,1}));
```

```
o74 : Ideal of R
```

This ideal does have dimension 0 and degree 12, so we have recovered the zeroes of Section 3.2.

```
i75 : dim Proj(R/Lines), degree(Lines)
```

```
o75 = (0, 12)
```

```
o75 : Sequence
```

We investigate the rest of the zeroes, which we obtain by taking the ideal quotient of I and the ideal of lines. As computed above, this has dimension 1 and degree 4.

```
i76 : Junk = I : Lines;
```

```
o76 : Ideal of R
```

```
i77 : dim Proj(R/Junk), degree Junk
```

```
o77 = (1, 4)
```

```
o77 : Sequence
```

We find the support of this extraneous component by taking its radical.

```
i78 : radical(Junk)
```

$$o78 = \text{ideal } (p_{\{0,\,3\}}\,,\; p_{\{0,\,2\}}\,,\; p_{\{0,\,1\}}\,,\; p_{\{1,\,2\}}^{2} + p_{\{1,\,3\}}^{2} + p_{\{2,\,3\}}^{2})$$

```
o78 : Ideal of R
```

From this, we see that the extraneous component is supported on an imaginary conic in the $\mathbb{P}^2$ of lines at infinity.

To understand the geometry behind this computation, observe that the sphere with radius r and center (a, b, c) has homogeneous equation

$$(x - wa)^2 + (y - wb)^2 + (z - wc)^2 = r^2 w^2.$$

At infinity, $w = 0$, this has equation

$$x^2 + y^2 + z^2 = 0.$$

The extraneous component is supported on the set of tangent lines to this imaginary conic. Aluffi and Fulton [1] studied this problem, using geometry to identify the extraneous ideal and the excess intersection formula [13] to obtain the answer of 12. Their techniques show that there will be 12 isolated lines tangent to 4 quadrics which have a smooth conic in common.

When the quadrics are spheres, the conic is the imaginary conic at infinity. Fulton asked the following question: Can all 12 lines be real if the (real) four quadrics share a real conic? We answer his question in the affirmative in the next section.

5.3   Lines Tangent to Real Quadrics Sharing a Real Conic

We consider four quadrics in $\mathbb{P}^3_{\mathbb{R}}$ sharing a non-singular conic, which we will take to be at infinity so that we may use local coordinates for $\mathbf{G}_{2,4}$ in our computations. The variety $\mathcal{V}(q) \subset \mathbb{P}^3_{\mathbb{R}}$ of a nondegenerate quadratic form q is determined up to isomorphism by the absolute value of the signature σ of the associated bilinear form. Thus there are three possibilities, 0, 2, or 4, for $|\sigma|$.

When $|\sigma| = 4$, the real quadric $\mathcal{V}(q)$ is empty. The associated symmetric matrix M is conjugate to the identity matrix, so $\wedge^2 M$ is also conjugate to the identity matrix. Hence $\mathcal{V}(\wedge^2 q)$ contains no real points. Thus we need not consider quadrics with $|\sigma| = 4$.

When $|\sigma| = 2$, we have $\mathcal{V}(q) \simeq S^2$, the 2-sphere. If the conic at infinity is imaginary, then $\mathcal{V}(q) \subset \mathbb{R}^3$ is an ellipsoid. If the conic at infinity is real, then $\mathcal{V}(q) \subset \mathbb{R}^3$ is a hyperboloid of two sheets. When $\sigma = 0$, we have $\mathcal{V}(q) \simeq S^1 \times S^1$, a torus. In this case, $\mathcal{V}(q) \subset \mathbb{R}^3$ is a hyperboloid of one sheet and the conic at infinity is real.

Thus either we have 4 ellipsoids sharing an imaginary conic at infinity, which we studied in Section 3.2; or else we have four hyperboloids sharing a real conic at infinity, and there are five possible combinations of hyperboloids of one or two sheets sharing a real conic at infinity. This gives six topologically distinct possibilities in all.

Theorem 5.2. *For each of the six topologically distinct possibilities of four real quadrics sharing a smooth conic at infinity, there exist four quadrics having the property that each of the 12 lines in $\mathbb{C}^3$ simultaneously tangent to the four quadrics is real.*

Proof. By the computation in Section 3.2, we need only check the five possibilities for hyperboloids. We fix the conic at infinity to be $x^2 + y^2 - z^2 = 0$. Then the general hyperboloid of two sheets containing this conic has equation in $\mathbb{R}^3$

$$(x - a)^2 + (y - b)^2 - (z - c)^2 + r = 0, \tag{11}$$

(with $r > 0$). The command `Two(a,b,c,r)` generates the associated symmetric matrix.

```
i79 : Two = (a, b, c, r) -> (
          matrix{{a^2 + b^2 - c^2 + r ,-a ,-b , c },
                 {        -a          , 1 , 0 , 0 },
                 {        -b          , 0 , 1 , 0 },
                 {         c          , 0 , 0 ,-1 }}
      );
```

The general hyperboloid of one sheet containing the conic $x^2 + y^2 - z^2 = 0$ at infinity has equation in $\mathbb{R}^3$

$$(x - a)^2 + (y - b)^2 - (z - c)^2 - r \; = \; 0, \tag{12}$$

(with $r > 0$). The command $\mathtt{One(a,b,c,r)}$ generates the associated symmetric matrix.

```
i80 : One = (a, b, c, r) -> (
          matrix{{a^2 + b^2 - c^2 - r ,-a ,-b , c },
                 {        -a          , 1 , 0 , 0 },
                 {        -b          , 0 , 1 , 0 },
                 {         c          , 0 , 0 ,-1 }}
      );
```

We consider i quadrics of two sheets (11) and $4 - i$ quadrics of one sheet (12). For each of these cases, the table below displays four 4-tuples of data (a, b, c, r) which give 12 real lines. (The data for the hyperboloids of one sheet are listed first.)

i	Data			
0	$(5, 3, 3, 16)$,	$(5, -4, 2, 1)$,	$(-3, -1, 1, 1)$,	$(2, -7, 0, 1)$
1	$(3, -2, -3, 6)$,	$(-3, -7, -6, 7)$,	$(-6, 3, -5, 2)$,	$(1, 6, -2, 5)$
2	$(6, 4, 6, 4)$,	$(-1, 3, 3, 6)$,	$(-7, -2, 3, 3)$,	$(-6, 7, -2, 5)$
3	$(-1, -4, -1, 1)$,	$(-3, 3, -1, 1)$,	$(-7, 6, 2, 9)$,	$(5, 6, -1, 12)$
4	$(5, 2, -1, 25)$,	$(6, -6, 2, 25)$,	$(-7, 1, 6, 1)$,	$(3, 1, 0, 1)$

We test each of these, using the formulation in local coordinates of Section 3.2.

```
i81 : R = QQ[y11, y12, y21, y22];

i82 : I = ideal (tangentTo(One( 5, 3, 3,16)),
                 tangentTo(One( 5,-4, 2, 1)),
                 tangentTo(One(-3,-1, 1, 1)),
                 tangentTo(One( 2,-7, 0, 1)));

o082 : Ideal of R

i83 : numRealSturm(charPoly(promote(y22, R/I), Z))

o083 = 12

i84 : I = ideal (tangentTo(One( 3,-2,-3, 6)),
                 tangentTo(One(-3,-7,-6, 7)),
                 tangentTo(One(-6, 3,-5, 2)),
                 tangentTo(Two( 1, 6,-2, 5)));

o084 : Ideal of R
```

```
i85 : numRealSturm(charPoly(promote(y22, R/I), Z))

o85 = 12

i86 : I = ideal (tangentTo(One( 6, 4, 6, 4)),
                 tangentTo(One(-1, 3, 3, 6)),
                 tangentTo(Two(-7,-2, 3, 3)),
                 tangentTo(Two(-6, 7,-2, 5)));

o86 : Ideal of R

i87 : numRealSturm(charPoly(promote(y22, R/I), Z))

o87 = 12

i88 : I = ideal (tangentTo(One(-1,-4,-1, 1)),
                 tangentTo(Two(-3, 3,-1, 1)),
                 tangentTo(Two(-7, 6, 2, 9)),
                 tangentTo(Two( 5, 6,-1,12)));

o88 : Ideal of R

i89 : numRealSturm(charPoly(promote(y22, R/I), Z))

o89 = 12

i90 : I = ideal (tangentTo(Two( 5, 2,-1,25)),
                 tangentTo(Two( 6,-6, 2,25)),
                 tangentTo(Two(-7, 1, 6, 1)),
                 tangentTo(Two( 3, 1, 0, 1)));

o90 : Ideal of R

i91 : numRealSturm(charPoly(promote(y22, R/I), Z))

o91 = 12
```

$\square$

In each of these enumerative problems there are 12 complex solutions. For each, we have done other computations showing that every possible number of real solutions (0, 2, 4, 6, 8, 10, or 12) can occur.

5.4   Generalization to Higher Dimensions

We consider lines tangent to quadrics in higher dimensions. First, we reinterpret the action of $\wedge^r V^*$ on $\wedge^r V$ described in (10) as follows. The vectors $\mathbf{x}_1, \ldots, \mathbf{x}_r$ and $\mathbf{y}_1, \ldots, \mathbf{y}_r$ define maps $\alpha : k^r \to V^*$ and $\beta : k^r \to V$. The matrix $[\langle \mathbf{x}_i, \mathbf{y}_j \rangle]$ is the matrix of the bilinear form on k^r given by $\langle \mathbf{u}, \mathbf{v} \rangle := (\alpha(\mathbf{u}), \beta(\mathbf{v}))$. Thus (10) vanishes when the bilinear form $\langle \cdot, \cdot \rangle$ on k^r is degenerate.

Now suppose that we have a quadratic form q on V given by a symmetric map $\varphi : V \to V^*$. This induces a quadratic form and hence a quadric on any r-plane H in V (with $H \not\subset \mathcal{V}(q)$). This induced quadric is singular when H is tangent to $\mathcal{V}(q)$. Since a quadratic form is degenerate only when the associated projective quadric is singular, we see that H is tangent to the quadric $\mathcal{V}(q)$ if and only if $(\wedge^r \varphi(\wedge^r H), \wedge^r H) = 0$. (This includes the case $H \subset \mathcal{V}(q)$.) We summarize this argument.

Theorem 5.3. *Let $\varphi\colon V \to V^*$ be a linear map with resulting bilinear form $(\varphi(\mathbf{u}), \mathbf{v})$. Then the locus of r-planes in V for which the restriction of this form is degenerate is the set of r-planes H whose Plücker coordinates are isotropic, $(\wedge^r \varphi(\wedge^r H), \wedge^r H) = 0$, with respect to the induced form on $\wedge^r V$.*

When φ is symmetric, this is the locus of r-planes tangent to the associated quadric in $\mathbb{P}(V)$.

We explore the problem of lines tangent to quadrics in $\mathbb{P}^n$. From the calculations of Section 5.1, we do not expect this to be interesting if the quadrics are general. (This is borne out for $\mathbb{P}^4$: we find 320 lines in $\mathbb{P}^4$ tangent to 6 general quadrics. This is the Bézout number, as $\deg \mathbf{G}_{2,5} = 5$ and the condition to be tangent to a quadric has degree 2.) This problem is interesting if the quadrics in $\mathbb{P}^n$ share a quadric in a $\mathbb{P}^{n-1}$. We propose studying such enumerative problems, both determining the number of solutions for general such quadrics, and investigating whether or not it is possible to have all solutions be real.

We use *Macaulay 2* to compute the expected number of solutions to this problem when $r = 2$ and $n = 4$. We first define some functions for this computation, which will involve counting the degree of the ideal of lines in $\mathbb{P}^4$ tangent to 6 general spheres. Here, X gives local coordinates for the Grassmannian, M is a symmetric matrix, `tanQuad` gives the equation in X for the lines tangent to the quadric given by M.

```
i92 : tanQuad = (M, X) -> (
           u := X^{0};
           v := X^{1};
           (u * M * transpose v)^2 -
           (u * M * transpose u) * (v * M * transpose v)
           );
```

`nSphere` gives the matrix M for a sphere with center V and squared radius r, and V and r give random data for a sphere.

```
i93 : nSphere = (V, r) ->
              (matrix {{r + V * transpose V}} || transpose V ) |
              ( V || id_((ring r)^n)
              );

i94 : V = () -> matrix table(1, n, (i,j) -> random(0, R));

i95 : r = () -> random(0, R);
```

We construct the ambient ring, local coordinates, and the ideal of the enumerative problem of lines in $\mathbb{P}^4$ tangent to 6 random spheres.

```
i96 : n = 4;

i97 : R = ZZ/1009[flatten(table(2, n-1, (i,j) -> z_(i,j)))];

i98 : X = 1 | matrix table(2, n-1, (i,j) -> z_(i,j))

o98 = | 1 0 z_(0,0) z_(0,1) z_(0,2) |
      | 0 1 z_(1,0) z_(1,1) z_(1,2) |

                2       5
o98 : Matrix R  <--- R
```

```
i99 : I = ideal (apply(1..(2*n-2),
                 i -> tanQuad(nSphere(V(), r()), X)));

o99 : Ideal of R
```

We find there are 24 lines in $\mathbb{P}^4$ tangent to 6 general spheres.

```
i100 : dim I, degree I

o100 = (0, 24)

o100 : Sequence
```

The expected numbers of solutions we have obtained in this way are displayed in the table below. The numbers in boldface are those which are proven.[4]

n	2	3	4	5	6
# expected	**4**	**12**	24	48	96

Acknowledgments. We thank Dan Grayson and Bernd Sturmfels: some of the procedures in this chapter were written by Dan Grayson and the calculation in Section 5.2 is due to Bernd Sturmfels.

References

1. P. Aluffi and W. Fulton: Lines tangent to four surfaces containing a curve. 2001.
2. E. Becker, M. G. Marinari, T. Mora, and C. Traverso: The shape of the Shape Lemma. In *Proceedings ISSAC-94*, pages 129–133, 1993.
3. E. Becker and Th. Wöermann: On the trace formula for quadratic forms. In *Recent advances in real algebraic geometry and quadratic forms (Berkeley, CA, 1990/1991; San Francisco, CA, 1991)*, pages 271–291. Amer. Math. Soc., Providence, RI, 1994.
4. D. N. Bernstein: The number of roots of a system of equations. *Funct. Anal. Appl.*, 9:183–185, 1975.
5. O. Bottema and G.R. Veldkamp: On the lines in space with equal distances to n given points. *Geometrie Dedicata*, 6:121–129, 1977.
6. A. M. Cohen, H. Cuypers, and H. Sterk, editors. *Some Tapas of Computer Algebra*. Springer-Varlag, 1999.
7. D. Cox, J. Little, and D. O'Shea: *Ideals, Varieties, Algorithms: An Introduction to Computational Algebraic Geometry and Commutative Algebra*. UTM. Springer-Verlag, New York, 1992.
8. A. I. Degtyarev and V. M. Kharlamov: Topological properties of real algebraic varieties: Rokhlin's way. *Uspekhi Mat. Nauk*, 55(4(334)):129–212, 2000.
9. D. Eisenbud: *Commutative Algebra With a View Towards Algebraic Geometry*. Number 150 in GTM. Springer-Verlag, 1995.
10. A. Eremenko and A. Gabrielov: Rational functions with real critical points and B. and M. Shapiro conjecture in real enumerative geometry. MSRI preprint 2000-002, 2000.

---

[4] As this was going to press, the obvious pattern was proven: There are $3 \cdot 2^{n-1}$ complex lines tangent to $2n - 2$ general spheres in $\mathbb{R}^n$, and all may be real [27].

11. A. Eremenko and A. Gabrielov: New counterexamples to pole placement by static output feedback. Linear Algebra and its Applications, to appear, 2001.
12. W. Fulton: *Intersection Theory*. Number 2 in Ergebnisse der Mathematik und ihrer Grenzgebiete. Springer-Verlag, 1984.
13. W. Fulton and R. MacPherson: Intersecting cycles on an algebraic variety. In P. Holm, editor, *Real and Complex Singularities*, pages 179–197. Oslo, 1976, Sijthoff and Noordhoff, 1977.
14. B. Huber, F. Sottile, and B. Sturmfels: Numerical Schubert calculus. *J. Symb. Comp.*, 26(6):767–788, 1998.
15. S. Kleiman: The transversality of a general translate. *Compositio Math.*, 28:287–297, 1974.
16. S. Kleiman and D. Laksov: Schubert calculus. *Amer. Math. Monthly*, 79:1061–1082, 1972.
17. D. Lichtblau: Finding cylinders through 5 points in $\mathbb{R}^3$. Mss., email address: danl@wolfram.com, 2001.
18. I.G. Macdonald, J. Pach, and T. Theobald: Common tangents to four unit balls in $\mathbb{R}^3$. To appear in *Discrete and Computational Geometry* **26**:1 (2001).
19. P. Pedersen, M.-F. Roy, and A. Szpirglas: Counting real zeros in the multivariate case. In *Computational Algebraic Geometry (Nice, 1992)*, pages 203–224. Birkhäuser Boston, Boston, MA, 1993.
20. J. Rosenthal and F. Sottile: Some remarks on real and complex output feedback. *Systems Control Lett.*, 33(2):73–80, 1998. For a description of the computational aspects, see http://www.nd.edu/~rosen/pole/.
21. F. Sottile: Enumerative geometry for the real Grassmannian of lines in projective space. *Duke Math. J.*, 87(1):59–85, 1997.
22. F. Sottile: The special Schubert calculus is real. *ERA of the AMS*, 5:35–39, 1999.
23. F. Sottile: The conjecture of Shapiro and Shapiro. An archive of computations and computer algebra scripts, http://www.expmath.org/extra/9.2/sottile/, 2000.
24. F. Sottile: Elementary transversality in the schubert calculus in any characteristic. math.AG/0010319, 2000.
25. F. Sottile: Real Schubert calculus: Polynomial systems and a conjecture of Shapiro and Shapiro. *Exper. Math.*, 9:161–182, 2000.
26. F. Sottile: Some real and unreal enumerative geometry for flag manifolds. *Mich. Math. J*, 48:573–592, 2000.
27. F. Sottile and T. Theobald: Lines tangent to $2n - 2$ spheres in $\mathbb{R}^n$. math.AG/0105180, 2001.
28. B. Sturmfels: *Gröbner Bases and Convex Polytopes*, volume 8 of *University Lecture Series*. American Math. Soc., Providence, RI, 1996.
29. J. Verschelde: Polynomial homotopies for dense, sparse, and determinantal systems. MSRI preprint 1999-041, 1999.
30. J. Verschelde: Numerical evidence of a conjecture in real algebraic geometry. *Exper. Math.*, 9:183–196, 2000.

Resolutions and Cohomology over Complete Intersections

Luchezar L. Avramov and Daniel R. Grayson*

This chapter contains a new proof and new applications of a theorem of Shamash and Eisenbud, providing a construction of projective resolutions of modules over a complete intersection. The duals of these infinite projective resolutions are finitely generated differential graded modules over a graded polynomial ring, so they can be represented in the computer, and can be used to compute Ext modules simultaneously in all homological degrees. It is shown how to write *Macaulay 2* code to implement the construction, and how to use the computer to determine invariants of modules over complete intersections that are difficult to obtain otherwise.

Introduction

Let $A = K[x_1, \ldots, x_e]$ be a polynomial ring with variables of positive degree over a field K, and $B = A/J$ a quotient ring modulo a homogeneous ideal.

In this paper we consider the case when B is a *graded complete intersection*, that is, when the defining ideal J is generated by a homogeneous A-regular sequence. We set up, describe, and illustrate a routine Ext, now implemented in *Macaulay 2*. For any two finitely generated graded B-modules M and N it yields a presentation of $\operatorname{Ext}_B^\bullet(M, N)$ as a bigraded module over an appropriately bigraded polynomial ring $S = A[X_1, \ldots, X_c]$.

A novel feature of our routine is that it computes the modules $\operatorname{Ext}_B^n(M, N)$ *simultaneously in all cohomological degrees* $n \geq 0$. This is made possible by the use of *cohomology operations*, a technique usually confined to theoretical considerations. Another aspect worth noticing is that, although the result is over a ring B with nontrivial relations, all the computations are made over the *polynomial ring S*; this may account for the effectiveness of the algorithm.

To explain the role of the complete intersection hypothesis, we cast it into the broader context of homological algebra over graded rings.

Numerous results indicate that the high syzygy modules of M exhibit 'similar' properties. For an outrageous example, assume that M has finite projective dimension. Its distant syzygies are then all equal to 0, and so—for trivial reasons—display an extremely uniform behavior. However, even this case has a highly nontrivial aspect: due to the Auslander-Buchsbaum Equality asymptotic information is available after at most $(e+1)$ steps. This accounts for the effectiveness of computer constructions of *finite* free resolutions.

---
* Authors supported by the NSF, grants DMS 99-70375 and DMS 99-70085.

Problems that computers are not well equipped to handle arise unavoidably when studying asymptotic behavior of *infinite* resolutions. We describe some, using graded Betti numbers $\beta_{ns}^B(M) = \dim_K \mathrm{Ext}_B^n(M, k)_{-s}$, where $k = B/(x_1, \ldots, x_e)B$, and regularity $\mathrm{reg}_B(M) = \sup_{n,s}\{s - n \mid \beta_{ns}^B(M) \neq 0\}$.

- *Irrationality*. There are rings B for which no recurrent relation with constant coefficients exists among the numbers $\beta_n^B(k) = \sum_s \beta_{ns}^B(k)$, see [1].
- *Irregularity*. For each $r \geq 2$ there exists a ring $B(r)$ with $\beta_{ns}^{B(r)}(k) = 0$ for $s \neq n$ and $0 \leq n \leq r$, but with $\beta_{r,r+1}^{B(r)}(k) \neq 0$, see [14].
- *Span*. If B is generated over K by elements of degree one and $\mathrm{reg}_B(k) \neq 0$, then $\mathrm{reg}_B(k) = \infty$, see [7].
- *Size*. There are inequalities $\beta_n^B(k) \geq \beta^n$ for all $n \geq 0$ and for some constant $\beta > 1$, unless B is a complete intersection, see [3].

These obstructions vanish miraculously when B is a graded complete intersection: For each M and all $n \gg 0$ the number $\beta_{n+1}(M)$ is a linear combination with constant coefficients of $\beta_{n-2c}^B(M), \ldots, \beta_n^B(M)$. If B is generated in degree one, then $\mathrm{reg}_B(k) = 0$ if and only if the ideal J is generated by quadratic forms. There are inequalities $\beta_n^B(M) \leq \beta(M)n^{c-1}$ for all $n \geq 1$ and for some constant $\beta(M) > 0$.

The algebra behind the miracle is a theorem of Gulliksen [12], who proves that $\mathrm{Ext}_B^\bullet(M, N)$ is a finitely generated bigraded module over a polynomial *ring of cohomology operators* $S = A[X_1, \ldots, X_c]$, where each variable X_i has cohomological degree 2. As a consequence of this result, problems in Homological Algebra can be answered in terms of Commutative Algebra.

Gulliksen's definition of the operators X_i as iterated connecting homomorphisms is badly suited for use by a computer. Other definitions have been given subsequently by several authors, see Remark 4.6. We take the approach of Eisenbud [11], who derives the operators from a specific B-free resolution of M, obtained by extending a construction of Shamash [15].

The resolution of Shamash and Eisenbud, and Gulliksen's Finiteness Theorem, are presented with detailed proofs in Section 4. They are obtained through a new construction—that of an intermediate resolution of M over the polynomial ring—that encodes C and all the null-homotopies of C corresponding to multiplication with elements of J; this material is contained in Section 3. It needs standard multilinear algebra, developed *ad hoc* in Section 2. Rules for juggling several gradings are discussed in an Appendix.

In Section 5 we present and illustrate the code for the routine Ext, which runs remarkably close to the proofs in Sections 3 and 4. Section 6 contains numerous computations of popular numerical invariants of a graded module, like its complexity, Poincaré series, and Bass series. They are extracted from knowledge of the bigraded modules $\mathrm{Ext}_B^\bullet(M, k)$ and $\mathrm{Ext}_B^\bullet(k, M)$, whose computation is also illustrated by examples, and is further used to obtain explicit equations for the cohomology variety $V_B^*(M)$ defined in [2]. For most invariants we include some short code that automates their computation. In Section 7 we extend these procedures to invariants of pairs of modules.

1   Matrix Factorizations

We start the discussion of homological algebra over a complete intersection with a very special case, that can be packaged attractively in matrix terms.

Let f be a non-zero-divisor in a commutative ring A.

Following Eisenbud [11, Sect. 5] we say that a pair (U, V) of matrices with entries in A, of sizes $k \times \ell$ and $\ell \times k$, is a *matrix factorization* of $-f$ if

$$U \cdot V = -f \cdot I_k \qquad \text{and} \qquad V \cdot U = -f \cdot I_\ell$$

where I_m denotes the $m \times m$ unit matrix. Localizing at f, one sees that $-f^{-1} \cdot U$ and V are inverse matrices over A_f; as a consequence $\ell = k$, and each equality above implies the other, for instance:

$$V \cdot U = \left(-f^{-1} \cdot U \right)^{-1} \cdot U = -f \cdot U^{-1} \cdot U = -f \cdot I_k$$

Here is a familiar example of matrix factorization, with $f = xy - wz$:

$$\begin{pmatrix} w & x \\ y & z \end{pmatrix} \cdot \begin{pmatrix} z & -y \\ -x & w \end{pmatrix} = -(xy - wz) \cdot \begin{pmatrix} 1 & 0 \\ 0 & 1 \end{pmatrix} = \begin{pmatrix} z & -y \\ -x & w \end{pmatrix} \cdot \begin{pmatrix} w & x \\ y & z \end{pmatrix}$$

Let now C_1 and C_0 be free A-modules of rank r, and let

$$d_1 : C_1 \to C_0 \qquad \text{and} \qquad s_0 : C_0 \to C_1$$

be A-linear homomorphisms defined by the matrices U and V, respectively, after bases have been tacitly chosen.

The second condition on the matrices U and V implies that d_1 is injective, while the first condition on these matrices shows that fC_0 is contained in $\mathrm{Im}(d_1)$. Setting $L = \mathrm{Coker}(d_1)$, one sees that the chosen matrix factorization defines a commutative diagram with exact rows

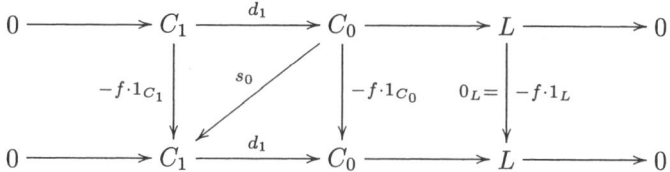

which expresses the following facts: $C = 0 \to C_1 \xrightarrow{d_1} C_0 \to 0$ is a free resolution of the A-module L, this module is annihilated by f, and s_0 is a homotopy between the maps $-f \cdot 1_C$ and 0_C, both of which lift $-f \cdot 1_L$.

Conversely, if an A-module L annihilated by f has a free resolution (C, d_1) of length 1, then $\mathrm{rank}_A C_1 = \mathrm{rank}_A C_0$, and any choice of homotopy s_0 between $-f \cdot 1_C$ and 0_C provides a matrix factorization of $-f$.

When we already have an A-module L with a presentation matrix U that defines an injective A-linear map, we can use *Macaulay 2* to create a matrix factorization (U, V) of $-f$.

Example 1.1. We revisit the familiar example from a higher perspective.

```
i1 : A = QQ[w,x,y,z]

o1 = A

o1 : PolynomialRing

i2 : U = matrix {{w,x},{y,z}}

o2 = | w x |
     | y z |

            2       2
o2 : Matrix A  <--- A

i3 : C = chainComplex U

       2       2
o3 = A   <-- A

       0       1

o3 : ChainComplex

i4 : L = HH_0 C

o4 = cokernel | w x |
              | y z |

                              2
o4 : A-module, quotient of A

i5 : f = -det U

o5 = x*y - w*z

o5 : A
```

Let's verify that f annihilates L.

```
i6 : f * L == 0

o6 = true
```

We use the **nullhomotopy** function.

```
i7 : s = nullhomotopy (-f * id_C)

       2                           2 ·
o7 = 1 : A  <---------------- A    : 0
                 {1} | z  -x |
                 {1} | -y  w |

o7 : ChainComplexMap
```

Let's verify that s is a null-homotopy for $-f$, using C.dd to obtain the differential of the chain complex C as a map of graded modules.

```
i8 : s * C.dd + C.dd * s == -f

o8 = true
```

We extract the matrix V from the null-homotopy to get our factorization.

```
i9 : V = s_0

o9 = {1} | z  -x |
     {1} | -y  w |
```

```
              2         2
o9 : Matrix A  <--- A
```

For every f and every $r \geq 1$ there exists a trivial matrix factorization of $-f$, namely, $(f \cdot I_k, -I_k)$; it can be obtained from the A-module $L = A^k/fA^k$. In general, it may not be clear how to find an A-module L with the properties necessary to obtain an 'interesting' matrix factorization of $-f$.

However, in some cases the supply is plentiful.

Remark 1.2. Let A be a graded polynomial ring in e variables of positive degree over a field K, let f be a homogeneous polynomial in A, and set $B = A/(f)$. Every B-module M of infinite projective dimension *generates* a matrix factorization (U, V) of $-f$, as follows.

Let (F, d_F) be a minimal graded free resolution of M over B, and set $L = \operatorname{Coker}(d_F \colon F_{e+1} \to F_e)$. Since M has infinite projective dimension, we have $L \neq 0$. By the Depth Lemma, $\operatorname{depth}_B L = \operatorname{depth} B$. On the other hand, $\operatorname{depth}_B L = \operatorname{depth}_A L$ and $\operatorname{depth} B = \operatorname{depth} A - 1$. By Hilbert's Syzygy Theorem, the minimal graded free resolution (C, d_C) of L over A is finite. By the Auslander-Buchsbaum Equality, $C_n = 0$ for $n > \operatorname{depth} A - \operatorname{depth}_A L = 1$.

The minimality of F ensures that all nonzero entries of the presentation matrix U of L are forms of positive degree. On the other hand, by [11, Sect. 0] the module L has no direct summand isomorphic to B: it follows that all nonzero entries of the homotopy matrix V are forms of positive degree (this is the reason for choosing L as above—stopping one step earlier in the resolution F could have produced a module L with a non-zero free direct summand).

Our reader would have noticed that *Macaulay 2* can read all the data and perform all the operations needed to construct a module L by the procedure described in the preceding remark. Here is how it does it.

Example 1.3. We produce a matrix factorization of $-f$, where

$$f = x^3 + 3y^3 - 2yz^2 + 5z^3 \in \mathbb{Q}[x, y, z] = A$$

generated by the module $M = B/\mathfrak{m}^2$, where $B = A/(f)$ and $\mathfrak{m} = (x, y, z)B$.

```
i10 : A = QQ[x,y,z];

i11 : f = x^3 + 3*y^3 - 2*y*z^2 + 5*z^3;

i12 : B = A/f;

i13 : m = ideal(x,y,z)

o13 = ideal (x, y, z)

o13 : Ideal of B
```

Let's take the B-module M and compute its minimal B-free resolution.

```
i14 : M = B^1/m^2;
```

```
i15 : F = resolution(M, LengthLimit=>8)
```

```
          1       6       9       9       9       9       9       9       9
o15 = B  <-- B  <-- B  <-- B  <-- B  <-- B  <-- B  <-- B  <-- B

          0       1       2       3       4       5       6       7       8
```

```
o15 : ChainComplex
```

We introduce a function **restrict1** N which accepts a B-module N and restricts scalars to produce an A-module.

```
i16 : restrict1 = N -> coker(lift(presentation N,A) | f);
```

Now make L as described in Remark 1.2

```
i17 : L = restrict1 cokernel F.dd_4;
```

We proceed as in Example 1.1 to get a matrix factorization.

```
i18 : C = res L;
```

```
i19 : U = C.dd_1;
```

```
              9               9
o19 : Matrix A  <--- A
```

```
i20 : print U
{4} | 0   xy  x2      y2      0           0           0         yz-5/2z2 0       |
{4} | 0   x2  -3y2    xy      yz-5/2z2 0              yz-5/2z2 0          0       |
{4} | x2  0   -2yz+5z2 0      y2-5/2yz yz-5/2z2 -5/2yz    0          0       |
{5} | 0   0   0       1/3z    0           0           0         1/2y      x       |
{5} | 0   0   -z      0       1/2y        0           1/2y      -1/2x     0       |
{5} | 0   -z  0       0       -1/2x       0           -1/2x     0         3y      |
{5} | 0   0   0       -1/3x   0           1/2y        -1/3z     0         0       |
{5} | -z  y   x       0       0           -1/2x       0         0         0       |
{5} | y   0   0       0       0           0           1/3x      0         -2y+5z  |
```

```
i21 : s = nullhomotopy (-f * id_C);
```

```
i22 : V = s_0;
```

```
              9               9
o22 : Matrix A  <--- A
```

```
i23 : print V
{6} | 0    0    -x   0        0           -2y2+5yz 0         -2yz+5z2 -3y2 |
{6} | 0    -x   0    0        0           -2yz+5z2 -3xy      -3y2     -3yz |
{6} | -x   y    0    0        -2yz+5z2 0              0         0        0    |
{6} | -3y  0    0    6yz-15z2 0              0         3x2       3xy      3xz  |
{6} | 0    2z   -3y  -15xz    -15yz       2x2         6yz-15z2 0        3x2  |
{6} | -2x  0    2z   0        -4yz+10z2 0              -6y2      2x2      0    |
{6} | 0    0    3y   -6xy+15xz -6y2+15yz 0              -6yz+15z2 0        -3x2 |
{6} | 2z   0    0    -6y2     2x2         2xy         0         0        0    |
{6} | 0    0    0    -x2      -xy         -y2         -xz       -yz      -z2  |
```

```
i24 : U*V+f==0
```

```
o24 = true
```

```
i25 : V*U+f==0
```

```
o25 = true
```

The procedure described above can be automated for more pleasant usage.

Code 1.4. The function `matrixFactorization` M produces a matrix factorization (U, V) of $-f$ generated by a module M over $B = A/(f)$.

```
i26 : matrixFactorization = M -> (
        B := ring M;
        f := (ideal B)_0;
        e := numgens B;
        F := resolution(M, LengthLimit => e+1);
        L := restrict1 cokernel F.dd_(e+1);
        C := res L;
        U := C.dd_1;
        s := nullhomotopy (-f * id_C);
        V := s_0;
        assert( U*V + f == 0 );
        assert( V*U + f == 0 );
        return (U,V));
```

We use the `assert` command to signal an error in case the matrices found don't satisfy our requirement for a matrix factorization.

Let's illustrate the new code with a slightly bigger module M than before.

Example 1.5. With the same A, f, B, and $\mathfrak{m}$ as in Example 1.3, we produce a matrix factorization generated by the B-module $M = B/\mathfrak{m}^3$.

```
i27 : time (U,V) = matrixFactorization(B^1/m^3);
      -- used 0.21 seconds
```

The parallel assignment statement above provides both variables U and V with matrix values. We examine their shapes without viewing the matrices themselves by appending a semicolon to the appropriate command.

```
i28 : U;

                15        15
o28 : Matrix A    <--- A

i29 : V;

                15        15
o29 : Matrix A    <--- A
```

Matrix factorizations were introduced to construct resolutions over the the residue ring $B = A/(f)$, using the following observation.

Remark 1.6. If (U, V) is a factorization of $-f$ by $k \times k$ matrices and the maps $d_1: C_1 \to C_0$ and $s_0: C_0 \to C_1$ are homomorphisms of free A-modules defined by U and V, respectively, then the sequence

$$\cdots \to C_1 \otimes_A B \xrightarrow{d_1 \otimes 1_B} C_0 \otimes_A B \xrightarrow{s_0 \otimes 1_B} C_1 \otimes_A B \xrightarrow{d_1 \otimes 1_B} C_0 \otimes_A B \to 0$$

of B-linear maps is a free resolution of the B-module $L = \mathrm{Coker}(d_1)$.

Indeed, freeness is clear, and we have a complex because $d_1 s_0 = -f \cdot 1_{C_0}$ and $s_0 d_1 = -f \cdot 1_{C_1}$. If $x \in C_1$ satisfies $\left(d_1 \otimes 1_B\right)(x \otimes 1) = 0$, then $d_1(x) = fy$ for some $y \in C_0$, hence $d_1 x = d_1 s_0(y)$. As d_1 is injective, we get $x = s_0(y)$, so $\mathrm{Ker}\left(d_1 \otimes 1_B\right) \subseteq \mathrm{Im}\left(s_0 \otimes 1_B\right)$; the reverse inclusion follows by symmetry.

Pooling Remarks 1.2 and 1.6 we recover Eisenbud's result [11, Sect. 6].

Theorem 1.7. *Let A be a graded polynomial ring in e variables of positive degree over a field K, and f a homogeneous polynomial in A. The minimal graded free resolution of every finitely generated graded module over $B = A/(f)$ becomes periodic of period 2 after at most e steps. The periodic part of the resolution is given by a matrix factorization of $-f$ generated by M.*

We illustrate the theorem on an already computed example.

Example 1.8. Let A, f, B, M, and F be as in Example 1.3.

To verify the periodicity of F we subtract pairs of differentials and compare the result with 0: direct comparison of the differentials would not work, because the free modules involved have different degrees.

```
i30 : F.dd_3 - F.dd_5 == 0

o30 = false

i31 : F.dd_4 - F.dd_6 == 0

o31 = false

i32 : F.dd_5 - F.dd_7 == 0

o32 = true
```

The first two answers above come as a surprise—and suggest a property of F that is weaker than the one we already know to be true!

There is an easy explanation: we checked the syzygy modules for *equality*, rather than for *isomorphism*. We do not know why *Macaulay 2* didn't produce an equality at the earliest possible stage, nor why it eventually produced one. The program has other strategies for computing resolutions, so let's try one.

```
i33 : M = B^1/m^2;

i34 : G = resolution(M, LengthLimit => 8, Strategy => 0)

            1       6       9       9       9       9       9       9       9
o34 = B <-- B <-- B <-- B <-- B <-- B <-- B <-- B <-- B

            0       1       2       3       4       5       6       7       8

o34 : ChainComplex

i35 : G.dd_3 - G.dd_5 == 0

o35 = true

i36 : G.dd_4 - G.dd_6 == 0

o36 = true

i37 : G.dd_5 - G.dd_7 == 0

o37 = true
```

The strategy paid off, revealing periodicity at the earliest possible stage. However, the algorithm used may be a lot slower that the default algorithm.

2 Graded Algebras

We describe some standard universal algebras over a commutative ring A.

Let Q denote a free A-module of rank c, and set $Q^* = \mathrm{Hom}_A(Q, A)$. We assign degree 2 to the elements of Q, and degree -2 to those of Q^*. We let $Q^\wedge$ denote a copy of Q whose elements are assigned degree 1; if x is an element of Q, then $x^\wedge$ denotes the corresponding element of $Q^\wedge$.

We use $\alpha = (\alpha_1, \ldots, \alpha_c) \in \mathbb{Z}^c$ as a multi-index, set $|\alpha| = \sum_i \alpha_i$, and order $\mathbb{Z}^c$ by the rule: $\alpha \geq \beta$ means $\alpha_i \geq \beta_i$ for each i. We let o denote the trivial element of $\mathbb{Z}^c$, and ϵ_i the i'th element of its standard basis.

Construction 2.1. For each integer $m \geq 0$ we form free A-modules

$$\mathsf{S}^m(Q^*) \quad \text{with basis} \quad \{X^\alpha : |\alpha| = m\}$$

$$\mathsf{D}^m(Q) \quad \text{with basis} \quad \{Y^{(\alpha)} : |\alpha| = m\}$$

$$\textstyle\bigwedge^m(Q^\wedge) \quad \text{with basis} \quad \{Y^{\wedge\alpha} : |\alpha| = m \quad \text{and} \quad \alpha \leq (\epsilon_1 + \cdots + \epsilon_c)\}$$

For $m < 0$ we declare the modules $\mathsf{S}^m(Q^*)$, $\mathsf{D}^m(Q)$, and $\bigwedge^m(Q^\wedge)$ to be equal to 0, and define the symbols X^α, $Y^{(\alpha)}$, and $Y^{\wedge\alpha}$ accordingly; in addition, we set $\bigwedge^m(Q^\wedge) = 0$ and $Y^{\wedge\alpha} = 0$ if $|\alpha| \not\leq (\epsilon_1 + \cdots + \epsilon_c)$, and we set

$$X_i = X^{\epsilon_i} \qquad Y_i = Y^{(\epsilon_i)} \qquad Y_i^\wedge = Y^{\wedge\epsilon_i} \qquad \text{for} \qquad i = 1, \ldots, c$$

Taking $\mathsf{S}^m(Q^*)$, $\mathsf{D}^m(Q)$, and $\bigwedge^m(Q^\wedge)$ as homogeneous components of degree $-2m$, $2m$, and m, respectively, we introduce graded algebras

$$S = \mathsf{S}(Q^*) \qquad D = \mathsf{D}(Q) \qquad E = \textstyle\bigwedge(Q^\wedge)$$

by defining products of basis elements by the formulas

$$X^\alpha \cdot X^\beta = X^{\alpha+\beta}$$

$$Y^{(\alpha)} \cdot Y^{(\beta)} = \prod_{i=1}^c \frac{(\alpha_i + \beta_i)!}{\alpha_i! \beta_i!} Y^{(\alpha+\beta)}$$

$$Y^{\wedge\alpha} \cdot Y^{\wedge\beta} = \mathrm{inv}(\alpha, \beta) Y^{\wedge\,\alpha+\beta}$$

where $\mathrm{inv}(\alpha, \beta)$ denotes the number of pairs (i, j) with $\alpha_i = \beta_j = 1$ and $i > j$. Thus, S is the *symmetric algebra* of Q^*, with $X^o = 1$, while D is the *divided powers algebra* of Q, with $Y^{(o)} = 1$, and E is the *exterior algebra* of $Q^\wedge$, with $Y^{\wedge o} = 1$. We identify S and the polynomial ring $A[X_1, \ldots, X_c]$.

A *homogeneous derivation* of a graded A-algebra W is a homogeneous A-linear map $d \colon W \to W$ such that the *Leibniz rule*

$$d(xy) = d(x)y + (-1)^{\deg x \cdot \deg d} x d(y)$$

holds for all homogeneous elements $x, y \in W$.

Construction 2.2. Each sequence $f_1, \ldots, f_c \in A$ yields a *Koszul map*

$$d_E \colon E \to E \qquad \text{defined by the formula}$$

$$d_E(Y^{\wedge \beta}) = \sum_{i=1}^{c} (-1)^{\beta_1 + \cdots + \beta_{i-1}} f_i Y^{\wedge \beta - \epsilon_i}$$

It is a derivation of degree -1 and satisfies $d_E^2 = 0$.

Construction 2.3. For every $X_i \in S^1(Q^*)$ and each $Y^{(\beta)} \in D^m(Q)$ we set

$$X_i \lrcorner Y^{(\beta)} = Y^{(\beta - \epsilon_i)} \in D^{m-1}(Q)$$

Extending this formula by A-bilinearity, we define $g \lrcorner y$ for all $g \in S^1(Q^*)$ and all $y \in D$. It is well known, and easily verified, that the map $g \lrcorner \colon y \mapsto g \lrcorner y$ is a graded derivation $D \to D$ of degree -2, and that the derivations associated with arbitrary g and g' commute. As a consequence, the formula

$$X^\alpha \lrcorner Y^{(\beta)} = (X_1 \lrcorner)^{\alpha_1} \cdots (X_c \lrcorner)^{\alpha_c} \left(Y^{(\beta)} \right) \in D^{|\beta - \alpha|}(Q)$$

extended A-linearly to all $u \in S$, defines on D a structure of graded S-module.

The usual products on $S \otimes_A E$ and $D \otimes_A E$ and the induced gradings

$$(S \otimes_A E)_n = \bigoplus_{\ell - 2k = n} S^k(Q^*) \otimes_A \textstyle\bigwedge^\ell(Q^\wedge)$$

$$(D \otimes_A E)_n = \bigoplus_{\ell + 2k = n} D^{(k)}(Q) \otimes_A \textstyle\bigwedge^\ell(Q^\wedge)$$

turn $S \otimes_A E$ and $D \otimes_A E$ into graded algebras. The second one is a graded module over the first, for the action $(u \otimes z) \cdot (y \otimes z') = (u \lrcorner v) \otimes (z \cdot z')$.

Construction 2.4. The element $w = \sum_{i=1}^{c} X_i \otimes Y_i^\wedge$ yields a *Cartan map*

$$d_{DE} \colon D \otimes_A E \to D \otimes_A E \qquad \text{defined by the formula}$$

$$d_{DE}(y \otimes z) = w \cdot (y \otimes z) = \sum_{i=1}^{c} (X_i \lrcorner y) \otimes (Y_i^\wedge \cdot z)$$

It is an E-linear derivation of degree -1, and $d_{DE}^2 = 0$ because $w^2 = 0$.

Lemma 2.5. *For each integer s define a complex G^s as follows:*

$$\cdots \to D^k(Q) \otimes_A \textstyle\bigwedge^{s-k}(Q^\wedge) \xrightarrow{w} D^{k-1}(Q) \otimes_A \textstyle\bigwedge^{s-k+1}(Q^\wedge) \to \cdots$$

with $D^0(Q) \otimes_A \bigwedge^s(Q^\wedge)$ in degree s. If $s > 0$, then G^s is split exact.

Proof. Note that for each $s \in \mathbb{Z}$ there exist isomorphisms of complexes $\bigoplus_{s=1}^{\infty} G^s \cong (D \otimes_A E)_{\geq 1} \cong \left(\bigotimes_{i=1}^{c} G(i) \right)_{\geq 1}$, where $G(i)$ is the complex

$$\cdots \to AY_i^{(k+1)} \otimes_A A \xrightarrow{w_i} (AY_i^{(k)}) \otimes (AY_i^\wedge) \xrightarrow{0} (AY_i^{(k)}) \otimes_A A \to \cdots$$

and w_i is left multiplication with $X_i \otimes Y_i^\wedge$. This map bijective, so each complex $G(i)_{\geq 1}$ is split exact. The assertion follows. $\square$

3   Universal Homotopies

This section contains the main new mathematical result of the paper.

We introduce a universal construction, that takes as input a projective resolution C of an A-module M and a finite set f of elements annihilating M; the output is a new projective resolution of M over A. If $f \neq \varnothing$, then the new resolution is infinite—even when C is finite—because it encodes additional data: the null-homotopies for $f \cdot 1_C$ for all $f \in f$, all compositions of such homotopies, and all relations between those compositions. This higher-order information tracks the transformation of the homological properties of M when its ring of operators is changed from A to $A/(f)$.

Our construction is motivated by, and is similar to, one due to Shamash [15] and Eisenbud [11]: assuming that the elements of f form an A-regular sequence, they produce a projective resolution of M over $A/(f)$. By contrast, we make no assumption whatsoever on f. With the additional hypothesis, in the next section we quickly recover the original result from the one below. As an added benefit, we eliminate the use of spectral sequences from the proof.

Theorem 3.1. *Let A be a commutative ring, let $f_1, \ldots, f_c$ be a sequence of elements of A, let M be an A-module annihilated by f_i for $i = 1, \ldots, c$, and let $r \colon C \to M$ be a resolution of M by projective (respectively, free) A-modules.*

There exists a family of homogeneous A-linear maps

$$\{d_\gamma \colon C \to C \mid \deg(d_\gamma) = 2|\gamma| - 1\}_{\gamma \in \mathbb{N}^c}$$

satisfying the following conditions

$$d_o = d_C \quad \text{is the differential of} \quad C$$

$$[d_o, d_\gamma] = \begin{cases} -f_i \cdot 1_C & \text{if} \quad \gamma = \epsilon_i \text{ for } i = 1, \ldots, c \\ -\sum^+_{\alpha + \beta = \gamma} d_\alpha d_\beta & \text{if} \quad |\gamma| \geq 2 \end{cases} \tag{1}$$

where $\sum^+$ denotes a summation restricted to indices in $\mathbb{N}^c \smallsetminus \{o\}$.

Any family $\{d_\gamma\}_{\gamma \in \mathbb{N}^c}$ as above defines an A-linear map of degree -1,

$$d_{CD} \colon C \otimes_A D \to C \otimes_A D \qquad \text{given by}$$

$$d_{CD}(x \otimes y) = \sum_{\gamma \in \mathbb{N}^c} d_\gamma(x) \otimes (X^\gamma \lrcorner y) \tag{2}$$

where D is the divided powers algebra defined in Construction 2.1, and the action of X^γ on D is defined in Construction 2.3.

With d_E and d_{DE} defined in Constructions 2.2 and 2.4 and the tensor product of maps of graded modules defined as in Remark 3.4, the map

$$d \colon C \otimes_A D \otimes_A E \to C \otimes_A D \otimes_A E \qquad \text{given by}$$

$$d = d_{CD} \otimes 1_E + 1_C \otimes d_{DE} + 1_C \otimes 1_D \otimes d_E \tag{3}$$

is an A-linear differential of degree −1, and the map

$$q: C \otimes_A D \otimes_A E \to M \qquad \text{given by}$$

$$q(x \otimes y \otimes z) = \begin{cases} yz \cdot r(x) & \text{if } \deg(y) = \deg(z) = 0 \\ 0 & \text{otherwise} \end{cases}$$

is a resolution of M by projective (respectively, free) A-modules.

For use in the proof, we bring up a few general homological points. A *bounded filtration* of a chain complex F is a sequence

$$0 = F^0 \subseteq F^1 \subseteq \cdots \subseteq F^{s-1} \subseteq F^s \subseteq \cdots$$

of subcomplexes such that for each n there exists an s with $F_n^s = F_n$. As usual, we let $\mathrm{gr}^s(F)$ denote the complex of A-modules F^s/F^{s-1}.

Lemma 3.2. *Let $q: F \to F'$ be a morphism of complexes with bounded filtrations, such that $q(F^s) \subseteq F'^s$ for all $s \in \mathbb{Z}$. If for each s the induced map $\mathrm{gr}^s(q): \mathrm{gr}^s(F) \to \mathrm{gr}^s(F')$ is a quasi-isomorphism, then so is q.*

Proof. Denoting q^s the restriction of q to F^s, we first show by induction on s that $H_n(q^s)$ is bijective for all n. The assertion is clear for $s = 0$, since $F^0 = 0$ and $F'^0 = 0$. For the inductive step, we assume that q^{s-1} is a quasi-isomorphism for some $s \geq 1$. We have a commutative diagram of complexes

$$\begin{array}{ccccccccc} 0 & \longrightarrow & F^{s-1} & \longrightarrow & F^s & \longrightarrow & \mathrm{gr}^s(F) & \longrightarrow & 0 \\ & & \downarrow{\scriptstyle q^{s-1}} & & \downarrow{\scriptstyle q^s} & & \downarrow{\scriptstyle \mathrm{gr}^s(q)} & & \\ 0 & \longrightarrow & F'^{s-1} & \longrightarrow & F'^s & \longrightarrow & \mathrm{gr}^s(F') & \longrightarrow & 0 \end{array}$$

By hypothesis and inductive assumption, in the induced diagram

$$\begin{array}{ccccccccc} H_{n+1}(\mathrm{gr}^s(F)) & \to & H_n(F^{s-1}) & \to & H_n(F^s) & \to & H_n(\mathrm{gr}^s(F)) & \to & H_{n-1}(F^{s-1}) \\ \downarrow{\scriptstyle H_{n+1}(\mathrm{gr}^s(q))}\cong & & \downarrow{\scriptstyle H_n(q^{s-1})}\cong & & \downarrow{\scriptstyle H_n(q^s)} & & \downarrow{\scriptstyle H_n(\mathrm{gr}^s(q))}\cong & & \downarrow{\scriptstyle H_{n-1}(q^{s-1})}\cong \\ H_{n+1}(\mathrm{gr}^s(F')) & \to & H_n(F'^{s-1}) & \to & H_n(F'^s) & \to & H_n(\mathrm{gr}^s F') & \to & H_{n-1}(F'^{s-1}) \end{array}$$

the four outer vertical maps are bijective. By the Five-Lemma, so is $H_n(q^s)$.

Now we fix an integer $n \in \mathbb{Z}$, and pick s so large that

$$F_k^s = F_k \qquad \text{and} \qquad F_k'^s = F_k' \qquad \text{hold for} \qquad k = n-1, n, n+1.$$

The choice of s implies that $H_n(F^s) = H_n(F)$, $H_n(F'^s) = H_n(F')$, and $H_n(q^s) = H_n(q)$. Since we have already proved that $H_n(q^s)$ is an isomorphism, we conclude that $H_n(q): H_n(F) \to H_n(F')$ is an isomorphism. $\qquad\square$

Remark 3.3. If (F, d_F) is a complex of A-modules, then $\text{Hom}_A^{\text{gr}}(F, F)$ denotes the graded module whose n'th component consists of the A-linear maps $g \colon F \to F$ with $g(F_i) \subseteq F_{i+n}$ for all $i \in \mathbb{Z}$.

If g, h are homogeneous A-linear maps, then their composition gh is homogeneous of degree $\deg(g) + \deg(h)$, and so is their graded commutator

$$[g, h] = gh - (-1)^{\deg g \cdot \deg h} hg$$

Commutation is a graded derivation: for each homogeneous map h' one has

$$[g, hh'] = [g, h]h' + (-1)^{\deg g \cdot \deg h} h[g, h']$$

The map $h \mapsto [d_F, h]$ has square 0, and transforms $\text{Hom}_A^{\text{gr}}(F, F)$ into a complex of A-modules; by definition, its cycles are the chain maps $F \to F$, and its boundaries are the null-homotopic maps.

Remark 3.4. If $p \colon F \to F'$ and $q \colon G \to G'$ are graded maps of graded modules, we define the tensor product $p \otimes q \colon F \otimes F' \to G \otimes G'$ by the formula $(p \otimes q)(f \otimes g) = (-1)^{\deg q \cdot \deg f}(p(f) \otimes q(g))$. With this convention, when $F = F'$ and $G = G'$, the graded commutator $[1_F \otimes q, p \otimes 1_G]$ vanishes.

Lemma 3.5. *Let M be an A-module and let $r \colon C \to M$ be a free resolution. If $g \colon C \to C$ is an A-linear map with $\deg(g) > 0$, and $[d_C, g] = 0$, then $g = [d_C, h]$ for some A-linear map $h \colon C \to C$ with $\deg(h) = \deg(g) + 1$.*

Proof. The augmentation $r \colon C \to M$ defines a chain map of degree zero

$$\text{Hom}_A^{\text{gr}}(C, r) \colon \text{Hom}_A^{\text{gr}}(C, C) \to \text{Hom}_A^{\text{gr}}(C, M)$$

The map induced in homology is an isomorphism: to see this, apply the 'comparison theorem for projective resolutions'. Since A-linear maps $C \to M$ of positive degree are trivial, the conclusion follows from Remark 3.3. □

Proof (of Theorem 3.1). Recall that D is the divided powers algebra of a free A-module Q with basis $Y_1, \ldots, Y_c$, that $X_1, \ldots, X_c$ is the dual basis of the free A-module Q^*, and S for the symmetric algebra of Q^*, see Construction 2.1 for details. We set $f = \sum_{i=1}^c f_i X_i \in S^1(Q^*)$.

We first construct the maps d_γ by induction on $|\gamma|$.

If $|\gamma| = 0$, then $\gamma = o$, so $d_o = d_C$ is predefined. If $|\gamma| = 1$, then $\gamma = \epsilon_i$ for some i with $1 \leq i \leq c$. Since f_i annihilates the B-module M, the map $-f_i \cdot 1_C$ lifts the zero map on M, hence is null-homotopic. For each i we take d_{ϵ_i} to be a null-homotopy, that is, $[d_o, d_{\epsilon_i}] = -f_i \cdot 1_C$. With these choices, the desired formulas hold for all γ with $|\gamma| \leq 1$.

Assume by induction that maps d_γ satisfying the conclusion of the lemma have been chosen for all $\gamma \in \mathbb{N}^c$ with $|\gamma| < n$, for some $n \geq 2$. Fix $\gamma \in \mathbb{N}^c$

with $|\gamma| = n$. Using Remark 3.3 and the induction hypothesis, we obtain

$$
\left[d_o, \sum_{\alpha+\beta=\gamma}^{+} d_\alpha d_\beta \right] = \sum_{\alpha+\beta=\gamma}^{+} \left([d_o, d_\alpha] d_\beta - d_\alpha [d_o, d_\beta] \right)
$$

$$
= \sum_{\alpha+\beta=\gamma}^{+} \left(\left(\sum_{\alpha'+\alpha''=\alpha}^{+} d_{\alpha'} d_{\alpha''} \right) d_\beta - d_\alpha \left(\sum_{\beta'+\beta''=\beta}^{+} d_{\beta'} d_{\beta''} \right) \right)
$$

$$
= \sum_{\alpha'+\alpha''+\beta=\gamma}^{+} d_{\alpha'} d_{\alpha''} d_\beta - \sum_{\alpha+\beta'+\beta''=\gamma}^{+} d_\alpha d_{\beta'} d_{\beta''}
$$

$$
= 0
$$

The map $-\sum_{\alpha+\beta=\gamma}^{+} d_\alpha d_\beta$ has degree $2|\gamma| - 2$, so by Lemma 3.5 it is equal to $[d_o, d_\gamma]$ for some A-linear map $d_\gamma : C \to C$ of degree $2|\gamma| - 1$. Choosing such a d_γ for each $\gamma \in \mathbb{N}^c$ with $|\gamma| = n$, we complete the step of the induction.

From the definition of d we obtain an expression

$$
d^2 = d_{CD}^2 \otimes 1_E + 1_C \otimes [d_{DE}, 1_D \otimes d_E] + [d_{CD} \otimes 1_E, 1_C \otimes d_{DE}] +
$$
$$
1_C \otimes d_{DE}^2 + 1_C \otimes 1_D \otimes d_E^2 + [d_{CD} \otimes 1_E, 1_C \otimes 1_D \otimes d_E]
$$

Constructions 2.4, 2.2, and Remark 3.4 show that the maps in the second row are equal to 0, so to prove that $d^2 = 0$ it suffices to establish the equalities

$$
d_{CD}^2 = -f \cdot 1_{C \otimes D} \tag{4}
$$
$$
[d_{DE}, 1_D \otimes d_E] = f \cdot 1_{D \otimes E} \tag{5}
$$
$$
[d_{CD} \otimes 1_E, 1_C \otimes d_{DE}] = 0 \tag{6}
$$

A direct computation with formula (1) proves equality (4) above:

$$
d_{CD}^2(x \otimes y) = d_{CD} \left(\sum_{\beta \in \mathbb{N}^c} d_\beta(x) \otimes \left(X^\beta \lrcorner y \right) \right)
$$

$$
= \sum_{\beta \in \mathbb{N}^c} \left(\sum_{\alpha \in \mathbb{N}^c} d_\alpha d_\beta(x) \otimes \left(X^\alpha \lrcorner \left(X^\beta \lrcorner y \right) \right) \right)
$$

$$
= \sum_{\alpha+\beta \in \mathbb{N}^c} d_\alpha d_\beta(x) \otimes \left(X^{\alpha+\beta} \lrcorner y \right)
$$

$$
= \sum_{i=1}^{c} -f_i x \otimes \left(X_i \lrcorner y \right)
$$

$$
= -f \cdot (x \otimes y)
$$

By Constructions 2.2 and 2.4, the maps $1_D \otimes d_E$ and d_{DE} are derivations of degree -1, so the commutator $[d_{DE}, 1_D \otimes d_E]$ is a derivation of degree -2. Every element of $D \otimes_A E$ is a product of elements $1 \otimes Y_i^\wedge$ of degree 1 and $Y_j^{(k)} \otimes 1$ of degree $2k$, so it suffices to check that the map on either side of

(6) takes the same value on those elements. For degree reasons, both sides vanish on $1 \otimes Y_i^{\wedge}$. We now complete the proof of equality (5) as follows:

$$
\begin{aligned}
[d_{DE}, 1_E &\otimes d_E](Y_j^{(k)} \otimes 1) \\
&= d_{DE}\big((1_D \otimes d_E)(Y_j^{(k)} \otimes 1)\big) + (1_E \otimes d_E)\big(d_{DE}(Y_j^{(k)} \otimes 1)\big) \\
&= (1_E \otimes d_E)(Y_j^{(k-1)} \otimes Y_j^{\wedge}) \\
&= Y_j^{(k-1)} \otimes f_j \\
&= f \cdot (Y_j^{(k)} \otimes 1)
\end{aligned}
$$

To derive equation (6) we use Constructions 2.2 and 2.4 once again:

$$
\begin{aligned}
\big((d_{CD} \otimes 1_E)&(1_C \otimes d_{DE})\big)(x \otimes y \otimes z) \\
&= (-1)^{\deg x} d_{CD}\Big(\sum_{i=1}^{c} x \otimes (X_i \lrcorner y) \otimes (Y_i^{\wedge} \cdot z)\Big) \\
&= (-1)^{\deg x} \sum_{i=1}^{c} \sum_{\gamma \in \mathbb{N}^c} d_{\gamma}(x) \otimes \big(X^{\gamma} \lrcorner (X_i \lrcorner y)\big) \otimes (Y_i^{\wedge} \cdot z) \\
&= - \sum_{\gamma \in \mathbb{N}^c} \sum_{i=1}^{c} (-1)^{\deg(d_{\gamma}(x))} d_{\gamma}(x) \otimes \big(X_i \lrcorner (X^{\gamma} \lrcorner y)\big) \otimes (Y_i^{\wedge} \cdot z) \\
&= - (1_C \otimes d_{DE})\Big(\sum_{\gamma \in \mathbb{N}^c} d_{\gamma}(x) \otimes (X^{\gamma} \lrcorner y) \otimes z\Big) \\
&= - \big((1_C \otimes d_{DE})(d_{CD} \otimes 1_E)\big)(x \otimes y \otimes z)
\end{aligned}
$$

It remains to show q is a quasi-isomorphism. Setting

$$
F^s = \bigoplus_{k+\ell \leq s} C \otimes_A \mathsf{D}^k(Q) \otimes_A \textstyle\bigwedge^{\ell}(Q^{\wedge}) \qquad \text{for} \qquad s \in \mathbb{Z}
$$

we obtain a bounded filtration of the complex $F = (C \otimes_A D \otimes_A E, d)$. On the other hand, we let F' denote the complex with $F_0' = M$ and $F_n' = 0$ for $n \neq 0$; the filtration defined by $F'^0 = 0$ and $F'^s = F'$ for $s \geq 1$ is obviously bounded, and $q(F^s) \subseteq F'^s$ holds for all $s \geq 0$. By Lemma 3.2 it suffices to show that the induced map $\mathrm{gr}^s(q) \colon \mathrm{gr}^s(F) \to \mathrm{gr}^s(F')$ is bijective for all s.

Inspection of the differential d of F shows that $\mathrm{gr}^s(F)$ is isomorphic to the tensor product of complexes $C \otimes_A G^s$, where G^s is the complex defined in Lemma 2.5. It is established there that G^s is split exact for $s > 0$, hence $\mathrm{H}_n(C \otimes_A G^s) = 0$ for all $n \in \mathbb{Z}$. As $G^0 = A$ and $\mathrm{gr}^0(q) = r$, we are done. $\quad\square$

4   Cohomology Operators

We present a new approach to the procedure of Shamash [15] and Eisenbud [11] for building projective resolutions over a complete intersection. We then use this resolution to prove a fundamental result of Gulliksen [12] on the structure of Ext modules over complete intersections.

A set $f = \{f_1, \ldots, f_c\} \subseteq A$ is *Koszul-regular* if the complex (E, d_E) of Construction 2.2, has $H_n(E) = 0$ for $n > 0$. A sufficient condition for Koszul-regularity is that the elements of f, in some order, form a regular sequence.

Theorem 4.1. *Let A be a commutative ring, $f = \{f_1, \ldots, f_c\} \subseteq A$ a subset, $B = A/(f)$ the residue ring, M a B-module, and $r: C \to M$ a resolution of M by projective (respectively, free) A-modules.*

Let $\{d_\gamma: C \to C\}_{\gamma \in \mathbb{N}^c}$ be a family of A-linear maps provided by Theorem 3.1, set $D' = D \otimes_A B$, and $y' = y \otimes 1$ for $y \in D$. The map

$$\partial: C \otimes_A D' \to C \otimes_A D' \qquad given\ by$$

$$\partial(x \otimes y') = \sum_{\gamma \in \mathbb{N}^c} d_\gamma(x) \otimes (X^\gamma \lrcorner y)' \tag{7}$$

is a B-linear differential of degree -1. If f is Koszul-regular, then the map

$$q': C \otimes_A D' \to M \qquad given\ by$$

$$q'(x \otimes y') = \begin{cases} y \cdot r(x) & if \deg(y') = 0 \\ 0 & otherwise \end{cases}$$

is a resolution of M by projective (respectively, free) A-modules.

Remark 4.2. Assume that in the theorem $f = \{f_1\}$. The module D_ℓ is then trivial if ℓ is odd, and is free with basis consisting of a single element $Y_1^{(\ell/2)}$ if ℓ is even. Thus, the resolution $C \otimes_A D'$ has the form

$$\cdots \xrightarrow{\partial_{2n+1}} \bigoplus_{j=0}^{\infty} C_{2j} \otimes_A BY_1^{(n-j)} \xrightarrow{\partial_{2n}} \bigoplus_{j=1}^{\infty} C_{2j-1} \otimes_A BY_1^{(n-j)} \xrightarrow{\partial_{2n-1}} \cdots$$

The simplest situation occurs when, in addition, C is a free resolution with $C_n = 0$ for $n \geq 2$. In this case the differential d_o has a single non-zero component, $d_1: C_1 \to C_0$, the homotopy d_{ϵ_1} between $-f \cdot 1_C$ and 0_C has a single non-zero component, $s_0: C_0 \to C_1$, and all the maps d_γ with $\gamma \in \mathbb{N}^1 \setminus \{o, \epsilon_1\}$ are trivial for degree reasons. It is now easy to see that the complex above coincides with the one constructed, *ad hoc*, in Remark 1.6.

Proof (of the theorem). In the notation of Theorem 3.1, we have equalities

$$C \otimes_A D' = (C \otimes_A D \otimes_A E) \otimes_E B \qquad and \qquad \partial = d \otimes 1_B$$

It follows that $\partial^2 = 0$. For each $s \geq 0$ consider the subcomplexes

$$F^s = \bigoplus_{k+\ell \leq s} C_k \otimes_A D_\ell \otimes_A E \qquad of \qquad F = C \otimes_A D \otimes_A E$$

$$F'^s = \bigoplus_{k+\ell \leq s} C_k \otimes_A D'_\ell \qquad of \qquad F' = C \otimes_A D'$$

They provide bounded filtrations of the complexes F and F', respectively, such that the map $p' = 1_C \otimes 1_D \otimes p \colon F \to F'$ satisfies $p'(F^s) \subseteq F'^s$ for all $s \geq 0$. Setting $G_s = \bigoplus_{k+\ell=s}(C_k \otimes_A D_\ell)$, we obtain equalities $\mathrm{gr}^s(F) = (G_s \otimes_A E, 1_{G_s} \otimes d_E)$ and $\mathrm{gr}^s(F') = (G_s \otimes_A B, 0)$ of complexes of A-modules.

If f is Koszul regular, then $p \colon E \to B$ is a quasi-isomorphism, hence so is $1_{G_s} \otimes p = \mathrm{gr}^s(p')$ for each $s \geq 0$. Lemma 3.2 then shows that p' is a quasi-isomorphism. The quasi-isomorphism $q \colon F \to M$ of Theorem 3.1 factors as $q = q'(1_C \otimes 1_D \otimes p)$, so we see that q' is a quasi-isomorphism, as desired. $\square$

Let M and N be B-modules, and let $\mathrm{Ext}_B^\bullet(M, N)$ denote the graded B-module having $\mathrm{Ext}_B^n(M, N)$ as component of degree $-n$. To avoid negative numbers, it is customary to regrade $\mathrm{Ext}_B^\bullet(M, N)$ by cohomological degree, under which the elements of $\mathrm{Ext}_B^n(M, N)$ are assigned degree n; we do not do it here, in order not to confuse Macaulay 2. Of course, these modules can be computed from any projective resolution of M over B.

The next couple of remarks collect a few innocuous observations. In hindsight, they provide some of the basic tools for studying cohomology of modules over complete intersections: see Remark 4.6 for some related material.

Remark 4.3. The resolution $(C \otimes_A D', \partial)$ provided by Theorem 4.1 is a graded module over the graded algebra S, with action defined by the formula

$$u \cdot (x \otimes y') = x \otimes (u \lrcorner y)'$$

and this action commutes with the differential ∂. The induced action provides a structure of graded S-module on the complex $\mathrm{Hom}_B(C \otimes_A D', N)$.

The action of S commutes with the differential $\partial^* = \mathrm{Hom}_B(\partial, N)$ of this complex, hence passes to its homology, making it a graded a S-module. Thus, each element $u \in S_{-2k} = \mathsf{S}^k(Q)$ determines homomorphisms

$$\mathrm{Ext}_B^n(M, N) \xrightarrow{\ u\ } \mathrm{Ext}_B^{n+2k}(M, N) \qquad \text{for all} \qquad n \in \mathbb{Z}$$

For this reason, from now on we refer to the graded ring S as the *ring of cohomology operators* determined by the Koszul-regular set f.

Remark 4.4. The canonical isomorphisms of complexes of A-modules

$$\mathrm{Hom}_B(C \otimes_A D', N) = S \otimes_A \mathrm{Hom}_A(C, N) = S \otimes_A \mathrm{Hom}_A(C, A) \otimes_A N$$

commute with the actions of S.

The following fundamental result shows that in many important cases the action of the cohomology operators is highly nontrivial.

Theorem 4.5. *Let A be a commutative ring, let f be a Koszul regular subset of A, and let S be the graded ring of cohomology operators defined by f.*

If M and N are finitely generated modules over $B = A/(f)$, and M has finite projective dimension over A (in particular, if A is regular), then the graded S-module $\mathrm{Ext}_B^\bullet(M, N)$ is finitely generated.

Proof. Choose a resolution $r\colon C \to M$ with C_n a finite projective A-module for each n and $C_n = 0$ for all $n \gg 0$. By Remark (4.4), the graded S-module $\mathrm{Hom}_B(C \otimes_A D', N)$ is finitely generated. Since S is noetherian, so is the submodule $\mathrm{Ker}(\partial^*)$, and hence the homology module, $\mathrm{Ext}_B^{\bullet}(M, N)$.   □

Remark 4.6. The resolution of Remark 4.2 was constructed by Shamash [15, Sect. 3], that of Theorem 4.1 by Eisenbud [11, Sect. 7]. The new aspect of our approach is indicated at the beginning of Section 3.

As introduced in Remark 4.3, the S-module structure on Ext may appear *ad hoc*. In fact, it is independent of all choices of resolutions and maps, it can be computed from *any* projective resolution of M over B, it is natural in both module arguments and—in an appropriate sense—in the ring argument, and it commutes with Yoneda products from either side. These properties were proved by Gulliksen [12, Sect. 2], Mehta [13, Ch. 2], Eisenbud [11, Sect. 4], and Avramov [2, Sect. 2]. However, each author used a different construction of cohomology operators, and comparison of the different approaches has turned to be an unexpectedly delicate problem. It was finally resolved in [8], where complete proofs of the main properties of the operators can be found.

Gulliksen [12, Sect. 3] established a stronger form of Theorem 4.5, without finiteness hypotheses on the ring A: If the A-module $\mathrm{Ext}_A^{\bullet}(M, N)$ is noetherian, then the S-module $\mathrm{Ext}_B^{\bullet}(M, N)$ is noetherian; this can be obtained from the complexes of Remark (4.4) by means of a spectral sequence, cf. [4, Sect. 6]. The converse of Gulliksen's theorem was proved in [6, Sect. 4].

For the rest of the paper we place ourselves in a situation where *Macaulay 2* operates best—graded modules over positively graded rings. This grading is inherited by the various Ext modules, and we keep careful track of it. Our conventions and bookkeeping procedures are discussed in detail in an Appendix, which the reader is invited to consult as needed.

For ease of reference, we collect some notation.

Notation 4.7. The following is assumed for the rest of the paper.

- K is a field.
- $\{x_h \mid \deg'(x_h) > 0\}_{h=1,\ldots,e}$ is a set of indeterminates over K.
- $A = K[x_1, \ldots, x_e]$, graded by $\deg'(a) = 0$ for $a \in K$.
- $f_1, \ldots, f_c$ is a homogeneous A-regular sequence in $(x_1, \ldots, x_e)^2$.
- $r_i = \deg'(f_i)$ for $i = 1, \ldots, c$.
- $\{X_i \mid \mathrm{Deg}\, X_i = (-2, -r_i)\}_{i=1,\ldots,c}$ is a set of indeterminates over A.
- $S = A[X_1, \ldots, X_c]$, bigraded by $\mathrm{Deg}(a) = (0, \deg'(a))$.
- $B = A/(\boldsymbol{f})$, with degree induced by $\deg'$.
- M and N are finitely generated graded B-modules.
- S acts as bigraded ring of cohomology operators on $\mathrm{Ext}_B^{\bullet}(M, N)$.
- $k = B/(x_1, \ldots, x_e)B$, with degree induced by $\deg'$.
- $R = S \otimes_A k \cong K[X_1, \ldots, X_c]$, with bidegree induced by Deg.

Remark 4.8. Under the conditions above, it is reasonable to ask when the B-free resolution G of Theorem 4.1, obtained from a *minimal* A-free resolution C of M, will itself be minimal. Shamash [15, Sect. 3] proves that G is minimal if $f_i \in (x_1, \ldots, x_e)\mathrm{ann}_A(M)$ for $i = 1, \ldots, c$. An obvious example with non-minimal G occurs when M has finite projective dimension over B: if $c > 0$ then G is infinite. A more interesting failure of minimality follows.

Example 4.9. Let A, f, B, and M be as in Example 1.5.

```
i38 : M = B^1/m^3;

i39 : F = resolution(M, LengthLimit=>8)

        1       10      16      15      15      15      15      15      15
o39 = B  <-- B  <-- B  <-- B  <-- B  <-- B  <-- B  <-- B  <-- B

        0       1       2       3       4       5       6       7       8

o39 : ChainComplex
```

Thus, the sequence of Betti numbers $\beta_n^B(M)$ is $(1, 10, 16, 15, 15, 15, \ldots)$.

```
i40 : M' = restrict1 M;

i41 : C = res M'

        1       10      15      6
o41 = A  <-- A  <-- A  <-- A  <-- 0

        0       1       2       3       4

o41 : ChainComplex
```

By Remark 4.2, the sequence $\mathrm{rank}_B F_n$ is $(1, 10, 16, 16, 16, 16, \ldots)$.

In a graded context, all cohomological entities discussed so far in the text acquire an extra grading, discussed in detail in the Appendix. The notions below are used, but not named, in [5] in a local situation.

Remark 4.10. We define the *reduced Ext module* for M and N over B by

$$\mathrm{ext}_B^\bullet(M, N) = \mathrm{Ext}_B^\bullet(M, N) \otimes_A k$$

With the induced bigrading and action, it is a bigraded module over the bigraded ring R, that we call the *reduced ring of cohomology operators*.

The dimension of the K-vector space $\mathrm{ext}_B^n(M, N)_s$ is equal to the number of generators of bidegree $(-n, s)$ in any minimal set of generators of the graded B-module $\mathrm{Ext}_B^n(M, N)$. We define the graded (respectively, ungraded) *Ext-generator series* of M and N to be the formal power series

$$G_B^{M,N}(t, u) = \sum_{n \in \mathbb{N}, s \in \mathbb{Z}} \dim_K \mathrm{ext}_B^n(M, N)_s \, t^n u^{-s} \in \mathbb{Z}[u, u^{-1}][[t]]$$

$$G_B^{M,N}(t) = \sum_{n=0}^{\infty} \dim_K \mathrm{ext}_B^n(M, N) \, t^n \in \mathbb{Z}[[t]]$$

There is a simple relation between these series: $G_B^{M,N}(t) = G_B^{M,N}(t, 1)$.

Corollary 4.11. *In the notation above,* $\mathrm{ext}_B^{\scriptscriptstyle\bullet}(M,N)$ *is a finitely generated bigraded* R-*module, and* $G_B^{M,N}(t,u)$ *represents a rational function of the form*

$$\frac{g_B^{M,N}(t,u)}{(1-t^2u^{r_1})\cdots(1-t^2u^{r_c})} \qquad \text{with} \qquad g_B^{M,N}(t,u)\in\mathbb{Z}[t,u,u^{-1}]$$

Proof. The assertion on finite generation results from Theorem 4.5 and the one on bigradings from Remark A.2. The form of the power series is then given by the Hilbert-Serre Theorem. □

5   Computation of Ext Modules

This section contains the main new computational result of the paper.

We discuss, apply, and present an algorithm that computes, for graded modules M and N over a graded complete intersection ring B, the graded B-modules $\mathrm{Ext}_B^n(M,N)$ simultaneously in all degrees n, along with all the cohomology operators defined in Remark 4.3.

More precisely, the input consists of a field K, a polynomial ring $A = K[x_1,\dots,x_e]$ with $\deg'(x_h) > 0$, a sequence $f_1,\dots,f_c$ of elements of A, and finitely generated modules M, N over $B = A/(f_1,\dots,f_c)$. The program checks whether the sequence consists of homogeneous elements, whether it is regular, and whether the modules M, N are graded, sending the appropriate error message if any one of these conditions is violated. If the input data pass those tests, then the program produces a presentation of the bigraded module $H = \mathrm{Ext}_B^{\scriptscriptstyle\bullet}(M,N)$, where the elements of $\mathrm{Ext}_B^n(M,N)$ have homological degree $-n$, over the polynomial ring $A[X_1,\dots,X_c]$, bigraded by $\mathrm{Deg}(a) = (0,\deg'(a))$ and $\mathrm{Deg}(X_i) = (-2,-\deg'(f_i))$.

The algorithm is based on the proofs of Theorems 4.1 and 4.5, and is presented in Code 5.4 below. We start with an informal discussion.

Remark 5.1. The routine `resolution` of *Macaulay 2* finds the matrices $d_{o,n}\colon C_n \to C_{n-1}$ of the differential d_C of a minimal free resolution C of M over A. Matrices $d_{\gamma,n}\colon C_n \to C_{n+2|\gamma|-1}$ satisfying equation (1) for $\gamma \in \mathbb{N}^c$ with $|\gamma| > 0$ are computed using the routine `nullhomotopy` of the *Macaulay 2* language in a loop that follows the first part of the proof of Theorem 4.1.

The transposed matrix $d_{\gamma,n}^*$ yields an endomorphism of the free bigraded A-module $C^* = \bigoplus_{n=0}^e \mathrm{Hom}_A(C_n, A)$ of rank m, where $m = \sum_{n=0}^e \mathrm{rank}_A C_n$. The $m \times m$ matrix $\widetilde{d}_{\gamma,n}^*$ describing this endomorphism is formed using the routines `transpose` and `sum`. The $m \times m$ matrix

$$\Delta = \sum_{n=0}^e (-1)^{n+2|\gamma|-1} \sum_{\substack{\gamma\in\mathbb{N}^c \\ |\gamma|\le(e-n+1)/2}} X^\gamma \cdot \widetilde{d}_{\gamma,n}^*$$

with entries in $S = A[X_1,\dots,X_c]$ defines an endomorphism of the free bigraded S-module $S \otimes_A C^*$. It induces an endomorphism $\overline{\Delta}$ of the bigraded

S-module $S \otimes_A C^* \otimes_A N$. The bigraded S-module $H = \mathrm{Ext}_B^{\bullet}(M, N)$ is computed as $H = \mathrm{Ker}(\overline{\Delta}) / \mathrm{Im}(\overline{\Delta})$ using the routine homology.

In the computations we let H denote the bigraded S-module $\mathrm{Ext}_B^{\bullet}(M, N)$. As the graded ring S is zero in odd homological degrees, there is a canonical direct sum decomposition $H = H^{\mathrm{even}} \oplus H^{\mathrm{odd}}$ of bigraded S-modules, where 'even' or 'odd' refers to the parity of the *first* degree in each pair $\mathrm{Deg}(x)$.

We begin with an example in codimension 1, where it is possible to construct the infinite resolution and the action of S on it by hand.

Example 5.2. Consider the ring $A = K[x]$ where the variable x is assigned degree 5, and set $B = A/(x^3)$. The bigraded ring of cohomology operators then is $S = A[X, x]$, where $\mathrm{Deg}(X) = (-2, -15)$ and $\mathrm{Deg}\, x = (0, 5)$.

For the B-modules $M = B/(x^2)$ and $N = B/(x)$, the bigraded S-module $H = \mathrm{Ext}_B^{\bullet}(M, N)$ is described by the isomorphism

$$H \cong (S/(x)) \oplus (S/(x))[1, 10]$$

A minimal free resolution of M over B is displayed below.

$$F = \quad \cdots \longrightarrow B[-30] \xrightarrow{-x} B[-25] \xrightarrow{x^2} B[-15] \xrightarrow{-x} B[-10] \xrightarrow{x^2} B \longrightarrow 0$$

This resolution is actually isomorphic to the resolution $C \otimes_A D'$ described in Remark 4.2, formed from the free resolution

$$C = \quad 0 \longrightarrow A[-10] \xrightarrow{x^2} A \longrightarrow 0$$

of M over A and the nullhomotopy d_{ϵ_1} displayed in the diagram

$$
\begin{array}{ccccccccc}
0 & \longrightarrow & A[-10] & \xrightarrow{x^2} & A & \longrightarrow & 0 \\
 & & \Big\downarrow{-x^3} & \swarrow{-x} & \Big\downarrow{-x^3} & & \\
0 & \longrightarrow & A[-10] & \xrightarrow{x^2} & A & \longrightarrow & 0
\end{array}
$$

The isomorphism of F with $C \otimes_A D'$ endows F with a structure of bigraded module over S, where the action of X on F is the chain map $F \to F$ of homological degree -2 and internal degree -15 that corresponds to the identity map of B in each component.

The bigraded S-module $H = \mathrm{Ext}_B^{\bullet}(M, N)$ is the homology of the complex

$$\mathrm{Hom}_R(F, N) = \quad 0 \to N \xrightarrow{0} N[10] \xrightarrow{0} N[15] \xrightarrow{0} N[25] \xrightarrow{0} N[30] \longrightarrow \cdots$$

where multiplication by $x \in S$ is the zero map, and for each $i \geq 0$ multiplication by $X \in S$ sends $N[15i]$ to $N[15i + 15]$ (respectively, $N[10i + 10]$ to $N[10i + 25]$) by the identity map. This description provides the desired isomorphisms of bigraded S-modules.

Here is how to compute H with Macaulay 2.

Create the rings and modules.

```
i42 : K = ZZ/103;

i43 : A = K[x,Degrees=>{5}];

i44 : B = A/(x^3);

i45 : M = B^1/(x^2);

i46 : N = B^1/(x);
```

Use the function Ext to compute $H = \mathrm{Ext}_B^\bullet(M, N)$ (the semicolon at the end of the line will suppress printing until we have assigned the name S to the ring of cohomology operators constructed by *Macaulay 2*.)

```
i47 : H = Ext(M,N);
```

We may look at the ring.

```
i48 : ring H

o48 = K [$X , x, Degrees => {{-2, -15}, {0, 5}}]
           1

o48 : PolynomialRing
```

Macaulay 2 has assigned the name $X_1 to the variable X. The dollar sign indicates an internal name that cannot be entered from the keyboard: if necessary, obtain the variable by entering S_0; notice that indexing in *Macaulay 2* starts with 0 rather than 1. Notice also the appearance of braces rather than parentheses in *Macaulay 2*'s notation for bidegrees.

```
i49 : degree \ gens ring H

o49 = {{-2, -15}, {0, 5}}

o49 : List
```

Assign the ring a name.

```
i50 : S = ring H;
```

We can now look at the S-module H.

```
i51 : H

o51 = cokernel {0, 0}    | 0 x |
               {-1, -10} | x 0 |

                              2
o51 : S-module, quotient of S
```

Each row in the display above is labeled with the bidegree of the corresponding generator of H. This presentation gives the isomorphisms of bigraded S-modules, already computed by hand earlier.

Let's try an example with a complete intersection of codimension 2. It is not so easy to do by hand, but can be checked using the theory in [4].

Example 5.3. Begin by constructing a polynomial ring $A = K[x, y]$.

```
i52 : A = K[x,y];
```

Now we produce a complete intersection quotient ring $B = A/(x^3, y^2)$.

```
i53 : J = ideal(x^3,y^2);

o53 : Ideal of A

i54 : B = A/J;
```

We take N to be the B-module $B/(x^2, xy)$.

```
i55 : N = cokernel matrix{{x^2,x*y}}

o55 = cokernel | x2 xy |

                      1
o55 : B-module, quotient of B
```

Remark A.1 shows that $H = \mathrm{Ext}_B^{\bullet}(N, N)$ is a bigraded module over the bigraded ring $S = A[X_1, X_2] = K[X_1, X_2, x, y]$ where

$$\mathrm{Deg}(X_1) = (-2, -3) \quad \mathrm{Deg}(X_2) = (-2, -2)$$
$$\mathrm{Deg}(x) = (0, 1) \quad\quad \mathrm{Deg}(y) = (0, 1)$$

Using *Macaulay 2* (below) we obtain an isomorphism of bigraded S-modules

$$H^{\mathrm{even}} \cong \frac{S}{(x^2, xy, y^2, xX_1, yX_1)} \oplus \frac{S}{(x, y)}[2, 2]$$

$$H^{\mathrm{odd}} \cong \left(\frac{S}{(x, y, X_1)} \oplus \frac{S}{(x, y)} \right)^2 [1, 1]$$

These isomorphisms also yield expressions for the graded B-modules:

$$\mathrm{Ext}_B^{2i}(N, N) \cong N \cdot X_2^i \oplus \bigoplus_{h=1}^{i} k \cdot X_1^h X_2^{i-h} \oplus \bigoplus_{h=0}^{i-1} k[2] \cdot X_1^h X_2^{i-1-h}$$

$$\mathrm{Ext}_B^{2i+1}(N, N) \cong \left(k[1] \cdot X_2^i \oplus \bigoplus_{h=0}^{i} k[1] \cdot X_1^{i-h} X_2^h \right)^2$$

Now we follow in detail the computation of the bigraded S-module H.

```
i56 : time H = Ext(N,N);
        -- used 0.2 seconds

i57 : ring H

o57 = K [$X , $X , x, y, Degrees => {{-2, -2}, {-2, -3}, {0, 1}, {0, 1}}]
          1     2

o57 : PolynomialRing

i58 : S = ring H;
```

One might wish to have a better view of the bidegrees of the variables of the ring S. An easy way to achieve this, with signs reversed, is to display the transpose of the matrix of variables.

```
i59 : transpose vars S
```

154   L. L. Avramov and D. R. Grayson

```
o59 = {2, 2}  | $X_1 |
      {2, 3}  | $X_2 |
      {0, -1} | x    |
      {0, -1} | y    |

                 4      1
o59 : Matrix S  <--- S
```

The internal degrees displayed for the cohomology operators may come as a surprise. To understand what is going on, recall that these degrees are determined by a choice of minimal generators for J. At this point we do not know what is the sequence of generators that *Macaulay 2* used, so let's compute those generators the way the program did.

```
i60 : trim J

               2   3
o60 = ideal (y , x )

o60 : Ideal of A
```

Notice that *Macaulay 2* has reordered the original sequence of generators. Now we see that our variable X_1, which corresponds to x^3, is denoted X_2 by *Macaulay 2*, and that X_2, which corresponds to y^2 is denoted X_1. This explains the bidegrees used by the program.

Display H.

```
i61 : H

o61 = cokernel {-2, -2} | 0 0 0 0 0 0 0 0 0  0  0  y x 0   0     0      ...
               {-1, -1} | y 0 0 0 0 x 0 0 0  0  0  0 0 $X_1 0     0      ...
               {-1, -1} | 0 0 0 y 0 0 0 x 0  0  0  0 0 0    $X_1 0      ...
               {-1, -1} | 0 y 0 0 x 0 0 0 0  0  0  0 0 0    0     0      ...
               {-1, -1} | 0 0 y 0 0 0 x 0 0  0  0  0 0 0    0     0      ...
               {0, 0}   | 0 0 0 0 0 0 0 0 y2 xy x2 0 0 0    0     $X_1y ...

                          6
o61 : S-module, quotient of S
```

That's a bit large, so we want to look at the even and odd parts separately.

We may compute the even and odd parts of H as the span of the generators of H with the appropriate parity. Since the two desired functions differ only in the predicate to be applied, we can generate them both by writing a function that accepts the predicate as its argument and returns a function.

```
i62 : partSelector = predicate -> H -> (
           R  := ring H;
           H' := prune image matrix {
               select(
                   apply(numgens H, i -> H_{i}),
                   f -> predicate first first degrees source f
                   )
               };
           H');

i63 : evenPart = partSelector even; oddPart = partSelector odd;
```

Now to obtain the even part, H^{even}, simply type

```
i65 : evenPart H
```

```
o65 = cokernel {-2, -2} | 0  0  0  y x 0    0    |
               {0, 0}   | y2 xy x2 0 0 $X_1y $X_1x |

                                2
o65 : S-module, quotient of S
```

Do the same thing for the odd part, H^{odd}.

```
i66 : oddPart H

o66 = cokernel {-1, -1} | 0 0 y 0 0 0 x 0 0    0   |
               {-1, -1} | 0 y 0 0 x 0 0 0 0    0   |
               {-1, -1} | 0 0 0 y 0 0 0 x 0  $X_1 |
               {-1, -1} | y 0 0 0 0 x 0 0 $X_1 0   |

                                4
o66 : S-module, quotient of S
```

These presentations yield the desired isomorphism of bigraded S-modules.

Here is the source code which implements the routine Ext. It is incorporated into *Macaulay 2*.

Code 5.4. The function Ext(M,N) computes $\mathrm{Ext}_B^{\bullet}(M, N)$ for graded modules M, N over a graded complete intersection ring B. The function code can be used to obtain a copy of the source code.

```
i67 : print code(Ext,Module,Module)
-- ../../../m2/ext.m2:82-171
Ext(Module,Module) := Module => (M,N) -> (
  cacheModule := youngest(M,N);
  cacheKey := (Ext,M,N);
  if cacheModule#?cacheKey then return cacheModule#cacheKey;
  B := ring M;
  if B =!= ring N
  then error "expected modules over the same ring";
  if not isCommutative B
  then error "'Ext' not implemented yet for noncommutative rings.";
  if not isHomogeneous B
  then error "'Ext' received modules over an inhomogeneous ring";
  if not isHomogeneous N or not isHomogeneous M
  then error "'Ext' received an inhomogeneous module";
  if N == 0 then B^0
  else if M == 0 then B^0
  else (
    p := presentation B;
    A := ring p;
    I := ideal mingens ideal p;
    n := numgens A;
    c := numgens I;
    if c =!= codim B
    then error "total Ext available only for complete intersections";
    f := apply(c, i -> I_i);
    pM := lift(presentation M,A);
    pN := lift(presentation N,A);
    M' := cokernel ( pM | p ** id_(target pM) );
    N' := cokernel ( pN | p ** id_(target pN) );
    C := complete resolution M';
    X := local X;
    K := coefficientRing A;
    -- compute the fudge factor for the adjustment of bidegrees
```

```
fudge := if #f > 0 then 1 + max(first \ degree \ f) // 2 else 0;
S := K(monoid [X_1 .. X_c, toSequence A.generatorSymbols,
  Degrees => {
    apply(0 .. c-1, i -> {-2, - first degree f_i}),
    apply(0 .. n-1, j -> { 0,   first degree A_j})
    },
  Adjust => v -> {- fudge * v#0 + v#1, - v#0},
  Repair => w -> {- w#1, - fudge * w#1 + w#0}
  ]);
-- make a monoid whose monomials can be used as indices
Rmon := monoid [X_1 .. X_c,Degrees=>{c:{2}}];
-- make group ring, so 'basis' can enumerate the monomials
R := K Rmon;
-- make a hash table to store the blocks of the matrix
blks := new MutableHashTable;
blks#(exponents 1_Rmon) = C.dd;
scan(0 .. c-1, i ->
    blks#(exponents Rmon_i) = nullhomotopy (- f_i*id_C));
-- a helper function to list the factorizations of a monomial
factorizations := (gamma) -> (
  -- Input: gamma is the list of exponents for a monomial
  -- Return a list of pairs of lists of exponents showing the
  -- possible factorizations of gamma.
  if gamma === {} then { ({}, {}) }
  else (
    i := gamma#-1;
    splice apply(factorizations drop(gamma,-1),
      (alpha,beta) -> apply (0..i,
          j -> (append(alpha,j), append(beta,i-j))))));
scan(4 .. length C + 1,
  d -> if even d then (
    scan( exponents \ leadMonomial \ first entries basis(d,R),
      gamma -> (
        s := - sum(factorizations gamma,
          (alpha,beta) -> (
            if blks#?alpha and blks#?beta
            then blks#alpha * blks#beta
            else 0));
        -- compute and save the nonzero nullhomotopies
        if s != 0 then blks#gamma = nullhomotopy s;
        ))));
-- make a free module whose basis elements have the right degrees
spots := C -> sort select(keys C, i -> class i === ZZ);
Cstar := S^(apply(spots C,
  i -> toSequence apply(degrees C_i, d -> {i,first d})));
-- assemble the matrix from its blocks.
-- We omit the sign (-1)^(n+1) which would ordinarily be used,
-- which does not affect the homology.
toS := map(S,A,apply(toList(c .. c+n-1), i -> S_i),
  DegreeMap => prepend_0);
Delta := map(Cstar, Cstar,
  transpose sum(keys blks, m -> S_m * toS sum blks#m),
  Degree => {-1,0});
DeltaBar := Delta ** (toS ** N');
assert isHomogeneous DeltaBar;
assert(DeltaBar * DeltaBar == 0);
-- now compute the total Ext as a single homology module
cacheModule#cacheKey = prune homology(DeltaBar,DeltaBar)))
```

Remark 5.5. The bigraded module $\text{Tor}^B_\bullet(M, N)$ is the homology of the complex $(C \otimes_A D') \otimes_B N$, where $C \otimes_A D'$ is the complex from Theorem 4.1. Observations parallel to Remarks 4.3 and A.1 show that $\text{Tor}^B_\bullet(M, N)$ inherits from D' a structure of bigraded S-module.

It would be desirable also to have algorithms to compute $\text{Tor}^B_\bullet(M, N)$ in the spirit of the algorithm presented above for $\text{Ext}^\bullet_B(M, N)$. If one of the modules has finite length, then each $\text{Tor}^B_n(M, N)$ is a B-module of finite length, and the computation of $\text{Tor}^B_\bullet(M, N)$ can be reduced to a computation of Ext by means of Matlis duality, which here can be realized as vector space duality over the field K. However, in homology there is no equivalent for the finiteness property described in Remark 4.4; it is an **open problem** to devise algorithms that would compute $\text{Tor}^B_\bullet(M, N)$ in general.

6 Invariants of Modules

In this section we apply our techniques to develop effective methods for computation (for graded modules over a graded complete intersection) of invariants such as cohomology modules, Poincaré series, Bass series, complexity, critical degree, and support varieties. For each invariant we produce code that computes it, and illustrate the action of the code on some explicit example.

Whenever appropriate, we describe **open problems** on which the computational power of *Macaulay 2* could be unleashed.

Notation 4.7 is used consistently throughout the section.

6.1 Cohomology Modules

We call the bigraded R-module $P = \text{Ext}^\bullet_B(M, k)$ the *contravariant cohomology module* of M over B, and the bigraded R-module $I = \text{Ext}^\bullet_B(k, M)$ the *covariant cohomology module* of M. Codes that display presentations of the cohomology modules are presented after a detailed discussion of an example.

Example 6.1.1. Let us create the ring $B = K[x, y, z]/(x^3, y^4, z^5)$.

```
i68 : A = K[x,y,z];

i69 : J = trim ideal(x^3,y^4,z^5)

         3   4   5
o69 = ideal (x , y , z )

o69 : Ideal of A

i70 : B = A/J;
```

We trimmed the ideal, so that we know the generators *Macaulay 2* will use.

This time we want a graded B-module M about whose homology we know nothing a priori. One way to proceed is to create M as the cokernel of some random matrix of forms; let's try a 3 by 2 matrix of quadratic forms.

```
i71 : f = random (B^3, B^{-2,-3})
```

```
o71 =  | 27x2+49xy-14y2-23xz-6yz-19z2 38x2y-34xy2+4y3+x2z+16xyz-y2z-5xz ···
       | -5x2+44xy+38y2+40xz+15yz+4z2 -37x2y+51xy2-36y3+26z2z-38xyz-17y ···
       | 21x2-30xy+32y2-47xz+7yz-50z2 -6x2y-14xy2-26y3-7x2z+41xyz+50y2z ···
```

```
            3      2
o71 : Matrix B  <--- B
```

We can't read the second column of that matrix, so let's display it separately.

```
i72 : f_{1}
```

```
o72 =  | 38x2y-34xy2+4y3+x2z+16xyz-y2z-5xz2-6yz2+47z3          |
       | -37x2y+51xy2-36y3+26z2z-38xyz-17y2z+17xz2-11yz2+8z3 |
       | -6x2y-14xy2-26y3-7x2z+41xyz+50y2z+26xz2+46yz2-44z3   |
```

```
            3      1
o72 : Matrix B  <--- B
```

Now let's make the module M.

```
i73 : M = cokernel f;
```

We are going to produce isomorphisms of bigraded modules

$$P^{\text{even}} \cong R[4,10] \oplus (X_1, X_2)[2,7] \oplus \left(\frac{R}{(X_1, X_2, X_3)}\right)^3 \oplus R^4[2,7]$$

$$P^{\text{odd}} \cong \frac{R}{(X_1, X_2, X_3)}[1,2] \oplus \left(\frac{R}{(X_1)}\right)^3 [3,9] \oplus \frac{R}{(X_1, X_2)}[1,3] \oplus R^6[3,9]$$

over the polynomial ring $R = K[X_1, X_2, X_3]$ over K, bigraded by

$$\text{Deg}(X_1) = (-2, -3) \qquad \text{Deg}(X_2) = (-2, -4) \qquad \text{Deg}(X_3) = (-2, -5)$$

Let's compute $\text{Ext}_B^\bullet(M, B/(x, y, z))$ by the routine from Section 5.

```
i74 : time P = Ext(M,B^1/(x,y,z));
        -- used 1.64 seconds
```

```
i75 : S = ring P;
```

Examine the variables of S; due to transposing, their bidegrees are displayed with the *opposite* signs.

```
i76 : transpose vars S
```

```
o76 = {2, 3}  | $X_1 |
      {2, 4}  | $X_2 |
      {2, 5}  | $X_3 |
      {0, -1} | x    |
      {0, -1} | y    |
      {0, -1} | z    |
```

```
            6      1
o76 : Matrix S  <--- S
```

The variables x, y, and z of A annihilate P, and so appear in many places in a presentation of P. To reduce the size of such a presentation, we pass to a ring which eliminates those variables.

```
i77 : R = K[X_1..X_3,Degrees => {{-2,-3},{-2,-4},{-2,-5}},
              Adjust => S.Adjust, Repair => S.Repair];

i78 : phi = map(R,S,{X_1,X_2,X_3,0,0,0})

o78 = map(R,S,{X , X , X , 0, 0, 0})
               1   2   3

o78 : RingMap R <--- S

i79 : P = prune (phi ** P);

i80 : transpose vars ring P

o80 = {2, 3} | X_1 |
      {2, 4} | X_2 |
      {2, 5} | X_3 |

                3     1
o80 : Matrix R  <--- R
```

As we planned, the original variables x, y, z, which act trivially on the cohomology, are no longer present in the ring. Next we compute presentations

```
i81 : evenPart P

o81 = cokernel {-4, -10} | 0   0    0    0   0    0   0    0   0   0    |
               {-4, -10} | 0   0    0    0   0    0   0    0   0   -X_2 |
               {-4, -11} | 0   0    0    0   0    0   0    0   0   X_1  |
               {0, 0}    | 0   0    X_3  0   0    X_2 0    X_1 0   0    |
               {0, 0}    | 0   X_3  0    0   X_2  0   X_1  0   0   0    |
               {0, 0}    | X_3 0    0    X_2 0    0   0    0   X_1 0    |
               {-2, -7}  | 0   0    0    0   0    0   0    0   0   0    |
               {-2, -7}  | 0   0    0    0   0    0   0    0   0   0    |
               {-2, -7}  | 0   0    0    0   0    0   0    0   0   0    |
               {-2, -7}  | 0   0    0    0   0    0   0    0   0   0    |

                           10
o81 : R-module, quotient of R

i82 : oddPart P

o82 = cokernel {-1, -2} | X_3 0    X_2 0   0    0   0    X_1 |
               {-3, -9} | 0   0    0   0   0    0   X_1 0    |
               {-3, -9} | 0   0    0   0   0    X_1 0   0    |
               {-3, -9} | 0   0    0   0   X_1  0   0   0    |
               {-1, -3} | 0   X_2  0   X_1 0    0   0   0    |
               {-3, -9} | 0   0    0   0   0    0   0   0    |
               {-3, -9} | 0   0    0   0   0    0   0   0    |
               {-3, -9} | 0   0    0   0   0    0   0   0    |
               {-3, -9} | 0   0    0   0   0    0   0   0    |
               {-3, -9} | 0   0    0   0   0    0   0   0    |
               {-3, -9} | 0   0    0   0   0    0   0   0    |

                          11
o82 : R-module, quotient of R
```

These presentations yield the desired isomorphisms of bigraded R-modules.

The procedure above can be automated by installing a method that will be run when Ext is presented with a module M and the residue field k. It displays a presentation of $\mathrm{Ext}_B^{\bullet}(M, k)$ as a bigraded R-module.

Code 6.1.2. The function `changeRing` H takes an S-module H and tensors it with R. It does this by constructing R and a ring homomorphism

$$\varphi \colon A[X_1, \ldots, X_c] = S \to R = K[X_1, \ldots, X_c]$$

```
i83 : changeRing = H -> (
        S := ring H;
        K := coefficientRing S;
        degs := select(degrees source vars S,
            d -> 0 != first d);
        R := K[X_1 .. X_#degs, Degrees => degs,
            Repair => S.Repair, Adjust => S.Adjust];
        phi := map(R,S,join(gens R,(numgens S - numgens R):0));
        prune (phi ** H)
        );
```

Code 6.1.3. The function `Ext(M,k)` computes $\text{Ext}^{\bullet}_B(M, k)$ when B is a graded complete intersection, M a graded B-module, and k is the residue field of B. The result is presented as a module over the ring $k[X_1, \ldots, X_c]$.

```
i84 : Ext(Module,Ring) := (M,k) -> (
        B := ring M;
        if ideal k != ideal vars B
        then error "expected the residue field of the module";
        changeRing Ext(M,coker vars B)
        );
```

Example 6.1.4. For a test, we run again the computation for P^{odd}.

```
i85 : use B;

i86 : k = B/(x,y,z);

i87 : use B;

i88 : P = Ext(M,k);

i89 : time oddPart P
        -- used 0.09 seconds
```

```
o89 = cokernel {-1, -2} | X_3 0    X_2 0    0    0    0    X_1 |
               {-3, -9} | 0   0    0   0    0    0    X_1 0   |
               {-3, -9} | 0   0    0   0    0    X_1 0   0   |
               {-3, -9} | 0   0    0   0    X_1 0   0   0   |
               {-1, -3} | 0   X_2 0   X_1 0    0    0    0   |
               {-3, -9} | 0   0    0   0    0    0    0    0   |
               {-3, -9} | 0   0    0   0    0    0    0    0   |
               {-3, -9} | 0   0    0   0    0    0    0    0   |
               {-3, -9} | 0   0    0   0    0    0    0    0   |
               {-3, -9} | 0   0    0   0    0    0    0    0   |
               {-3, -9} | 0   0    0   0    0    0    0    0   |
```

                                                                   . . .
```
o89 : K [X , X , X , Degrees => {{-2, -3}, {-2, -4}, {-2, -5}}]-module · · ·
          1   2   3                                                . . .
```

We also introduce code for computing the covariant cohomology modules.

Code 6.1.5. The function `Ext(k,M)` computes $\text{Ext}^{\bullet}_B(k, M)$ when B is a graded complete intersection, M a graded B-module, and k is the residue field of B. The result is presented as a module over the ring $k[X_1, \ldots, X_c]$.

```
i90 : Ext(Ring,Module) := (k,M) -> (
        B := ring M;
        if ideal k != ideal vars B
        then error "expected the residue field of the module";
        changeRing Ext(coker vars B,M)
        );
```

Let's see the last code in action.

Example 6.1.6. For B and M from Example 6.1.1 we compute the odd part of the covariant cohomology module $\mathrm{Ext}_B^\bullet(k, M)$.

```
i91 : time I = Ext(k,M);
      -- used 14.81 seconds

i92 : evenPart I

o92 = cokernel {0, 6} | 37X_2   37X_1  |
               {0, 6} | -18X_2  -18X_1 |
               {0, 6} | -13X_2  -13X_1 |
               {0, 6} | -37X_2  -37X_1 |
               {0, 6} | 22X_2   22X_1  |
               {0, 6} | 0       0      |
               {0, 6} | X_2     X_1    |
                                                             . . .
o92 : K [X , X , X , Degrees => {{-2, -3}, {-2, -4}, {-2, -5}}]-module  · · ·
           1   2   3                                           . . .

i93 : oddPart I

o93 = cokernel {-1, 5} | -48X_3 13X_3  34X_3  3X_3   0    0      0       · · ·
               {-1, 5} | 3X_3   -40X_3 8X_3   8X_3   0    0      0       · · ·
               {-1, 5} | -X_3   37X_3  -13X_3 -35X_3 0    0      0       · · ·
               {-1, 4} | 4X_2   20X_2  3X_2   -47X_2 4X_1 20X_1  3X_1  · · ·
               {-1, 4} | 0      51X_2  0      -30X_2 0    51X_1  0       · · ·
               {-1, 4} | 0      12X_2  0      -3X_2  0    12X_1  0       · · ·
               {-1, 4} | 42X_2  12X_2  46X_2  25X_2  42X_1 12X_1 46X_   · · ·
               {-1, 4} | 45X_2  24X_2  -14X_2 -35X_2 45X_1 24X_1 -14X  · · ·
               {-1, 4} | 0      0      X_2    0      0    0      X_1   · · ·
               {-1, 4} | X_2    0      0      0      X_1  0      0       · · ·
               {-1, 4} | 0      -40X_2 0      10X_2  0    -40X_1 0       · · ·
               {-1, 4} | 0      X_2    0      0      0    X_1    0       · · ·
               {-1, 3} | 0      0      0      X_1    0    0      0       · · ·
                                                             . . .
o93 : K [X , X , X , Degrees => {{-2, -3}, {-2, -4}, {-2, -5}}]-module  · · ·
           1   2   3                                           . . .
```

6.2   Poincaré Series

The *graded Betti number* of M over B is the number $\beta_{ns}^B(M)$ of direct summands isomorphic to the free module $B[-s]$ in the n'th module of a minimal free resolution of M over B. It can be computed from the equality

$$\beta_{ns}^B(M) = \dim_K \mathrm{Ext}_B^n(M, k)_s$$

The *graded Poincaré series* of M over B is the generating function

$$P_M^B(t, u) = \sum_{n \in \mathbb{N},\, s \in \mathbb{Z}} \beta_{ns}^B(M)\, t^n u^{-s} \in \mathbb{Z}[u, u^{-1}][[t]]$$

It is easily computable with *Macaulay 2* from the contravariant cohomology module, by using the `hilbertSeries` routine.

Code 6.2.1. The function `poincareSeries2` M computes the graded Poincaré series of a graded module M over a graded complete intersection B.

First we set up a ring whose elements can serve as Poincaré series.

```
i94 : T = ZZ[t,u,Inverses=>true,MonomialOrder=>RevLex];
```

```
i95 : poincareSeries2 = M -> (
        B := ring M;
        k := B/ideal vars B;
        P := Ext(M,k);
        h := hilbertSeries P;
        T':= degreesRing P;
        substitute(h, {T'_0=>t^-1,T'_1=>u^-1})
      );
```

The last line in the code above replaces the variables in the Poincaré series provided by the `hilbertSeries` function with the variables in our ring T.

The *n*th *Betti number* $\beta_n^B(M)$ of M over B is the rank of the nth module in a minimal resolution of M by free B-modules. The *Poincaré series* $P_M^B(t)$ is the generating function of the Betti numbers. There are expressions

$$\beta_n^B(M) = \sum_{s=0}^{\infty} \beta_{ns}^B(M) \qquad \text{and} \qquad P_M^B(t) = P_M^B(t, 1)$$

Accordingly, the code for $P_M^B(t)$ just replaces in $P_M^B(t, u)$ the variable u by 1.

Code 6.2.2. The function `poincareSeries1` M computes the Poincaré series of a graded module M over a graded complete intersection B.

```
i96 : poincareSeries1 = M -> (
        substitute(poincareSeries2 M, {u=>1_T})
      );
```

Now let's use these codes in computations.

Example 6.2.3. To get a module whose Betti sequence initially decreases, we form an artinian complete intersection B' and take M' to be a cosyzygy in a minimal injective resolution of the residue field k. Since B' is self-injective, this can be achieved by taking a syzygy of k, then transposing its presentation matrix. Of course, we ask *Macaulay 2* to carry out these steps.

```
i97 : A' = K[x,y,z];
```

```
i98 : B' = A'/(x^2,y^2,z^3);
```

```
i99 : C' = res(B'^1/(x,y,z), LengthLimit => 6)

          1        3        6        10        15        21        28
o99 = B'  <-- B'  <-- B'  <-- B'    <-- B'    <-- B'    <-- B'

      0        1        2        3         4         5         6

o99 : ChainComplex

i100 : M' = coker transpose C'.dd_5

o100 = cokernel {-5} | -y  0    0   0   z   0 0 0   0 0   0   0 0   0 0 |
                {-5} | -x -y    0   0   0   z 0 0   0 0   0   0 0   0 0 |
                {-5} | 0   x   -y   0   0   0 z 0   0 0   0   0 0   0 0 |
                {-5} | 0   0    x  -y   0   0 0 z   0 0   0   0 0   0 0 |
                {-5} | 0   0    0  -x   0   0 0 0   z 0   0   0 0   0 0 |
                {-5} | 0   0    0   0   0   y 0 0   0 0   0   0 0   0 0 |
                {-5} | 0   0    0   0   0  -x y 0   0 0   0   0 0   0 0 |
                {-5} | 0   0    0   0   0   0 x y   0 0   0   0 0   0 0 |
                {-5} | 0   0    0   0   0   0 0 x   y 0   0   0 0   0 0 |
                {-5} | 0   0    0   0   0   0 0 0  -x y   0   0 0   0 0 |
                {-5} | 0   0    0   0   0   0 0 0   x 0   0   0 0   0 0 |
                {-6} | 0   0    0   0   0   0 0 0   0 -y  0   z 0   0 0 |
                {-6} | 0   0    0   0   0   0 0 0   0 x  -y   0 z   0 0 |
                {-6} | 0   0    0   0   0   0 0 0   0 0  -x   0 0   z 0 |
                {-6} | z2  0    0   0   0   0 0 0   0 0   0   y 0   0 0 |
                {-6} | 0  -z2   0   0   0   0 0 0   0 0   0   x y   0 0 |
                {-6} | 0   0   z2   0   0   0 0 0   0 0   0   0 -x  y 0 |
                {-6} | 0   0    0  z2   0   0 0 0   0 0   0   0 0   x 0 |
                {-7} | 0   0    0   0   0   0 0 0   0 0   0   0 0   0 z |
                {-7} | 0   0    0   0   0   0 0 0   0 z2  0   0 0   0 y |
                {-7} | 0   0    0   0   0   0 0 0   0 0  z2   0 0   0 x |
                                            21
o100 : B'-module, quotient of B'
```

Compute the Poincaré series in two variables $P_{M'}^{B'}(t, u)$.

```
i101 : poincareSeries2 M'

        -7       -6        -5        -6        -5        -4    2 -5        2 - ···
        3u    + 7u    + 11u    + t*u    + 5t*u    + 9t*u   - 6t u   - 14t u  ···
o101 = ------------------------------------------------------------------------ - ···
                                                                          ···
                                                                          ···

o101 : Divide
```

Example 6.2.4. We compute $P_M^B(t)$ for the module M from Example 6.1.1.

```
i102 : p = poincareSeries1 M

                  2      3      4       5      6     7
         3 + 2t - 5t  + 4t  + 12t  + t  - 4t  - t
o102 = ---------------------------------------------
                      2        2        2
              (1 - t )(1 - t )(1 - t )

o102 : Divide
```

We have written some rather naïve code for simplifying rational functions as
above. It locates factors of the form $1 - t^n$ in the denominator, factors out

$1-t$, and factors out $1+t$ if n is even. Keeping the factors of the denominator separate, it then cancels as many of them as it can with the numerator.

```
i103 : load "simplify.m2"
```

```
i104 : simplify p
```

$$o104 = \frac{3 - t - 4t^2 + 8t^3 + 4t^4 - 3t^5 - t^6}{(1 + t)^2 (1 - t)^3}$$

```
o104 : Divide
```

In this case, it succeeded in canceling a factor of $1 + t$.

Example 6.2.5. We compute some Betti numbers for M. We use the division operation in the Euclidean domain $T' = \mathbb{Q}[t, t^{-1}]$ with the reverse monomial ordering to compute power series expansions.

```
i105 : T' = QQ[t,Inverses=>true,MonomialOrder=>RevLex];
```

```
i106 : expansion = (n,q) -> (
            t := T'_0;
            rho := map(T',T,{t,1});
            num := rho value numerator q;
            den := rho value denominator q;
            n = n + first degree den;
            n = max(n, first degree num + 1);
            (num + t^n) // den
            );
```

Now let's expand the Poincaré series up to t^{20}.

```
i107 : expansion(20,p)
```

$$o107 = 3 + 2t + 4t^2 + 10t^3 + 15t^4 + 25t^5 + 32t^6 + 46t^7 + 55t^8 + 73t^9 + \cdots$$

```
o107 : T'
```

Just to make sure, let's compare the first few coefficients with the more pedestrian way of doing the computation, one Ext module at a time.

```
i108 : psi = map(K,B)
```

```
o108 = map(K,B,{0, 0, 0})
```

```
o108 : RingMap K <--- B
```

```
i109 : apply(10, i -> rank (psi ** Ext^i(M,coker vars B)))
```

```
o109 = {3, 2, 4, 10, 15, 25, 32, 46, 55, 73}
```

```
o109 : List
```

Now we restore t to its original use.

```
i110 : use T;
```

6.3   Complexity

The *complexity* of M is the least $d \in \mathbb{N}$ such that the function

$$n \mapsto \dim_K \mathrm{Ext}_B^n(M, k)$$

is bounded above by a polynomial of degree $d - 1$ (with the convention that the zero polynomial has degree -1). This number, denoted $\mathrm{cx}_B(M)$, was introduced in [2] to measures on a polynomial scale the rate of growth of the Betti numbers of M. It is calibrated so that $\mathrm{cx}_B(M) = 0$ if and only if M has finite projective dimension. Corollary 4.11 yields

$$P_M^B(t) = \frac{p_M^B(t)}{(1 - t^2)^c} \qquad \text{for some} \qquad p_M^B(t) \in \mathbb{Z}[t]$$

Decomposing the right hand side into partial fractions, one sees that $\mathrm{cx}_R(M)$ equals the order of the pole of $P_M^B(t)$ at $t = 1$; in particular, $\mathrm{cx}_R(M, N) \leq c$. However, since we get $P_M^B(t)$ from a computation of the R-module $P = \mathrm{Ext}_B^{\bullet}(M, k)$, it is natural to obtain $\mathrm{cx}_R(M)$ as the Krull dimension of P.

Code 6.3.1. The function `complexity` M yields the complexity of a graded module M over a graded complete intersection ring B.

```
i111 : complexity = M -> dim Ext(M,coker vars ring M);
```

Example 6.3.2. We compute $\mathrm{cx}_B(M)$ for M from Example 6.1.1.

```
i112 : complexity M

o112 = 3
```

6.4   Critical Degree

The *critical degree* of M is the least integer ℓ for which the minimal resolution F of M admits a chain map $g: F \to F$ of degree $m < 0$, such that $g_{m+n}: F_{m+n} \to F_n$ is surjective for all $n > \ell$. This number, introduced in [6] and denoted $\mathrm{crdeg}_B M$, is meaningful over every graded ring B. It is equal to the projective dimension whenever the latter is finite.

When B is a complete intersection it is proved in [6, Sect. 7] that $\mathrm{crdeg}_B M$ is finite and yields important information on the Betti sequence:

- if $\mathrm{cx}_B M \leq 1$, then $\beta_n^B(M) = \beta_{n+1}^B(M)$ for all $n > \mathrm{crdeg}_B M$.
- if $\mathrm{cx}_B M \geq 2$, then $\beta_n^B(M) < \beta_{n+1}^B(M)$ for all $n > \mathrm{crdeg}_B M$.

Thus, it is interesting to know $\mathrm{crdeg}_B M$, or at least to have a good upper bound. Here is what is known, in terms of $h = \mathrm{depth}\, B - \mathrm{depth}_B M$.

- if $\mathrm{cx}_B M = 0$, then $\mathrm{crdeg}_B M = h$.
- if $\mathrm{cx}_B M = 1$, then $\mathrm{crdeg}_B M \leq h$.
- if $\mathrm{cx}_B M = 2$, then $\mathrm{crdeg}_B M \leq h + 1 + \max\{2\beta_h^B(M) - 1, 2\beta_{h+1}^B(M)\}$.

The first part is the Auslander-Buchsbaum Equality, the second part is proved in [11, Sect. 6], the third is established in [4, Sect. 7].

These upper bounds are realistic: there exist examples in complexity 1 when they are reached, and examples in complexity 2 when they are not more than twice the actual value of the critical degree. If $cx_R M \geq 3$, then it is an **open problem** whether the critical degree of M can be bounded in terms that do not depend on the action of the cohomology operators.

However, in every concrete case $crdeg_R M$ can be computed explicitly by using *Macaulay 2*. Indeed, it is proved in [6, Sect. 7] that $crdeg_R M$ is equal to the highest degree of a non-zero element in the socle of the R-module $Ext_B^\bullet(M, k)$, that is, the submodule consisting of elements annihilated by $(X_1, \ldots, X_c)$. The socle is naturally isomorphic to $Hom_B(k, Ext_B^\bullet(M, k))$, so it can be obtained by standard *Macaulay 2* routines.

For instance, for the module M from Example 6.1.1, we get

```
i113 : k = coker vars ring H;
```

```
i114 : prune Hom(k,H)
```

```
o114 = 0
```

```
o114 : K [$X , $X , x, y, Degrees => {{-2, -2}, {-2, -3}, {0, 1}, {0, ···
              1    2
```

The degrees displayed above show that $crdeg_R M = 1$.

Of course, one might prefer to see the number $crdeg_B M$ directly.

Code 6.4.1. The function `criticalDegree` M computes the critical degree of a graded module M over a graded complete intersection ring B.

```
i115 : criticalDegree = M -> (
           B := ring M;
           k := B / ideal vars B;
           P := Ext(M,k);
           k = coker vars ring P;
           - min ( first \ degrees source gens prune Hom(k,P))
           );
```

Let's test the new code in a couple of cases.

Example 6.4.2. For the module M of Example 6.1.1 we have

```
i116 : criticalDegree M
```

```
o116 = 1
```

in accordance with what was already observed above.

For the module M' of Example 6.2.3 we obtain

```
i117 : criticalDegree M'
```

```
o117 = 5
```

6.5   Support Variety

Let $\overline{K}$ denote an algebraic closure of K. The *support variety* $V_B^*(M)$ is the algebraic set in $\overline{K}^c$ defined by the annihilator of $\operatorname{Ext}_B^*(M, k)$ over $R = K[X_1, \ldots, X_c]$. This 'geometric image' of the contravariant cohomology module was introduced in [2] and used to study the minimal free resolution of M. The dimension of the support variety is equal to the complexity $\operatorname{cx}_R(M)$, that we can already compute. There is no need to associate a variety to the covariant cohomology module, see 7.4.

Since $V_B^*(M)$ is defined by homogeneous equations, it is a cone in $\overline{K}^c$. An important **open problem** is whether every cone in $\overline{K}^c$ that can be defined over K is the variety of some B-module M. By [2, Sect. 6] all linear subspaces and all hypersurfaces arise in this way, but little more is known in general.

Feeding our computation of $\operatorname{Ext}_B^*(M, k)$ to standard *Macaulay 2* routines we write code for determining a set of equations defining $V_B^*(M)$.

Code 6.5.1. The function `supportVarietyIdeal` M yields a set of polynomial equations with coefficients in K, defining the support variety $V_B^*(M)$ in $\overline{K}^c$ for a graded module M over a graded complete intersection B.

```
i118 : supportVarietyIdeal = M -> (
           B := ring M;
           k := B/ideal vars B;
           ann Ext(M,k)
           );
```

As before, we illustrate the code with explicit computations. In view of the open problem mentioned above, we fix a ring and a type of presentation, then change randomly the presentation matrix in the hope of finding an 'interesting' variety. The result of the experiment is assessed in Remark 6.5.3.

Example 6.5.2. Let $\mathbb{F}_7$ denote the prime field with 7 elements, and form the zero-dimensional complete intersection $B'' = \mathbb{F}_7[x, y, z]/(x^7, y^7, z^7)$.

```
i119 : K'' = ZZ/7;

i120 : A'' = K''[x,y,z];

i121 : J'' = ideal(x^7,y^7,z^7);

o121 : Ideal of A''

i122 : B'' = A''/J'';
```

We apply the code above to search, randomly, for some varieties. Using `scan` we print the results from several runs with one command.

```
i123 : scan((1,1) .. (3,3), (r,d) -> (
           V := cokernel random (B''^r,B''^{-d});
           << "--------------------------------------------------- ...
           << endl
           << "V = " << V << endl
           << "support variety ideal = "
           << timing supportVarietyIdeal V
           << endl))
```

```
-------------------------------------------------------------------
V = cokernel | -2x+3y+2z |
support variety ideal = ideal (X  - 2X , X  + X )
                               2    3   1    3
                    -- 0.7 seconds
-------------------------------------------------------------------
V = cokernel | 3x2-2xy+xz-3yz |
support variety ideal = ideal(X  + 3X  + 2X )
                              1     2     3
                    -- 0.48 seconds
-------------------------------------------------------------------
V = cokernel | -2x3+3x2y+y3-x2z-3y2z-xz2-3z3 |
support variety ideal = 0
                    -- 1.54 seconds
-------------------------------------------------------------------
V = cokernel | -3y+3z |
             | -2x-2y |
support variety ideal = ideal(X  + X  - X )
                              1    2    3
                    -- 0.86 seconds
-------------------------------------------------------------------
V = cokernel | -x2+2y2-xz+yz+3z2 |
             | 2xy-3xz-3yz-2z2   |
support variety ideal = 0
                    -- 1.31 seconds
-------------------------------------------------------------------
V = cokernel | -x3-2x2y-xy2-2xyz+3y2z+2xz2-yz2-2z3 |
             | 2xy2+3y3-3x2z-2y2z+2xz2+2yz2        |
support variety ideal = 0
                    -- 2.21 seconds
-------------------------------------------------------------------
V = cokernel | 3x-y-z    |
             | -3x-y+2z  |
             | x-2y+3z   |
support variety ideal = 0
                    -- 1.1 seconds
-------------------------------------------------------------------
V = cokernel | 2x2-2xy+2y2+2xz-3z2  |
             | -x2+2xy+y2+3xz+3yz-z2 |
             | -2xz+2yz+2z2          |
support variety ideal = 0
                    -- 1.67 seconds
-------------------------------------------------------------------
V = cokernel | 2x3-x2y+2xy2-y3-2xyz+3y2z+xz2+3yz2+z3  |
             | -3x3-3x2y+3xy2+2x2z+3xyz-3y2z-xz2       |
             | -3x3-2x2y-xy2-2y3-2xyz+y2z+xz2+3yz2-z3  |
support variety ideal = 0
                    -- 1.92 seconds
-------------------------------------------------------------------
```

Remark 6.5.3. The (admittedly short) search above did not turn up any non-linear variety. This should be contrasted with the known result that *every* cone in $\overline{\mathbb{F}}_7{}^3$ is the support variety of some B''-module.

Indeed, B'' is isomorphic to the group algebra $\mathbb{F}_7[G]$ of the elementary abelian group $G = \mathbb{C}_7 \times \mathbb{C}_7 \times \mathbb{C}_7$, where $\mathbb{C}_7$ is a cyclic group of order 7. It is shown in [2, Sect. 7] that $V^*_{B''}(V)$ is equal to a variety $V^*_G(V)$, defined in a different way in [9] by Carlson. He proves in [10] that if K is a field of characteristic $p > 0$, and G is an elementary abelian p-group of rank c, then every cone in $\overline{K}^c$ is the rank variety of a finitely generated module over $K[G]$.

6.6 Bass Series

The *graded Bass number* $\mu_B^{ns}(M)$ of M over B is the number of direct summands isomorphic to $U[s]$ in the n'th module of a minimal graded injective resolution of M over B, where U is the injective envelope of k. It satisfies

$$\mu_B^{ns}(M) = \dim_K \operatorname{Ext}_B^n(k, M)_s$$

The *graded Bass series* of M over B is the generating function

$$I_B^M(t, u) = \sum_{n \in \mathbb{N}, s \in \mathbb{Z}} \mu_B^{ns}(M)\, t^n u^s \in \mathbb{Z}[u, u^{-1}][[t]]$$

It is easily computable with *Macaulay 2* from the covariant cohomology module, by using the `hilbertSeries` routine.

Code 6.6.1. The function `bassSeries2` M computes the graded Bass series of a graded module M over a graded complete intersection B.

```
i124 : bassSeries2 = M -> (
          B := ring M;
          k := B/ideal vars B;
          I := Ext(k,M);
          h := hilbertSeries I;
          T':= degreesRing I;
          substitute(h, {T'_0=>t^-1, T'_1=>u})
          );
```

As with Betti numbers and Poincaré series, there are ungraded versions of Bass numbers and Bass series; they are given, respectively, by

$$\mu_B^n(M) = \sum_{s=0}^{\infty} \mu_B^{ns}(M) \qquad \text{and} \qquad I_B^M(t) = I_B^M(t, 1)$$

Code 6.6.2. The function `bassSeries1` M computes the Bass series of a graded module M over a graded complete intersection B.

```
i125 : bassSeries1 = M -> (
          substitute(bassSeries2 M, {u=>1_T})
          );
```

Now let's use these codes in computations.

Example 6.6.3. For k, the residue field of B, the contravariant and covariant cohomology modules coincide. For comparison, we compute side by side the Poincaré series and the Bass series of k, when $B = K[x, y, z]/(x^3, y^4, z^5)$ is the ring defined in Example 6.1.1.

```
i126 : use B;

i127 : L = B^1/(x,y,z);

i128 : p = poincareSeries2 L

                2 2      3 3
        1 + 3t*u + 3t u  + t u
o128 = ------------------------------
              2 3      2 4      2 5
        (1 - t u )(1 - t u )(1 - t u )

o128 : Divide

i129 : b = bassSeries2 L

                 -1     2 -2     3 -3
        1 + 3t*u    + 3t u   + t u
o129 = ----------------------------------
           2 -3      2 -4      2 -5
        (1 - t u )(1 - t u )(1 - t u )

o129 : Divide
```

The reader would have noticed that the two series are different, and that one
is obtained from the other by the substitution $u \mapsto u^{-1}$. This underscores
the different meanings of the graded Betti numbers and Bass numbers.

Example 6.6.4. Here we compute the graded and ungraded Bass series of
the B-module M of Example 6.1.1.

```
i130 : b2 = bassSeries2 M

         6      3       4       5     2 2     2 3     3     3       3 2   ...
        7u  + t*u + 9t*u + 3t*u  - t u  - t u  - 4t - 3t u - 3t u + ...
o130 = --------------------------------------------------------------------- ...
                                      2 -3      2 -4      2 -5
                               (1 - t u )(1 - t u )(1 - t u )

o130 : Divide

i131 : b1 = bassSeries1 M;

i132 : simplify b1

             2     3     4
        7 + 6t - 8t  - 2t  + 3t
o132 = --------------------------
              2       3
        (1 + t) (1 - t)

o132 : Divide
```

7   Invariants of Pairs of Modules

In this final section we compute invariants of a pair (M, N) of graded modules
over a graded complete intersection B, derived from the reduced Ext module
$\mathrm{ext}_B^{\bullet}(M, N)$ defined in Remark 4.10. The treatment here is parallel to that
in Section 6. When one of the modules M or N is equal to the residue field
k, the invariants discussed below reduce to those treated in that section.

7.1 Reduced Ext Module

The reduced Ext module $\operatorname{ext}_B^{\bullet}(M, N) = \operatorname{Ext}_B^{\bullet}(M, N) \otimes_A k$ defined in Remark 4.10 is computed from our basic routine $\mathtt{Ext(M,N)}$ by applying the function $\mathtt{changeRing}$ defined in Code 6.1.2.

Code 7.1.1. The function $\mathtt{ext(M,N)}$ computes $\operatorname{ext}_B^{\bullet}(M, N)$ when M and N are graded modules over a graded complete intersection B.

```
i133 : ext = (M,N) -> changeRing Ext(M,N);
```

Example 7.1.2. Using the ring $B = K[x, y, z]/(x^3, y^4, z^5)$ and the module M created in Example 6.1.1, we make new modules

$$N = B/(x^2 + z^2, \, y^3) \qquad \text{and} \qquad N' = B/(x^2 + z^2, \, y^3 - 2z^3)$$

```
i134 : use B;

i135 : N = B^1/(x^2 + z^2,y^3);

i136 : time rH = ext(M,N);
        -- used 15.91 seconds

i137 : evenPart rH
```

```
o137 = cokernel {-4, -9} | 0    0   0   0   0   0   0    0    0   0   0   · · ·
                {0, 2}   | 0    0   0   0   0   0   0    X_3  0   0   0   · · ·
                {0, 2}   | 0    0   0   0   0   X_3 0    0    0   0   0   · · ·
                {0, 2}   | 0    0   0   0   0   0   X_3  0    0   0   0   · · ·
                {0, 2}   | 0    0   0   0   X_3 0   0    0    0   0   0   · · ·
                {0, 2}   | 0    0   0   X_3 0   0   0    0    0   0   0   · · ·
                {0, 2}   | 0    0   X_3 0   0   0   0    0    0   0   0   · · ·
                {0, 2}   | 0    X_3 0   0   0   0   0    0    0   0   X_2 · · ·
                {0, 2}   | X_3  0   0   0   0   0   0    0    0   X_2 0   · · ·
                {-2, -4} | 0    0   0   0   0   0   0    0    0   0   0   · · ·
                {-2, -4} | 0    0   0   0   0   0   0    0    0   0   0   · · ·
                {-2, -4} | 0    0   0   0   0   0   0    0    0   0   0   · · ·
                {-2, -4} | 0    0   0   0   0   0   0    0    0   0   0   · · ·
                {-2, -4} | 0    0   0   0   0   0   0    0    0   0   0   · · ·
                {-2, -4} | 0    0   0   0   0   0   0    0    0   0   0   · · ·
                {0, 1}   | 0    0   0   0   0   0   0    0    X_2 0   0   · · ·

                                                                        · · ·
o137 : K [X , X , X , Degrees => {{-2, -3}, {-2, -4}, {-2, -5}}]-modul · · ·
           1   2   3                                                    · · ·
```

```
i138 : oddPart rH
```

```
o138 = cokernel {-3, -6} | 0        0   0   0   0   0   0    0    0   0   · · ·
                {-3, -6} | 0        0   0   0   0   0   0    0    0   0   · · ·
                {-3, -6} | 0        0   0   0   0   0   0    0    0   0   · · ·
                {-3, -6} | 0        0   0   0   0   0   0    0    0   0   · · ·
                {-3, -6} | 0        0   0   0   0   0   0    0    0   0   · · ·
                {-1, -1} | -39X_3 0  0   0   0   0   0   0    X_2  0   · · ·
                {-1, -1} | 31X_3  0  0   0   0   0   0   X_2  0    0   · · ·
                {-1, -1} | -34X_3 0  0   0   0   0   X_2 0    0    0   · · ·
                {-1, -1} | -35X_3 0  0   0   0   X_2 0   0    0    0   · · ·
```

```
          {-1, -1} | -29X_3 0   0    0    X_2 0   0   0   0   0    ...
          {-1, -1} | 12X_3  0   0    X_2 0   0   0   0   0   0    ...
          {-1, -1} | -8X_3  0   X_2 0    0   0   0   0   0   0    ...
          {-1, -1} | X_3    X_2 0   0    0   0   0   0   0   0    ...
          {-3, -7} | 0      0   0   0    0   0   0   0   0   X_1  ...

                                                                 ...
o138 : K [X , X , X , Degrees => {{-2, -3}, {-2, -4}, {-2, -5}}]-modul ...
         1   2   3                                               ...

i139 : N' = B^1/(x^2 + z^2,y^3 - 2*z^3);

i140 : time rH' = ext(M,N');
       -- used 20.26 seconds

i141 : evenPart rH'

o141 = cokernel {-4, -8} | 0  0   0   0   0   0   0   0   0   0   0  ...
                {-4, -8} | 0  0   0   0   0   0   0   0   0   0   0  ...
                {-4, -9} | 0  0   0   0   0   0   0   0   0   0   0  ...
                {-4, -9} | 0  0   0   0   0   0   0   0   0   0   0  ...
                {-4, -9} | 0  0   0   0   0   0   0   0   0   0   0  ...
                {-4, -9} | 0  0   0   0   0   0   0   0   0   0   0  ...
                {-4, -9} | 0  0   0   0   0   0   0   0   0   0   0  ...
                {-4, -9} | 0  0   0   0   0   0   0   0   0   0   0  ...
                {0, 2}   | 0  0   0   0   0   0   X_3 0   0   0   0  ...
                {0, 2}   | 0  0   0   0   0   X_3 0   0   0   0   0  ...
                {-2, -4} | 0  0   0   0   0   0   0   0   0   0   0  ...
                {-2, -4} | 0  0   0   0   0   0   0   0   0   0   0  ...
                {-2, -4} | 0  0   0   0   0   0   0   0   0   0   0  ...
                {-2, -4} | 0  0   0   0   0   0   0   0   0   0   0  ...
                {-2, -4} | 0  0   0   0   0   0   0   0   0   0   0  ...
                {-2, -4} | 0  0   0   0   0   0   0   0   0   0   0  ...
                {0, 2}   | 0  0   0   0   X_3 0   0   0   0   0   0  ...
                {0, 2}   | 0  0   X_3 0   0   0   0   0   0   0   X_2 ...
                {-2, -4} | 0  0   0   0   0   0   0   0   0   0   0  ...
                {-2, -4} | 0  0   0   0   0   0   0   0   0   0   0  ...
                {-2, -4} | 0  0   0   0   0   0   0   0   0   0   0  ...
                {0, 2}   | 0  0   0   X_3 0   0   0   0   0   0   0  ...
                {0, 2}   | 0  X_3 0   0   0   0   0   0   0   X_2 0  ...
                {-2, -4} | 0  0   0   0   0   0   0   0   0   0   0  ...
                {-2, -4} | 0  0   0   0   0   0   0   0   0   0   0  ...
                {-2, -4} | 0  0   0   0   0   0   0   0   0   0   0  ...
                {-2, -4} | 0  0   0   0   0   0   0   0   0   0   0  ...
                {-2, -4} | 0  0   0   0   0   0   0   0   0   0   0  ...
                {0, 2}   | X_3 0  0   0   0   0   0   0   0   0   0  ...
                {0, 1}   | 0  0   0   0   0   0   0   X_2 0   0   0  ...
                {0, 1}   | 0  0   0   0   0   0   0   0   X_2 0   0  ...

                                                                 ...
o141 : K [X , X , X , Degrees => {{-2, -3}, {-2, -4}, {-2, -5}}]-modul ...
         1   2   3                                               ...

i142 : oddPart rH'

o142 = cokernel {-3, -6} | 0  0   0   0   0   0   0   0   -42X_2 21X_ ...
                {-3, -6} | 0  0   0   0   0   0   0   0   -6X_2  -32X ...
                {-3, -6} | 0  0   0   0   0   0   0   0   -8X_2  12X_ ...
                {-3, -6} | 0  0   0   0   0   0   0   0   26X_2  -36X ...
                {-3, -6} | 0  0   0   0   0   0   0   0   50X_2  18X_ ...
```

```
{-3, -6} | 0    0    0    0   0   0    0   0    31X_2  7X_2 ···
{-3, -7} | 0    0    0    0   0   0    0   0    0      0    ···
{-3, -7} | 0    0    0    0   0   0    0   0    0      0    ···
{-3, -7} | 0    0    0    0   0   0    0   0    0      0    ···
{-3, -7} | 0    0    0    0   0   0    0   0    0      0    ···
{-3, -7} | 0    0    0    0   0   0    0   0    0      X_1  ···
{-3, -7} | 0    0    0    0   0   0    0   0    X_1    0    ···
{-1, -2} | 0    0    0    X_2 0   0    0   X_1  0      0    ···
{-1, -2} | 0    0    X_2  0   0   0    X_1 0    0      0    ···
{-1, -2} | 0    X_2  0    0   0   X_1  0   0    0      0    ···
{-1, -2} | X_2  0    0    0   X_1 0    0   0    0      0    ···

                                                            ···
o142 : K [X , X , X , Degrees => {{-2, -3}, {-2, -4}, {-2, -5}}]-modul ···
         1   2   3                                          ···
```

7.2   Ext-generator Series

The Ext-generator series $G_B^{M,N}(t,u)$ defined in Remark 4.10 generalizes both the Poincaré series of M and the Bass series of N, as seen from the formulas

$$P_M^B(t,u) = G_B^{M,k}(t,u) \quad \text{and} \quad I_B^N(t,u) = G_B^{k,N}(t,u^{-1})$$

Similar equalities hold for the corresponding series in one variable. Codes for computing Ext-generator series are easy to produce.

Code 7.2.1. The function extgenSeries2(M,N) computes $G_B^{M,N}(t,u)$ when M and N are graded modules over a graded complete intersection B, and presents it as a rational function with denominator $(1 - t^2 u^{r_1}) \cdots (1 - t^2 u^{r_c})$.

```
i143 : extgenSeries2 = (M,N) -> (
         H := ext(M,N);
         h := hilbertSeries H;
         T':= degreesRing H;
         substitute(h, {T'_0=>t^-1,T'_1=>u^-1})
         );
```

Code 7.2.2. The function extgenSeries1(M,N) computes the Ext-generator series in one variable for a pair (M, N) of graded modules over a graded complete intersection B.

```
i144 : extgenSeries1 = (M,N) -> (
         substitute(extgenSeries2(M,N), {u=>1_T})
         );
```

Example 7.2.3. For M, N, and N' as in Example 7.1.2 we obtain

```
i145 : time extgenSeries2(M,N)
       -- used 0.44 seconds

         -2    -1           2      2 2      2 3      2 4      3 4     3  ···
        8u   + u   + 8t*u - 8t u - 9t u  - 9t u  + 7t u  - 8t u  - 8t u ···
o145 = ---------------------------------------------------------------- ···
                                                                        ···
                                                                        ···

o145 : Divide
```

```
i146 : g=time extgenSeries1(M,N)
       -- used 0.13 seconds

              2     3      4     5     6    7
       9 + 8t - 19t  - 11t  + 17t + 4t  - 7t  - t
o146 = ---------------------------------------------
              2        2        2
        (1 - t )(1 - t )(1 - t )

o146 : Divide

i147 : simplify g

              2    3    4
       9 - t - 9t  + 6t + t
o147 = ---------------------
                   2
        (1 + t)(1 - t)

o147 : Divide

i148 : time extgenSeries2(M,N')

         -2     -1    2      2      2 2     2 3      2 4     3 5    ...
       7u  + 2u  + 4t*u - 7t u - 9t u - 9t u + 16t u - 4t u + 2 ...
o148 = ------------------------------------------------------------ ...
                                                                    ...
                                                                    ...

o148 : Divide

i149 : g'=time extgenSeries1(M,N')
       -- used 0.18 seconds

              2     3     4     5     6
       9 + 4t - 9t  + 4t + 8t  - 2t  - 2t
o149 = -----------------------------------
              2        2        2
        (1 - t )(1 - t )(1 - t )

o149 : Divide

i150 : simplify g'

              2    3     5
       9 - 5t - 4t  + 8t - 2t
o150 = ----------------------
               2        3
        (1 + t) (1 - t)

o150 : Divide
```

7.3   Complexity

The *complexity* of a pair of B-modules (M, N) is the least $d \in \mathbb{N}$ such that there exists a polynomial of degree $d - 1$ bounding above the function

$$n \mapsto \dim_K \operatorname{ext}_B^n(M, N)$$

It is denoted $cx_B(M, N)$ and measures on a polynomial scale the rate of growth of the minimal number of generators of $\operatorname{Ext}_B^n(M, N)$; it vanishes if and only if $\operatorname{Ext}_B^n(M, N) = 0$ for all $n \gg 0$. Corollary 4.11 yields

$$G_B^{M,N}(t) = \frac{h(t)}{(1 - t^2)^c} \qquad \text{for some} \qquad h(t) \in \mathbb{Z}[t]$$

so decomposition into partial fractions shows that $cx_R(M, N)$ equals the order of the pole of $G_B^{M,N}(t)$ at $t = 1$. Alternatively, $cx_R(M, N)$ can be obtained by computing the Krull dimension of a reduced Ext module over R.

Code 7.3.1. The function `complexityPair(M,N)` yields the complexity of a pair (M, N) of graded modules over a graded complete intersection ring B.

```
i151 : complexityPair = (M,N) -> dim ext(M,N);
```

Example 7.3.2. For M, N, and N' as in Example 7.1.2 we have

```
i152 : time complexityPair(M,N)
       -- used 0.39 seconds

o152 = 2

i153 : time complexityPair(M,N')
       -- used 0.12 seconds

o153 = 3
```

7.4   Support Variety

Let $\overline{K}$ be an algebraic closure of K. The *support variety* $V_B^*(M, N)$ is the algebraic set in $\overline{K}^c$ defined by the annihilator of $\operatorname{ext}_B^{\bullet}(M, N)$ over $R = K[X_1, \ldots, X_c]$. It is clear from the definition that $V_B^*(M, k)$ is equal to the variety $V_B^*(M)$ defined in 6.5. One of the main results of [5, Sect. 5] shows that $V_B^*(M, N) = V_B^*(M) \cap V_B^*(N)$, and, as a consequence, $V_B^*(M) = V_B^*(M, M) = V_B^*(k, M)$. The dimension of $V_B^*(M, N)$ is equal to the complexity $cx_R(M, N)$, already computed above.

Feeding our computation of $\operatorname{ext}_B^{\bullet}(M, N)$ to standard *Macaulay 2* routines we write code for determining a set of equations defining $V_B^*(M, N)$.

Code 7.4.1. The function `supportVarietyPairIdeal(M,N)` yields a set of polynomial equations with coefficients in K, defining the variety $V_B^*(M, N)$ in $\overline{K}^c$ for graded modules M, N over a graded complete intersection B.

```
i154 : supportVarietyPairIdeal = (M,N) -> ann ext(M,N);
```

Example 7.4.2. For M, N, and N' as in Example 7.1.2 we have

```
i155 : time supportVarietyPairIdeal(M,N)
       -- used 0.97 seconds

o155 = ideal X
             1

o155 : Ideal of K [X , X , X , Degrees => {{-2, -3}, {-2, -4}, {-2, -5}}]
                   1   2   3
```

```
i156 : time supportVarietyPairIdeal(M,N')
     -- used 1.73 seconds

o156 = 0

o156 : Ideal of K [X , X , X , Degrees => {{-2, -3}, {-2, -4}, {-2, -5}}]
                   1   2   3
```

Appendix A. Gradings

Our purpose here is to set up a context in which the theory of Sections 3 and 4 translates into data that *Macaulay 2* can operate with.

A first point is to develop a *flexible* and *consistent* scheme within which to handle the two kinds of degrees we deal with—the internal gradings of the input, and the homological degrees created during computations.

A purely technological difficulty arises when our data are presented to *Macaulay 2*: it only accepts multidegrees whose first component is positive, which is *not* the case for rings of cohomology operators.

A final point, mostly notational, tends to generate misunderstanding and errors if left unaddressed. On the printed page, the difference between homological and cohomological conventions is handled graphically by switching between sub- and super-indices, and reversing signs; both authors were used to it, but *Macaulay 2* has so far refused to read TeX printouts.

The *raison d'être* of the following remarks was to debug communications between the three of us.

Remark A.1. Only one degree, denoted deg, appears in Section 2, and anywhere in the main text before Notation 4.7; when needed, it will be referred to as *homological degree*.

Assume that $A = \bigoplus_{h \in \mathbb{Z}} A_h$ is a graded ring. Any element a of A_h is said to be homogeneous of *internal degree* h; the notation for this is $\deg' a = h$. Let $f = \{f_1, \ldots, f_c\}$ be a Koszul-regular set consisting of homogeneous elements. We give the ring $B = A/(f)$ the induced grading, and extend the notation for internal degree to all graded B-modules M.

Let M be a graded B-module. For any integer e, we let $M[e]$ denote the graded module with $M[e]_d = M_{d+e}$. We take a projective resolution C of M by graded A-modules, with differential d_C preserving internal degrees. Recall that we have been writing $\deg x = n$ to indicate that x is an element in C_n; we refer to this situation also by saying that x has *homological degree* x. We combine both degrees in a single *bidegree*, denoted Deg, as follows:

$$\mathrm{Deg}\, x = (\deg x, \deg' x)$$

For a bigraded module H and pair of integers (e, e'), we let $H[e, e']$ denote the bigraded module with $H[e, e']_{d,d'} = H_{d+e,d'+e'}$.

Because $\deg Y_i = 2$, the elements of the free B-module Q have homological degree 2. We introduce an internal grading $\deg'$ on Q by setting $\deg' Y_i = r_i$, where $r_i = \deg' f_i$ for $i = 1, \ldots, c$. With this choice, the homomorphism $f: Q \to A$ acquires internal degree 0 (of course, this was the reason behind our choice of grading in the first place). The internal grading on Q defines, in the usual way, internal gradings on all symmetric and exterior powers of Q and Q^*; in particular, $\deg' Y^{(\alpha)} = \sum \alpha_i r_i$ and $\deg' Y^{\wedge \beta} = \sum \beta_i r_i$. Thus, the ring $S = A[X_1, \ldots, X_c]$ acquires a bigrading defined by $\operatorname{Deg} a = (0, h)$ for all elements $a \in A_h$ and $\operatorname{Deg} X_i = (-2, -r_i)$ for $i = 1, \ldots, c$.

In this context, we call S the *bigraded ring of cohomology operators*.

Since the differential d_C has internal degree 0, a null-homotopic chain map $C \to C$ which is homogeneous of internal degree r will have a null-homotopy that is itself homogeneous of internal degree r. In the proof of Theorem 3.1 we construct maps d_γ as null-homotopies, so we may arrange for them to be homogeneous maps with $\deg' d_\gamma = \sum \gamma_i d_i$. Our grading assumptions guarantee that d is homogeneous with $\operatorname{Deg} d = (-1, 0)$.

With these data, the B-free resolution $C \otimes_A D'$ provided by Theorem 4.1 becomes one by graded B-modules, and its differential ∂ is homogeneous with $\operatorname{Deg} \partial = (-1, 0)$. For any graded B-module N, these properties are transferred to the complex $\operatorname{Hom}_B(C \otimes_A D', N)$ and its differential.

We sum up the contents of Remarks 4.3 and A.1.

Remark A.2. If A is a graded ring, $\{f_1, \ldots, f_c\}$ is a Koszul-regular set consisting of homogeneous elements, B is the residue ring $A/(\boldsymbol{f})$, and M, N are graded B-modules, then $\operatorname{Ext}_B^*(M, N)$ is a bigraded module over the ring $S = A[X_1, \ldots, X_c]$, itself bigraded by setting $\operatorname{Deg} a = (0, \deg'(a))$ for all homogeneous $a \in A$ and $\operatorname{Deg} X_i = (-2, -\deg'(f_i))$ for $i = 1, \ldots, c$.

Remark A.3. The core algorithms of the program can handle multi-graded rings and modules, but only if each variable in the ring has positive first component of its multi-degree. At the moment, a user who needs a multi-graded ring R which violates this requirement must provide two linear maps: R.Adjust, that transforms the desired multi-degrees into ones satisfying this requirement, as well as its inverse map, R.Repair. The routine Ext, discussed above, incorporates such adjustments for the rings of cohomology operators it creates. When we wish to create related rings with some of the same multi-degrees, we may use the same adjustment operator.

References

1. David J. Anick: A counterexample to a conjecture of Serre. *Ann. of Math. (2)*, 115(1):1–33, 1982 and 116(3):661, 1982.
2. Luchezar L. Avramov: Modules of finite virtual projective dimension. *Invent. Math.*, 96(1):71–101, 1989.
3. Luchezar L. Avramov: Infinite free resolutions. In *Six lectures on commutative algebra (Bellaterra, 1996)*, pages 1–118. Progress in Math., vol 166, Birkhäuser, Basel, 1998.

4. Luchezar L. Avramov and Ragnar-Olaf Buchweitz: Homological algebra modulo a regular sequence with special attention to codimension two. *J. Algebra*, 230(1):24–67, 2000.

5. Luchezar L. Avramov and Ragnar-Olaf Buchweitz: Support varieties and cohomology over complete intersections. *Invent. Math.*, 142(2):285–318, 2000.

6. Luchezar L. Avramov, Vesselin N. Gasharov, and Irena V. Peeva: Complete intersection dimension. *Inst. Hautes Études Sci. Publ. Math.*, 86:67–114, 1997.

7. Luchezar L. Avramov and Irena Peeva: Finite regularity and Koszul algebras. *Amer. J. Math*, 123(2):275–281, 2001.

8. Luchezar L. Avramov and Li-Chuan Sun: Cohomology operators defined by a deformation. *J. Algebra*, 204(2):684–710, 1998.

9. Jon F. Carlson: The varieties and the cohomology ring of a module. *J. Algebra*, 85(1):104–143, 1983.

10. Jon F. Carlson: The variety of an indecomposable module is connected. *Invent. Math.*, 77(2):291–299, 1984.

11. David Eisenbud: Homological algebra on a complete intersection, with an application to group representations. *Trans. Amer. Math. Soc.*, 260(1):35–64, 1980.

12. Tor H. Gulliksen: A change of ring theorem with applications to Poincaré series and intersection multiplicity. *Math. Scand.*, 34(1):167–183, 1974.

13. Vikram Mehta: *Endomorphisms of complexes and modules over Golod rings.* Thesis. University of California, Berkeley, CA, 1975.

14. Jan-Erik Roos: Commutative non-Koszul algebras having a linear resolution of arbitrarily high order. Applications to torsion in loop space homology. *C. R. Acad. Sci. Paris Sér. I Math.*, 316(11):1123–1128, 1993.

15. Jack Shamash: The Poincaré series of a local ring. *J. Algebra*, 12:453–470, 1969.

Algorithms for the Toric Hilbert Scheme

Michael Stillman, Bernd Sturmfels, and Rekha Thomas

The toric Hilbert scheme parametrizes all algebras isomorphic to a given semi-group algebra as a multigraded vector space. All components of the scheme are toric varieties, and among them, there is a fairly well understood co-herent component. It is unknown whether toric Hilbert schemes are always connected. In this chapter we illustrate the use of *Macaulay 2* for exploring the structure of toric Hilbert schemes. In the process we will encounter al-gorithms from commutative algebra, algebraic geometry, polyhedral theory and geometric combinatorics.

Introduction

Consider the multigrading of the polynomial ring $R = \mathbb{C}[x_1, \ldots, x_n]$ specified by a non-negative integer $d \times n$-matrix $A = (a_1, \ldots, a_n)$ such that degree $(x_i) = a_i \in \mathbb{N}^d$. This defines a decomposition $R = \bigoplus_{b \in \mathbb{N}A} R_b$, where $\mathbb{N}A$ is the subsemigroup of $\mathbb{N}^d$ spanned by $a_1, \ldots, a_n$, and R_b is the $\mathbb{C}$-span of all monomials $x^u = x_1^{u_1} \cdots x_n^{u_n}$ with degree $Au = a_1 u_1 + \cdots + a_n u_n = b$. The *toric Hilbert scheme* $Hilb_A$ parametrizes all A-homogeneous ideals $I \subset R$ (ideals that are homogeneous under the multigrading of R by $\mathbb{N}A$) with the property that $(R/I)_b$ is a 1-dimensional $\mathbb{C}$-vector space, for all $b \in \mathbb{N}A$. We call such an ideal I an A-*graded* ideal. Equivalently, I is A-graded if it is A-homogeneous and R/I is isomorphic as a multigraded vector space to the semigroup algebra $\mathbb{C}[\mathbb{N}A] = R/I_A$, where

$$I_A := \langle x^u - x^v \ : \ Au = Av \rangle \subset R$$

is the *toric ideal* of A. An A-graded ideal is generated by binomials and monomials in R since, by definition, any two monomials x^u and x^v of the same degree $Au = Av$ must be $\mathbb{C}$-linearly dependent modulo the ideal.

We recommend [22, §4, §10] as an introductory reference for the topics in this chapter. The study of toric Hilbert schemes for $d = 1$ goes back to Arnold [1] and Korkina et al.[13], and it was further developed by Sturmfels ([21] and [22, §10]). Peeva and Stillman [17] introduced the scheme structure that gives the toric Hilbert scheme its universal property, and from this they derive a formula for the tangent space of a point on $Hilb_A$. Maclagan recently showed that the quadratic binomials in [21, §5] define the same scheme as the determinantal equations in [17]. Both of these systems of global equations are generally much too big for practical computations. Instead, most of our algorithms are based on the local equations given by Peeva and Stillman in [16] and the combinatorial approach of Maclagan and Thomas in [14].

We begin with the computation of a toric ideal using *Macaulay 2*. Our running example throughout this chapter is the following 2×5-matrix:

$$A = \begin{pmatrix} 1 & 1 & 1 & 1 & 1 \\ 0 & 1 & 2 & 7 & 8 \end{pmatrix}, \tag{1}$$

which we input to *Macaulay 2* as a list of lists of integers.

```
i1 : A = {{1,1,1,1,1},{0,1,2,7,8}};
```

The toric ideal of A lives in the multigraded ring $R := \mathbb{C}[a,b,c,d,e]$.

```
i2 : R = QQ[a..e,Degrees=>transpose A];
```

```
i3 : describe R
```

```
o3 = QQ [a, b, c, d, e, Degrees => {{1, 0}, {1, 1}, {1, 2}, {1, 7}, {1 ···
```

We use Algorithm 12.3 in [22] to compute I_A. The first step is to find a matrix B whose rows generate the lattice $ker_{\mathbb{Z}}(A) := \{x \in \mathbb{Z}^n : Ax = 0\}$.

```
i4 : B = transpose syz matrix A
```

```
o4 = | 1 -2  1  0 0 |
     | 0  5 -6  1 0 |
     | 0  6 -7  0 1 |

               3         5
o4 : Matrix ZZ <--- ZZ
```

Although in theory any basis of $ker_{\mathbb{Z}}(A)$ will suffice, in practice it is more efficient to use a *reduced* basis [20, §6.2], which can be computed using the *basis reduction* package LLL.m2 in *Macaulay 2*. The command LLL when applied to the output of `syz matrix A` will return a matrix of the same size whose columns form a reduced lattice basis for $ker_{\mathbb{Z}}(A)$. The output appears in compressed form as follows:

```
i5 : load "LLL.m2";
```

```
i6 : LLL syz matrix A
```

```
o6 = | 0   1  2  |
     | 1  -1  0  |
     | -1  0 -3  |
     | -1 -1  2  |
     | 1   1 -1  |

               5         3
o6 : Matrix ZZ <--- ZZ
```

We recompute B using this package to get the following 3×5 matrix.

```
i7 : B = transpose LLL syz matrix A
```

```
o7 = | 0 1 -1 -1 1  |
     | 1 -1 0 -1 1  |
     | 2 0 -3 2 -1  |

               3         5
o7 : Matrix ZZ <--- ZZ
```

The advantage of a reduced basis may not be apparent in small examples. However, as the size of A increases, it becomes increasingly important for the termination of Algorithm 12.3 in [22]. (To appreciate this, consider the matrix (7) from Section 4.)

A row $b = b^+ - b^-$ of B is then coded as the binomial $x^{b^+} - x^{b^-} \in R$, and we let J be the ideal generated by all such binomials.

```
i8 : toBinomial = (b,R) -> (
          top := 1_R; bottom := 1_R;
          scan(#b, i -> if b_i > 0 then top = top * R_i^(b_i)
                  else if b_i < 0 then bottom = bottom * R_i^(-b_i));
          top - bottom);

i9 : J = ideal apply(entries B, b -> toBinomial(b,R))

                              2 2     3
o9 = ideal (- c*d + b*e, - b*d + a*e, a d  - c e)

o9 : Ideal of R
```

The toric ideal equals $(J : (x_1 \cdots x_n)^\infty)$, which is computed via n successive saturations as follows:

```
i10 : scan(gens ring J, f -> J = saturate(J,f))
```

Putting the above pieces of code together, we get the following procedure for computing the toric ideal of a matrix A.

```
i11 : toricIdeal = (A) -> (
          n := #(A_0);
          R = QQ[vars(0..n-1),Degrees=>transpose A,MonomialSize=>16];
          B := transpose LLL syz matrix A;
          J := ideal apply(entries B, b -> toBinomial(b,R));
          scan(gens ring J, f -> J = saturate(J,f));
          J
          );
```

See [2], [11] and [22, §4, §12] for other algorithms for computing toric ideals and various ideas for speeding up the computation.

In our example, $I_A = \langle cd - be, bd - ae, b^2 - ac, a^2d^2 - c^3e, c^4 - a^3e, bc^3 - a^3d, ad^4 - c^2e^3, d^6 - ce^5 \rangle$, which we now compute using this procedure.

```
i12 : I = toricIdeal A;

o12 : Ideal of R

i13 : transpose mingens I

o13 = {-2, -9}  | cd-be    |
      {-2, -8}  | bd-ae    |
      {-2, -2}  | b2-ac    |
      {-4, -14} | a2d2-c3e |
      {-4, -8}  | c4-a3e   |
      {-4, -7}  | bc3-a3d  |
      {-5, -28} | ad4-c2e3 |
      {-6, -42} | d6-ce5   |

                  8         1
o13 : Matrix R   <--- R
```

This ideal defines an embedding of $\mathbb{P}^1$ as a degree 8 curve into $\mathbb{P}^4$. We will see in Section 3 that its toric Hilbert scheme $Hilb_A$ has a non-reduced component.

This chapter is organized into four sections and two appendices as follows. The main goal in Section 1 is to describe an algorithm for generating all monomial A-graded ideals for a given A. These monomial ideals are the vertices of the *flip graph* of A whose connectivity is equivalent to the connectivity of $Hilb_A$. We describe how all neighbors of a given vertex of this graph can be calculated. In Section 2, we explain the role of polyhedral geometry in the study of $Hilb_A$. Our first algorithm tests for *coherence* in a monomial A-graded ideal. We then show how to compute the polyhedral complexes supporting A-graded ideals, which in turn relate the flip graph of A to the *Baues graph* of A. For unimodular matrices, these two graphs coincide and hence our method of computing the flip graph can be used to compute the Baues graph. Section 3 explores the components of $Hilb_A$ via local equations around the torus fixed points of the scheme. We include a combinatorial interpretation of these local equations from the point of view of integer programming. The scheme $Hilb_A$ has a *coherent* component, which is examined in detail in Section 4. We prove that this component is, in general, not normal and that its normalization is the toric variety of the Gröbner fan of I_A. We conclude the chapter with two appendices, each containing one large piece of *Macaulay 2* code that we use in this chapter. Appendix A displays code from the *Macaulay 2* file `polarCone.m2` that is used to convert a generator representation of a polyhedron to an inequality representation and vice versa. Appendix B displays code from the file `minPres.m2` used for computing minimal presentations of polynomial quotient rings. The main ingredient of this package is the subroutine `removeRedundantVariables`, which is what we use in this chapter.

1   Generating Monomial Ideals

We start out by computing the *Graver basis* Gr_A, which is the set of binomials in I_A that are minimal with respect to the partial order defined by

$$x^u - x^v \leq x^{u'} - x^{v'} \quad \Longleftrightarrow \quad x^u \text{ divides } x^{u'} \text{ and } x^v \text{ divides } x^{v'}.$$

The set Gr_A is a *universal Gröbner basis* of I_A and has its origins in the theory of integer programming [9]. It can be computed using [22, Algorithm 7.2], a *Macaulay 2* version of which is given below.

```
i14 : graver = (I) -> (
          R := ring I;
          k := coefficientRing R;
          n := numgens R;
          -- construct new ring S with 2n variables
          S := k[Variables=>2*n,MonomialSize=>16];
          toS := map(S,R,(vars S)_{0..n-1});
```

```
toR := map(R,S,vars R | matrix(R, {toList(n:1)}));
-- embed I in S
m := gens toS I;
-- construct the toric ideal of the Lawrence
-- lifting of A
i := 0;
while i < n do (
    wts := join(toList(i:0),{1},toList(n-i-1:0));
    wts = join(wts,wts);
    m = homogenize(m,S_(n+i),wts);
    i=i+1;
    );
J := ideal m;
scan(gens ring J, f -> J = saturate(J,f));
-- apply the map toR to the minimal generators of J
f := matrix entries toR mingens J;
p := sortColumns f;
f_p) ;
```

The above piece of code first constructs a new polynomial ring S in n more variables than R. Assume $S = \mathbb{C}[x_1, \ldots, x_n, y_1, \ldots, y_n]$. The inclusion map $\texttt{toS} : R \to S$ embeds the toric ideal I in S and collects its generators in the matrix m. A binomial $x^a - x^b$ lies in Gr_A if and only if $x^a y^b - x^b y^a$ is a minimal generator of the toric ideal in S of the $(d+n) \times 2n$ matrix

$$\Lambda(A) := \begin{pmatrix} A & 0 \\ I_n & I_n \end{pmatrix},$$

which is called the *Lawrence lifting* of A. Since $u \in ker_{\mathbb{Z}}(A) \Leftrightarrow (u, -u) \in ker_{\mathbb{Z}}(\Lambda(A))$, we use the while loop to homogenize the binomials in m with respect to $\Lambda(A)$, using the n new variables in S. This converts a binomial $x^a - x^b \in$ m to the binomial $x^a y^b - x^b y^a$. The ideal generated by these new binomials is labeled J. As before, we can now successively saturate J to get the toric ideal of $\Lambda(A)$ in S. The image of the minimal generators of this toric ideal under the map $\texttt{toR} : S \to R$ such that $x_i \mapsto x_i$ and $y_i \mapsto 1$ is precisely the Graver basis Gr_A. These binomials are the entries of the matrix f and is output by the program.

In our example Gr_A consists of 42 binomials.

```
i15 : Graver = graver I

o15 = | -cd+be -bd+ae -b2+ac -cd2+ae2 -a2d2+c3e -c4+a2bd -c4+a3e -bc3+ ···

            1        42
o15 : Matrix R  <--- R
```

Returning to the general case, an element b of $\mathbb{N}A$ is called a *Graver degree* if there exists a binomial $x^u - x^v$ in the Graver basis Gr_A such that $Au = Av = b$. If b is a Graver degree then the set of monomials in R_b is the corresponding *Graver fiber*. In our running example there are 37 distinct Graver fibers. We define the $\texttt{ProductIdeal}$ of A as $PI := \langle x^a x^b : x^a - x^b \in Gr_A \rangle$. This ideal is contained in every monomial ideal of $Hilb_A$ and hence no monomial in PI can be a standard monomial of a monomial A-graded ideal. Since our purpose in constructing Graver fibers is to use them to generate all monomial A-graded ideals, we will be content with listing just the monomials in each Graver fiber that do not lie in PI. Since R is multigraded by A, we can obtain such a presentation of a Graver fiber by simply asking for the basis of R in degree b modulo PI.

```
i16 : graverFibers = (Graver) -> (
          ProductIdeal := (I) -> ( trim ideal(
            apply(numgens I, a -> (
                f := I_a; leadTerm f * (leadTerm f - f)))));
          PI := ProductIdeal ideal Graver;
          R := ring Graver;
          new HashTable from apply(
                unique degrees source Graver,
                d -> d => compress (basis(d,R) % PI) ));

i17 : fibers = graverFibers Graver

o17 = HashTable{{2, 2} => | ac b2 |                                          }
                {2, 8} => | ae bd |
                {2, 9} => | be cd |
                {3, 16} => | ae2 bde cd2 |
                {4, 14} => | a2d2 c3e |
                {4, 7} => | a3d bc3 |
                {4, 8} => | a3e a2bd c4 |
                {5, 10} => | a3ce a2b2e a2bcd ab3d c5 |
                {5, 14} => | a3d2 ac3e b2c2e bc3d |
                {5, 16} => | a3e2 a2cd2 ab2d2 c4e |
                {5, 21} => | a2d3 bc2e2 c3de |
                {5, 22} => | a2d2e abd3 c3e2 |
                {5, 28} => | ad4 c2e3 |
                {5, 7} => | a4d abc3 b3c2 |
                {5, 8} => | a4e a3bd ac4 b2c3 |
                {6, 12} => | a3c2e a2bc2d ab4e b5d c6 |
                {6, 14} => | a4d2 a2c3e abc3d b4ce b3c2d |
                {6, 18} => | a3ce2 a2b2e2 a2c2d2 b4d2 c5e |
                {6, 21} => | a3d3 abc2e2 ac3de b3ce2 bc3d2 |
                {6, 24} => | a3e3 a2cd2e abcd3 b3d3 c4e2 |
                {6, 28} => | a2d4 ac2e3 b2ce3 c3d2e |
                {6, 30} => | a2d2e2 acd4 b2d4 c3e3 |
                {6, 35} => | ad5 bce4 c2de3 |
                {6, 36} => | ad4e bd5 c2e4 |
                {6, 42} => | ce5 d6 |
                {6, 7} => | a5d a2bc3 b5c |
                {6, 8} => | a5e a4bd a2c4 b4c2 |
                {7, 14} => | a5d2 a3c3e a2bc3d b6e b5cd c7 |
                {7, 21} => | a4d3 a2bc2e2 a2c3de abc3d2 b5e2 b3c2d2 |
                {7, 28} => | a3d4 a2c2e3 ac3d2e b4e3 bc3d3 |
                {7, 35} => | a2d5 abce4 ac2de3 b3e4 c3d3e |
                {7, 42} => | ace5 ad6 b2e5 c2d2e3 |
                {7, 49} => | be6 cde5 d7 |
                {7, 7} => | a6d a3bc3 b7 |
                {7, 8} => | a6e a5bd a3c4 b6c |
                {8, 56} => | ae7 bde6 cd2e5 d8 |
                {8, 8} => | a7e a6bd a4c4 b8 |

o17 : HashTable
```

For example, the Graver degree $(8, 8)$ corresponds to the Graver fiber

$$\{\, \underline{a^7 e},\ \underline{a^6 bd},\ \underline{a^4 c^4},\ a^3 b^2 c^3,\ a^2 b^4 c^2,\ ab^6 c,\ \underline{b^8}\,\}.$$

Our *Macaulay 2* code outputs only the four underlined monomials, in the format | a7e a6bd a4c4 b8 |. The three non-underlined monomials lie in the ProductIdeal. Graver degrees are important because of the following result.

Lemma 1.1 ([22, Lemma 10.5]). *The multidegree of any minimal generator of any ideal I in $Hilb_A$ is a Graver degree.*

The next step in constructing the toric Hilbert scheme is to compute all its fixed points with respect to the scaling action of the n-dimensional algebraic torus $(\mathbb{C}^*)^n$. (The torus $(\mathbb{C}^*)^n$ acts on R by scaling variables : $\lambda \mapsto \lambda \cdot x := (\lambda_1 x_1, \ldots, \lambda_n x_n)$.) These fixed points are the monomial ideals M lying on $Hilb_A$. Every term order $\prec$ on the polynomial ring R gives such a monomial ideal: $M = in_\prec(I_A)$, the initial ideal of the toric ideal I_A with respect to $\prec$. Two ideals J and J' are said to be *torus isomorphic* if $J = \lambda \cdot J'$ for some $\lambda \in (\mathbb{C}^*)^n$. Any monomial A-graded ideal that is torus isomorphic to an initial ideal of I_A is said to be *coherent*. In particular, the initial ideals of I_A are coherent and they can be computed by [22, Algorithm 3.6] applied to I_A. A refinement and fast implementation can be found in the software package TiGERS by Huber and Thomas [12].

Now we wish to compute all monomial ideals M on $Hilb_A$ regardless of whether M is coherent or not. For this we use the procedure generateAmonos given below. This procedure takes in the Graver basis Gr_A and records the numerator of the Hilbert series of I_A in trueHS. It then computes the Graver fibers of A, sorts them and calls the subroutine selectStandard to generate a candidate for a monomial ideal on $Hilb_A$.

```
i18 : generateAmonos = (Graver) -> (
            trueHS := poincare coker Graver;
            fibers := graverFibers Graver;
            fibers = apply(sort pairs fibers, last);
            monos = {};
            selectStandard := (fibers, J) -> (
            if #fibers == 0 then (
                if trueHS == poincare coker gens J
                then (monos = append(monos,flatten entries mingens J)));
            ) else (
                P := fibers_0;
                fibers = drop(fibers,1);
                P = compress(P % J);
                nP := numgens source P;
                -- nP is the number of monomials not in J.
                if nP > 0 then (
                    if nP == 1 then selectStandard(fibers,J)
                    else (--remove one monomial from P,take the rest.
                        P = flatten entries P;
                        scan(#P, i -> (
                            J1 := J + ideal drop(P,{i,i});
                            selectStandard(fibers, J1)))));
            ));
            selectStandard(fibers, ideal(0_(ring Graver)));
            ) ;
```

The arguments to the subroutine selectStandard are the Graver fibers given as a list of matrices and a monomial ideal J that should be included in every A-graded ideal that we generate. The subroutine then loops through each Graver fiber, and at each step selects a standard monomial from that fiber and updates the ideal J by adding the other monomials in this fiber to J. The final J output by the subroutine is the candidate ideal that is sent back to generateAmonos. It is stored by the program if its Hilbert series agrees with that of I_A. All the monomial A-graded ideals are stored in the list monos. Below, we ask *Macaulay 2* for the cardinality of monos and its first ten elements.

```
i19 : generateAmonos Graver;

i20 : #monos

o20 = 281

i21 : scan(0..9, i -> print toString monos#i)
{c*d, b*d, b^2, c^3*e, c^4, b*c^3, c^2*e^3, b*c^2*e^2, b*c*e^4, d^6}
{c*d, b*d, b^2, c^3*e, c^4, b*c^3, c^2*e^3, b*c^2*e^2, c*e^5, b*c*e^4, ···
{c*d, b*d, b^2, c^3*e, c^4, b*c^3, c^2*e^3, b*c^2*e^2, c*e^5, b*c*e^4, ···
{c*d, b*d, b^2, c^3*e, c^4, b*c^3, c^2*e^3, b*c^2*e^2, c*e^5, b*c*e^4, ···
{c*d, b*d, b^2, c^3*e, c^4, b*c^3, c^2*e^3, b*c^2*e^2, d^6, a*d^5}
{c*d, b*d, b^2, c^3*e, c^4, b*c^3, b*c^2*e^2, a*d^4, d^6}
{c*d, b*d, b^2, c^3*e, c^4, b*c^3, a*d^4, a^2*d^3, d^6}
{c*d, b*d, b^2, a^2*d^2, c^4, b*c^3, a*d^4, d^6}
{c*d, b*d, b^2, a^2*d^2, a^3*d, c^4, a*d^4, d^6}
{c*d, b*d, b^2, a^3*e, a^2*d^2, a^3*d, a*d^4, d^6}
```

The monomial ideals (torus-fixed points) on $Hilb_A$ form the vertices of the *flip graph* of A whose edges correspond to the torus-fixed curves on $Hilb_A$. This graph was introduced in [14] and provides structural information about $Hilb_A$. The edges emanating from a monomial ideal M can be constructed as follows: For any minimal generator x^u of M, let x^v be the unique monomial with $x^v \notin M$ and $Au = Av$. Form the *wall ideal*, which is generated by $x^u - x^v$ and all minimal generators of M other than x^u, and let M' be the initial monomial ideal of the wall ideal with respect to any term order $\succ$ for which $x^v \succ x^u$. It can be shown that M' is the unique initial monomial ideal of the wall ideal that contains x^v. If M' lies on $Hilb_A$ then $\{M, M'\}$ is an edge of the flip graph. We now illustrate the *Macaulay 2* procedure for computing all flip neighbors of a monomial A-graded ideal.

```
i22 : findPositiveVector = (m,s) -> (
            expvector := first exponents s - first exponents m;
            n := #expvector;
            i := first positions(0..n-1, j -> expvector_j > 0);
            splice {i:0, 1, (n-i-1):0}
            );

i23 : flips = (M) -> (
            R := ring M;
            -- store generators of M in monoms
            monoms := first entries generators M;
            result := {};
            -- test each generator of M to see if it leads to a neighbor
```

```
scan(#monoms, i -> (
  m := monoms_i;
  rest := drop(monoms,{i,i});
  b := basis(degree m, R);
  s := (compress (b % M))_(0,0);
  J := ideal(m-s) + ideal rest;
  if poincare coker gens J == poincare coker gens M then (
    w := findPositiveVector(m,s);
    R1 := (coefficientRing R)[generators R, Weights=>w];
    J = substitute(J,R1);
    J = trim ideal leadTerm J;
    result = append(result,J);
  )));
  result
);
```

The code above inputs a monomial A-graded ideal M whose minimal generators are stored in the list monoms. The flip neighbors of M will be stored in result. For each monomial x^u in monoms we need to test whether it yields a flip neighbor of M or not. At the i-th step of this loop, we let m be the i-th monomial in monoms. The list rest contains all monomials in monoms except m. We compute the standard monomial s of M of the same degree as m. The wall ideal of $m - s$ is the binomial ideal J generated by $m - s$ and the monomials in rest. We then check whether J is A-graded by comparing its Hilbert series with that of M. (Alternately, one could check whether M is the initial ideal of the wall ideal with respect to $m \succ s$.) If this is the case, we use the subroutine findPositiveVector to find a unit vector $w = (0, \ldots, 1, \ldots, 0)$ such that $w \cdot s > w \cdot m$. The flip neighbor is then the initial ideal of J with respect to w and it is stored in result. The program outputs the minimal generators of each flip neighbor. Here is an example.

```
i24 : R = QQ[a..e,Degrees=>transpose A];

i25 : M = ideal(a*e,c*d,a*c,a^2*d^2,a^2*b*d,a^3*d,c^2*e^3,
              c^3*e^2,c^4*e,c^5,c*e^5,a*d^5,b*e^6);

o25 : Ideal of R

i26 : F = flips M

                  2 2   3    4   2 3   3 2    5     5    ...
o26 = {ideal (a*e, c*d, a*c, a d , a d, c , c e , c e , a*d , c*e , b* ...

o26 : List

i27 : #F

o27 = 4

i28 : scan(#F, i -> print toString entries mingens F_i)
{{a*e, c*d, a*c, a^2*d^2, a^3*d, c^4, c^2*e^3, c^3*e^2, a*d^5, c*e^5, ...
{{c*d, a*e, a*c, a^2*d^2, a^2*b*d, a^3*d, c^3*e^2, c^4*e, c^5, a*d^4, ...
{{a*e, c*d, a*c, a^2*d^2, a^3*d, a^2*b*d, c^2*e^3, c^3*e^2, c^4*e, c^5 ...
{{a*e, a*c, c*d, a^2*b*d, a^3*d, a^2*d^2, c^2*e^3, c^3*e^2, c^4*e, c^5 ...
```

It is an open problem whether the toric Hilbert scheme $Hilb_A$ is connected. Recent work in geometric combinatorics [19] suggests that this is probably false for some A. This result and its implications for $Hilb_A$ will be discussed further in Section 2. The following theorem of Maclagan and Thomas [14] reduces the connectivity of $Hilb_A$ to a combinatorial problem.

Theorem 1.2. *The toric Hilbert scheme $Hilb_A$ is connected if and only if the flip graph of A is connected.*

We now have two algorithms for listing monomial ideals on $Hilb_A$. First, there is the *backtracking algorithm* whose *Macaulay 2* implementation was described above. Second, there is the *flip search algorithm*, which starts with any coherent monomial ideal M and then constructs the connected component of M in the flip graph of A by carrying out local flips as above. This procedure is also implemented in TiGERS [12]. Clearly, the two algorithms will produce the same answer if and only if $Hilb_A$ is connected. In other words, finding an example where $Hilb_A$ is disconnected is equivalent to finding a matrix A for which the flip search algorithm produces fewer monomial ideals than the backtracking algorithm.

2   Polyhedral Geometry

Algorithms from polyhedral geometry are essential in the study of the toric Hilbert scheme. Consider the problem of deciding whether or not a given monomial ideal M in $Hilb_A$ is coherent. This problem gives rise to a system of linear inequalities as follows: Let $x^{u_1}, \ldots, x^{u_r}$ be the minimal generators of M, and let x^{v_i} be the unique standard monomial with $Au_i = Av_i$. Then M is coherent if and only if there exists a vector $w \in \mathbb{R}^n$ such that $w \cdot (u_i - v_i) > 0$ for $i = 1, \ldots, r$. Thus the test for coherence amounts to solving a *feasibility problem of linear programming*, and there are many highly efficient algorithms (based on the simplex algorithms or interior point methods) available for this task. For our experimental purposes, it is convenient to use the code polarCone.m2, given in Appendix A, which is based on the (inefficient but easy-to-implement) *Fourier-Motzkin elimination* method (see [25] for a description). This code converts the generator representation of a polyhedron to its inequality representation and vice versa. A simple example is given in Appendix A. In particular, given a Gröbner basis $\mathcal{G}$ of I_A, the function polarCone will compute all the extreme rays of the *Gröbner cone* $\{w \in \mathbb{R}^n : w \cdot (u_i - v_i) \geq 0 \text{ for each } x^{u_i} - x^{v_i} \in \mathcal{G}\}$.

We now show how to use *Macaulay 2* to decide whether a monomial A-graded ideal M is coherent. The first step in this calculation is to compute all the standard monomials of M of the same degree as the minimal generators of M. We do this using the procedure stdMonomials.

```
i29 : stdMonomials = (M) -> (
            R := ring M;
            RM := R/M;
            apply(numgens M, i -> (
                    s := basis(degree(M_i),RM); lift(s_(0,0), R)))
            );
```

As an example, consider the following monomial A-graded ideal.

```
i30 : R = QQ[a..e,Degrees => transpose A ];

i31 : M = ideal(a^3*d, a^2*b*d, a^2*d^2, a*b^3*d, a*b^2*d^2, a*b*d^3,
                a*c, a*d^4, a*e, b^5*d, b^4*d^2, b^3*d^3, b^2*d^4,
                b*d^5, b*e, c*e^5);

o31 : Ideal of R

i32 : toString stdMonomials M

o32 = {b*c^3, c^4, c^3*e, c^5, c^4*e, c^3*e^2, b^2, c^2*e^3, b*d, c^6, ...
```

From the pairs x^u, x^v of minimal generators x^u and the corresponding standard monomials x^v, the function inequalities creates a matrix whose columns are the vectors $u - v$.

```
i33 : inequalities = (M) -> (
             stds := stdMonomials(M);
             transpose matrix apply(numgens M, i -> (
                  flatten exponents(M_i) -
                       flatten exponents(stds_i))));

i34 : inequalities M

o34 = | 3  2  2  1  1  1  1  1  1  0  0  0  0  0  0  0 |
      | -1 1  0  3  2  1 -2  0 -1  5  4  3  2  1  1  0 |
      | -3 -4 -3 -5 -4 -3 1 -2  0 -6 -5 -4 -3 -2 -1 1  |
      | 1  1  2  1  2  3  0  4 -1  1  2  3  4  5 -1 -6 |
      | 0  0 -1  0 -1 -2  0 -3  1  0 -1 -2 -3 -4  1  5 |

                5          16
o34 : Matrix ZZ  <--- ZZ
```

It is convenient to simplify the output of the next procedure using the following program to divide an integer vector by the g.c.d. of its components. We also load polarCone.m2, which is needed in decideCoherence below.

```
i35 : primitive := (L) -> (
             n := #L-1; g := L#n;
             while n > 0 do (n = n-1; g = gcd(g, L#n););
             if g === 1 then L else apply(L, i -> i // g));

i36 : load "polarCone.m2"

i37 : decideCoherence = (M) -> (
             ineqs := inequalities M;
             c := first polarCone ineqs;
             m := - sum(numgens source c, i -> c_{i});
             prods := (transpose m) * ineqs;
             if numgens source prods != numgens source compress prods
             then false else primitive (first entries transpose m));
```

Let K be the cone $\{x \in \mathbb{R}^n : g \cdot x \leq 0$, for all columns g of ineqs $\}$. The command polarCone ineqs computes a pair of matrices P and Q such that K is the sum of the cone generated by the columns of P and the subspace generated by the columns of Q. Let m be the negative of the sum of the columns of P. Then m lies in the cone $-K$. The entries in the matrix prods are the dot products $g \cdot m$ for each column g of ineqs. Since M is a monomial A-graded ideal, it is coherent if and only if K is full dimensional, which is the case if and only if no dot product $g \cdot m$ is zero. This is the conditional in the if .. then statement of decideCoherence. If M is coherent, the program outputs the primitive representative of m and otherwise returns the boolean false. Notice that if M is coherent, the cone $-K$ is the Gröbner cone corresponding to M and the vector m is a weight vector w such that $in_w(I_A) = M$. We now test whether the ideal M from line i29 is coherent.

```
i38 : decideCoherence M

o38 = {0, 0, 1, 15, 18}

o38 : List
```

Hence, M is coherent: it is the initial ideal with respect to the weight vector $w = (0, 0, 1, 15, 18)$ of the toric ideal in our running example (1). Here is one of the 55 noncoherent monomial A-graded ideals of this matrix.

```
i39 : N = ideal(a*e,c*d,a*c,c^3*e,a^3*d,c^4,a*d^4,a^2*d^3,c*e^5,
               c^2*e^4,d^7);

o39 : Ideal of R

i40 : decideCoherence N

o40 = false
```

In the rest of this section, we study the connection between A-graded ideals and polyhedral complexes defined on A. This study relates the flip graph of the toric Hilbert scheme to the Baues graph of the configuration A. (See [18] for a survey of the Baues problem and its relatives). Let $pos(A) := \{Au : u \in \mathbb{R}^n, u \geq 0\}$ be the cone generated by the columns of A in $\mathbb{R}^d$. A *polyhedral subdivision* Δ of A is a collection of full dimensional subcones $pos(A_\sigma)$ of $pos(A)$ such that the union of these subcones is $pos(A)$ and the intersection of any two subcones is a face of each. Here $A_\sigma := \{a_j : j \in \sigma \subseteq \{1, \ldots, n\}\}$. It is customary to identify Δ with the set of sets $\{\sigma : pos(A_\sigma) \in \Delta\}$. If every cone in the subdivision Δ is simplicial (the number of extreme rays of the cone equals the dimension of the cone), we say that Δ is a *triangulation* of A. The simplicial complex corresponding to a triangulation Δ is uniquely obtained by including in Δ all the subsets of every $\sigma \in \Delta$. We refer the reader to [22, §8] for more details.

For each $\sigma \in \Delta$, let I_σ be the prime ideal that is the sum of the toric ideal I_{A_σ} and the monomial ideal $\langle x_j : j \notin \sigma \rangle$. Recall that two ideals J and J' are said to be *torus isomorphic* if $J = \lambda \cdot J'$ for some $\lambda \in (\mathbb{C}^*)^n$. The following theorem shows that polyhedral subdivisions of A are related to A-graded ideals via their radicals.

Theorem 2.1 (Theorem 10.10 [22, §10]). *If I is an A-graded ideal, then there exists a polyhedral subdivision $\Delta(I)$ of A such that $\sqrt{I} = \cap_{\sigma \in \Delta(I)} J_\sigma$ where each component J_σ is a prime ideal that is torus isomorphic to I_σ.*

We say that $\Delta(I)$ supports the A-graded ideal I. When M is a monomial A-graded ideal, $\Delta(M)$ is a triangulation of A. In particular, if M is coherent (i.e, $M = in_w(I_A)$ for some weight vector w), then $\Delta(M)$ is the *regular* or *coherent* triangulation of A induced by w [22, §8]. The coherent triangulations of A are in bijection with the vertices of the *secondary polytope* of A [3], [8].

It is convenient to represent a triangulation Δ of A by its *Stanley-Reisner* ideal $I_\Delta := \langle x_{i_1} x_{i_2} \cdots x_{i_k} : \{i_1, i_2, \ldots, i_k\}$ is a non-face of $\Delta\rangle$. If M is a monomial A-graded ideal, Theorem 2.1 implies that $I_{\Delta(M)}$ is the radical of M. Hence we will represent triangulations of A by their Stanley-Reisner ideals. As seen below, the matrix in our running example has eight distinct triangulations corresponding to the eight distinct radicals of the 281 monomial A-graded ideals computed earlier. All eight are coherent.

$$
\begin{array}{rcl}
\{\{1,2\},\{2,3\},\{3,4\},\{4,5\}\} & \leftrightarrow & \langle ac, ad, ae, bd, be, ce\rangle \\
\{\{1,3\},\{3,4\},\{4,5\}\} & \leftrightarrow & \langle b, ad, ae, ce\rangle \\
\{\{1,2\},\{2,4\},\{4,5\}\} & \leftrightarrow & \langle c, ad, ae, be\rangle \\
\{\{1,2\},\{2,3\},\{3,5\}\} & \leftrightarrow & \langle d, ac, ae, be\rangle \\
\{\{1,3\},\{3,5\}\} & \leftrightarrow & \langle b, d, ae\rangle \\
\{\{1,4\},\{4,5\}\} & \leftrightarrow & \langle b, c, ae\rangle \\
\{\{1,2\},\{2,5\}\} & \leftrightarrow & \langle c, d, ae\rangle \\
\{\{1,5\}\} & \leftrightarrow & \langle b, c, d\rangle
\end{array}
$$

The Baues graph of A is a graph on all the triangulations of A in which two triangulations are adjacent if they differ by a single *bistellar flip* [18]. The *Baues problem* from discrete geometry asked whether the Baues graph of a point configuration can be disconnected for some A. Every edge of the secondary polytope of A corresponds to a bistellar flip, and hence the subgraph of the Baues graph that is induced by the coherent triangulations of A is indeed connected: it is precisely the edge graph of the secondary polytope of A. The Baues problem was recently settled by Santos [19] who gave an example of a six dimensional point configuration with 324 points for which there is an isolated (necessarily non-regular) triangulation.

Santos' configuration would also have a disconnected flip graph and hence a disconnected toric Hilbert scheme if it were true that *every* triangulation of A supports a monomial A-graded ideal. However, Peeva has shown that this need not be the case (Theorem 10.13 in [22, §10]). Hence, the map from the set of all monomial A-graded ideals to the set of all triangulations of A that sends $M \mapsto \Delta(M)$ is not always surjective, and it is unknown whether Santos' 6×324 configuration has a disconnected toric Hilbert scheme.

Thus, even though one cannot in general conclude that the existence of a disconnected Baues graph implies the existence of a disconnected flip graph, there is an important special situation in which such a conclusion is possible. We call an integer matrix A of full row rank *unimodular* if the absolute value of each of its non-zero maximal minors is the same constant. A matrix A is unimodular if and only if every monomial A-graded ideal is square-free. For a unimodular matrix A, the Baues graph of A coincides with the flip graph of A. As you might expect, Santos' configuration is not unimodular.

Theorem 2.2 (Lemma 10.14 [22, §10]). *If A is unimodular, then each triangulation of A supports a unique (square-free) monomial A-graded ideal. In this case, a monomial A-graded ideal is coherent if and only if the triangulation supporting it is coherent.*

Using Theorem 2.2 we can compute all the triangulations of a unimodular matrix since they are precisely the polyhedral complexes supporting monomial A-graded ideals. Then we could enumerate the connected component of a coherent monomial A-graded ideal in the flip graph of A to decide whether the Baues/flip graph is disconnected.

Let Δ_r be the standard r-simplex that is the convex hull of the $r+1$ unit vectors in $\mathbb{R}^{r+1}$, and let $A(r,s)$ be the $(r+s+2) \times (r+1)(s+1)$ matrix whose columns are the products of the vertices of Δ_r and Δ_s. All matrices of type $A(r,s)$ are unimodular. From the product of two triangles we get

$$A(2,2) := \begin{pmatrix} 1\,1\,1\,0\,0\,0\,0\,0\,0 \\ 0\,0\,0\,1\,1\,1\,0\,0\,0 \\ 0\,0\,0\,0\,0\,0\,1\,1\,1 \\ 1\,0\,0\,1\,0\,0\,1\,0\,0 \\ 0\,1\,0\,0\,1\,0\,0\,1\,0 \\ 0\,0\,1\,0\,0\,1\,0\,0\,1 \end{pmatrix}.$$

We can now use our algebraic algorithms to compute all the triangulations of $A(2,2)$. Since *Macaulay 2* requires the first entry of the degree of every variable in a ring to be positive, we use the following matrix with the same row space as $A(2,2)$ for our computation:

```
i41 : A22 =
        {{1,1,1,1,1,1,1,1,1},{0,0,0,1,1,1,0,0,0},{0,0,0,0,0,0,1,1,1},
        {1,0,0,1,0,0,1,0,0},{0,1,0,0,1,0,0,1,0},{0,0,1,0,0,1,0,0,1}};

i42 : I22 = toricIdeal A22

o42 = ideal (f*h - e*i, c*h - b*i, f*g - d*i, e*g - d*h, c*g - a*i, b* ···

o42 : Ideal of R
```

The ideal I22 is generated by the 2 by 2 minors of a 3 by 3 matrix of indeterminates. This is the ideal of $\mathbb{P}^2 \times \mathbb{P}^2$ embedded in $\mathbb{P}^8$ via the Segre embedding.

```
i43 : Graver22 = graver I22;

                 1        15
o43 : Matrix R   <--- R

i44 : generateAmonos(Graver22);

i45 : #monos

o45 = 108
```

```
i46 : scan(0..9,i->print toString monos#i)
{f*h, c*h, f*g, e*g, c*g, b*g, c*e, c*d, b*d}
{f*h, d*h, c*h, f*g, c*g, b*g, c*e, c*d, b*d}
{d*i, f*h, d*h, c*h, c*g, b*g, c*e, c*d, b*d}
{e*i, c*h, f*g, e*g, c*g, b*g, c*e, c*d, b*d}
{e*i, d*i, c*h, e*g, c*g, b*g, c*e, c*d, b*d}
{e*i, d*i, d*h, c*h, c*g, b*g, c*e, c*d, b*d}
{f*h, c*h, f*g, e*g, c*g, b*g, c*e, a*e, c*d}
{e*i, c*h, f*g, e*g, c*g, b*g, c*e, a*e, c*d, b*d*i}
{e*i, c*h, f*g, e*g, c*g, b*g, c*e, a*e, c*d, a*f*h}
{e*i, d*i, c*h, e*g, c*g, b*g, c*e, a*e, c*d}
```

Thus there are 108 monomial $A(2,2)$-graded ideals and decideCoherence will check that all of them are coherent. Since $A(2,2)$ is unimodular, each monomial $A(2,2)$-graded ideal is square-free and is hence radical. These 108 ideals represent the 108 triangulations of $A(2,2)$ and we have listed ten of them above. The flip graph (equivalently, Baues graph) of $A(2,2)$ is connected. However, it is unknown whether the Baues graph of $A(r,s)$ is connected for all values of (r,s).

3   Local Equations

Consider the reduced Gröbner basis of a toric ideal I_A for a term order w:

$$\left\{ x^{u_1} - x^{v_1},\; x^{u_2} - x^{v_2},\; \ldots,\; x^{u_r} - x^{v_r} \right\}. \tag{2}$$

The initial ideal $M = in_w(I_A) = \langle x^{u_1}, x^{u_2}, \ldots, x^{u_r} \rangle$ is a coherent monomial A-graded ideal. In particular, it is a $(\mathbb{C}^*)^n$-fixed point on the toric Hilbert scheme $Hilb_A$. We shall explain a method, due to Peeva and Stillman [16], for computing local equations of $Hilb_A$ around such a fixed point. A variant of this method also works for computing the local equations around a non-coherent monomial ideal M, but that variant involves local algebra, specifically Mora's tangent cone algorithm, which is not yet fully implemented in *Macaulay 2*. See [16] for details.

We saw how to compute the flip graph of A in Section 1. The vertices of this graph are the $(\mathbb{C}^*)^n$-fixed points M and its edges correspond to the $(\mathbb{C}^*)^n$-fixed curves. By computing and decomposing the local equations around each M, we get a complete description of the scheme $Hilb_A$.

The first step is to introduce a new variable z_i for each binomial in our Gröbner basis (2) and to consider the following r binomials:

$$x^{u_1} - z_1 \cdot x^{v_1},\; x^{u_2} - z_2 \cdot x^{v_2},\; \ldots,\; x^{u_r} - z_r \cdot x^{v_r} \tag{3}$$

in the polynomial ring $\mathbb{C}[x, z]$ in $n + r$ indeterminates. The term order w can be extended to an elimination term order in $\mathbb{C}[x, z]$ so that x^{u_i} is the leading term of $x^{u_i} - z_i \cdot x^{v_i}$ for all i. We compute the minimal first syzygies of the monomial ideal M, and form the corresponding S-pairs of binomials in (3). For each S-pair

$$\frac{lcm(x^{u_i}, x^{u_j})}{x^{u_i}} \cdot \left(x^{u_i} - z_i \cdot x^{v_i} \right) \;-\; \frac{lcm(x^{u_i}, x^{u_j})}{x^{u_j}} \cdot \left(x^{u_j} - z_j \cdot x^{v_j} \right)$$

we compute a normal form with respect to (3) using the extended term order w. The result is a binomial in $\mathbb{C}[x, z]$ that factors as

$$x^\alpha \cdot z^\beta \cdot (z^\gamma - z^\delta),$$

where $\alpha \in \mathbb{N}^n$ and $\beta, \gamma, \delta \in \mathbb{N}^r$. Note that this normal form is not unique but depends on our choice of a reduction path. Let J_M denote the ideal in $\mathbb{C}[z_1, \ldots, z_r]$ generated by all binomials $z^\beta \cdot (z^\gamma - z^\delta)$ gotten from normal forms of all the S-pairs considered above.

Proposition 3.1 ([16]). *The ideal J_M is independent of the reduction paths chosen. It defines a subscheme of $\mathbb{C}^r$ isomorphic to an affine open neighborhood of the point M on the toric Hilbert scheme $Hilb_A$.*

We apply this technique to compute a particularly interesting affine chart of $Hilb_A$ for our running example. Consider the following set of 13 binomials:

$$\{\, ae - z_1bd,\ cd - z_2be,\ ac - z_3b^2,\ a^2d^2 - z_4c^3e,\ a^2bd - z_5c^4,$$
$$a^3d - z_6bc^3,\ c^2e^3 - z_7ad^4,\ c^3e^2 - z_8abd^3,\ c^4e - z_9ab^2d^2,$$
$$c^5 - z_{10}ab^3d,\ ce^5 - z_{11}d^6,\ ad^5 - z_{12}bce^4,\ be^6 - z_{13}d^7 \,\}.$$

If we set $z_1 = z_2 = \cdots = z_{13} = 1$ then we get a generating set for the toric ideal I_A. The 13 monomials obtained by setting $z_1 = z_2 = \cdots = z_{13} = 0$ generate the initial monomial ideal $M = in_w(I_A)$ with respect to the weight vector $w = (9, 3, 5, 0, 0)$. Thus M is one of the 226 coherent monomial A-graded ideals of our running example. The above set of 13 binomials in $\mathbb{C}[x, z]$ give the universal family for $Hilb_A$ around this M.

The local chart of $Hilb_A$ around the point M is a subscheme of affine space $\mathbb{C}^{13}$ with coordinates $z_1, \ldots, z_{13}$, whose defining equations are obtained as follows: Extend the weight vector w by assigning weight zero to all variables z_i, so that the first term in each of the above 13 binomials is the leading term. For each pair of binomials corresponding to a minimal syzygy of M, form their S-pair and then reduce it to a normal form with respect to the 13 binomials above. For instance,

$$S\big(c^5 - z_{10}ab^3d,\ ce^5 - z_{11}d^6\big) = z_{11}c^4d^6 - z_{10}ab^3de^5 \longrightarrow b^4d^2e^4 \cdot (z_2^4z_{11} - z_1z_{10}).$$

Each such normal form is a monomial in a, b, c, d, e times a binomial in $z_1, \ldots, z_{13}$. The set of all these binomials, in the z-variables, generates the ideal J_M of local equations of $Hilb_A$ around M. In our example, J_M is generated by 27 nonzero binomials. This computation can be done in *Macaulay 2* using the procedure `localCoherentEquations`.

```
i47 : localCoherentEquations = (IA) -> (
              -- IA is the toric ideal of A living in a ring equipped
              -- with weight order w, if we are computing the local
              -- equations about the initial ideal of IA w.r.t. w.
              R := ring IA;
```

```
w := (monoid R).Options.Weights;
M := ideal leadTerm IA;
S := first entries ((gens M) % IA);
-- Make the universal family J in a new ring.
nv := numgens R; n := numgens M;
T = (coefficientRing R)[generators R, z_1 .. z_n,
                        Weights => flatten splice{w, n:0},
                        MonomialSize=>16];
M = substitute(generators M,T);
S = apply(S, s -> substitute(s,T));
J = ideal apply(n, i ->
            M_(0,i) - T_(nv + i) * S_i);
-- Find the ideal Ihilb of local equations about M:
spairs := (gens J) * (syz M);
g := forceGB gens J;
B = (coefficientRing R)[z_1 .. z_n,MonomialSize=>16];
Fones := map(B,T, matrix(B,{splice {nv:1}}) | vars B);
Ihilb := ideal Fones (spairs % g);
Ihilb
);
```

Suppose we wish to calculate the local equations about $M = in_w(I_A)$. The input to localCoherentEquations is the toric ideal I_A living in a polynomial ring equipped with the weight order specified by w. This is done as follows:

```
i48 : IA = toricIdeal A;

o48 : Ideal of R

i49 : Y = QQ[a..e, MonomialSize => 16,
            Degrees => transpose A, Weights => {9,3,5,0,0}];

i50 : IA = substitute(IA,Y);

o50 : Ideal of Y
```

The initial ideal M is calculated in the third line of the algorithm, and S stores the standard monomials of M of the same degrees as the minimal generators of M. We could have calculated S using our old procedure stdMonomials but this involves computing the monomials in R_b for various values of b, which can be slow on large examples. As by-products, localCoherentEquations also gets J, the ideal of the universal family for $Hilb_A$ about M, the ring T of this ideal, and the ring B of Ihilb, which is the ideal of the affine patch of $Hilb_A$ about M. The matrix spairs contains all the S-pairs between generators of J corresponding to the minimal first syzygies of M. The command forceGB is used to declare the generators of J to be a Gröbner basis, and Fones is the ring map from T to B that sends each of a, b, c, d, e to one and the z variables to themselves. The columns of the matrix (spairs % g) are the normal forms of the polynomials in spairs with respect to the forced Gröbner basis g and the ideal Ihilb of local equations is generated by the image of these normal forms in the ring B under the map Fones.

```
i51 : JM = localCoherentEquations(IA)

                                                        . . .
o51 = ideal (z z  - z , z z  - z , - z z  + z , - z z  + z , - z z  + ···
              1 2    3   1 2    3     4 7    2     5 8    2     1 5    ···
```

o51 : Ideal of B

Removing duplications among the generators:

$$J_M = \langle z_1 - z_{10}z_{11}, z_2 - z_4z_7, z_2 - z_5z_8, z_2 - z_{11}z_{12}, z_2 - z_1z_{11}z_{13},$$
$$z_3 - z_1z_2, z_3 - z_5z_9, z_4 - z_1z_5, z_6 - z_3z_5, z_6 - z_1z_2z_5, z_7 - z_1z_{10}, z_8 - z_1z_7,$$
$$z_9 - z_1z_8, z_{12} - z_1z_{13}, z_1z_2 - z_5z_9, z_1z_2 - z_1z_5z_8, z_1z_2 - z_1^2z_4z_{10}, z_1z_2 - z_1^2z_5z_7,$$
$$z_1z_2 - z_1z_{11}z_{12}, z_1z_2 - z_2z_{10}z_{11}, z_1^3z_4 - z_3z_{11}, z_1z_5z_8 - z_4z_8, z_2z_{10} - z_1z_{12},$$
$$z_3z_4 - z_1z_6, z_3z_7 - z_2z_8, z_3z_8 - z_2z_9, z_3z_{10} - z_2z_7\rangle.$$

Notice that there are many generators of J_M that have a single variable as one of its terms. Using these generators we can remove variables from other binomials. This is done in *Macaulay 2* using the subroutine removeRedundantVariables, which is the main ingredient of the package minPres.m2 for computing the minimal presentations of polynomial quotient rings. Both removeRedundantVariables and minPres.m2 are explained in Appendix B. The command removeRedundantVariables applied to an ideal in a polynomial ring (not quotient ring) creates a ring map from the ring to itself that sends the redundant variables to polynomials in the non-redundant variables and the non-redundant variables to themselves. Applying this to our ideal J_M we obtain the following simplifications.

```
i52 : load "minPres.m2";

i53 : G = removeRedundantVariables JM
```

$$o53 = map(B,B,\{z_5z_{10}^3z_{11}^2, z_5^4z_{10}^3z_{11}, z_5^4z_{10}^3z_{11}, z_5z_{10}^3z_{11}, z_5, z_5^2z_{10}^4z_{11}^3, z_{10}^2 \cdots$$

```
o53 : RingMap B <--- B

i54 : ideal gens gb(G JM)
```

$$o54 = ideal(z_5z_{10}^3z_{11}^2 - z_{10}z_{11}^2z_{13})$$

o54 : Ideal of B

Thus our affine patch of $Hilb_A$ has the coordinate ring

$$\mathbb{C}[z_1, z_2, \ldots, z_{13}]/J_M \simeq \frac{\mathbb{C}[z_5, z_{10}, z_{11}, z_{13}]}{\langle z_5z_{10}^3z_{11}^2 - z_{10}z_{11}^2z_{13}\rangle} = \frac{\mathbb{C}[z_5, z_{10}, z_{11}, z_{13}]}{\langle (z_5z_{10}^2 - z_{13})z_{10}z_{11}^2\rangle}.$$

Hence, we see immediately that there are three components through the point M on $Hilb_A$. The restriction of the coherent component to the affine neighborhood of M on $Hilb_A$ is defined by the ideal quotient $(J_M : (z_1z_2 \cdots z_{13})^\infty)$ and hence the first of the above components is an affine patch of the coherent component. Locally near M it is given by the single equation $z_5z_{10}^2 - z_{13} = 0$ in $\mathbb{A}^4$. It is smooth and, as expected, has dimension three. The second component, $z_{10} = 0$, is also of dimension three and is smooth at M. The third component, given by $z_{11}^2 = 0$ is more interesting. It has dimension three as well, but is not reduced. Thus we have proved the following result.

Proposition 3.2. *The toric Hilbert scheme $Hilb_A$ of the matrix*

$$A = \begin{pmatrix} 1 & 1 & 1 & 1 & 1 \\ 0 & 1 & 2 & 7 & 8 \end{pmatrix}$$

is not reduced.

We can use the ring map G from above to simplify J so as to involve only the four variables z_5, z_{10}, z_{11} and z_{13}.

```
i55 : CX = QQ[a..e, z_5,z_10,z_11,z_13, Weights =>
            {9,3,5,0,0,0,0,0,0}];
```

```
i56 : F = map(CX, ring J, matrix{{a,b,c,d,e}} |
            substitute(G.matrix,CX))
```

$$
\begin{array}{c}
\qquad\qquad\qquad\qquad\quad 3\ \ 2 \qquad\quad 4\ \ 3 \qquad\qquad \cdots \\
o56 = \text{map(CX,T,\{a, b, c, d, e, z\ z}\ \ , z\ z\ \ z\ \ , z\ z\ \ z\ \ , z \cdots \\
\qquad\qquad\qquad\qquad\quad 10\ 11\quad 5\ 10\ 11\quad 5\ 10\ 11\quad 5\ 10\ 11\quad \cdots
\end{array}
$$

```
o56 : RingMap CX <--- T
```

Applying this map to J we get the ideal J1,

```
i57 : J1 = F J
```

$$
\begin{array}{c}
\qquad\qquad\qquad\qquad\qquad\qquad\quad 3\ \ 2 \qquad\quad 2\ \ 4\ \ 3 \qquad \cdots \\
o57 = \text{ideal (c*d - b*e*z}\ \ z\ \ , a*e - b*d*z\ z\ \ z\ \ , a*c - b\ z\ z\ \ z\ \ , a \cdots \\
\qquad\qquad\qquad\qquad\quad 10\ 11\qquad\quad 5\ 10\ 11\qquad\quad 5\ 10\ 11\quad \cdots
\end{array}
$$

```
o57 : Ideal of CX
```

and adding the ideal $\langle z_{11}^2 \rangle$ to J1 we obtain the universal family for the non-reduced component of $Hilb_A$ about M.

```
i58 : substitute(ideal(z_11^2),CX) + J1
```

$$
\begin{array}{c}
\qquad\quad 2 \qquad\qquad\qquad\qquad\qquad\qquad 3\ \ 2 \qquad\quad 2\ \ 4\quad \cdots \\
o58 = \text{ideal (z}\ \ , c*d - b*e*z\ \ z\ \ , a*e - b*d*z\ z\ \ z\ \ , a*c - b\ z\ z\ \ z\ \cdots \\
\qquad\qquad 11\qquad\qquad 10\ 11\qquad\quad 5\ 10\ 11\qquad\quad 5\ 10\quad \cdots
\end{array}
$$

```
o58 : Ideal of CX
```

In the rest of this section, we present an interpretation of the ideal J_M in terms of the combinatorial theory of *integer programming*. See, for instance, [22, §4] or [24] for the relevant background. Our reduced Gröbner basis (2) is the *minimal test set* for the family of integer programs

$$\text{Minimize} \quad w \cdot u \quad \text{subject to } A \cdot u = b \text{ and } u \in \mathbb{N}^n, \qquad (4)$$

where $A \in \mathbb{N}^{d \times n}$ and $w \in \mathbb{Z}^n$ are fixed and b ranges over $\mathbb{N}^d$. If $u' \in \mathbb{N}^n$ is any feasible solution to (4), then the corresponding optimal solution $u \in \mathbb{N}^n$ is computed as follows: the monomial x^u is the unique normal form of $x^{u'}$ modulo the Gröbner basis (2).

Suppose we had reduced $x^{u'}$ modulo the binomials (3) instead of (2). Then the output has a z-factor that depends on our choice of reduction path. To be precise, suppose the reduction path has length m and at the j-th step we had used the reduction $x^{u_{\mu_j}} \to z_{\mu_j} \cdot x^{v_{\mu_j}}$. Then we would obtain the normal form

$$z_{\mu_1} z_{\mu_2} z_{\mu_3} \cdots z_{\mu_m} \cdot x^u.$$

Reduction paths can have different lengths. If we take another path that has length m' and uses $x^{u_{\nu_j}} \to z_{\nu_j} \cdot x^{v_{\nu_j}}$ at the j-th step, then the output would be

$$z_{\nu_1} z_{\nu_2} z_{\nu_3} \cdots z_{\nu_{m'}} \cdot x^u.$$

Theorem 3.3. *The ideal J_M of local equations on $Hilb_A$ is generated by the binomials*

$$z_{\mu_1} z_{\mu_2} z_{\mu_3} \cdots z_{\mu_m} - z_{\nu_1} z_{\nu_2} z_{\nu_3} \cdots z_{\nu_{m'}},$$

each encoding a pair of distinct reduction sequences from a feasible solution of an integer program of the type (4) to the corresponding optimal solution using the minimal test set in (2).

Proof. The given ideal is contained in J_M because its generators are differences of monomials arising from the possible reduction paths of $lcm(x^{u_i}, x^{u_j})$, for $1 \leq i, j \leq r$. Conversely, any reduction sequence can be transformed into an equivalent reduction sequence using S-pair reductions. This follows from standard arguments in the proof of Buchberger's criterion [5, §2.6, Theorem 6], and it implies that the binomials $z_{\mu_1} \cdots z_{\mu_m} - z_{\nu_1} \cdots z_{\nu_{m'}}$ are $\mathbb{C}[z]$-linear combinations of the generators of J_M.   □

A given feasible solution of an integer program (4) usually has many different reduction paths to the optimal solution using the reduced Gröbner basis (2). For our matrix 1 and cost vector $w = (9, 3, 5, 0, 0)$, the monomial $a^2 bde^6$ encodes the feasible solution $(2, 1, 0, 1, 6)$ of the integer program

$$\text{Minimize} \quad w \cdot u \quad \text{subject to} \quad A \cdot u = \begin{pmatrix} 10 \\ 56 \end{pmatrix} \quad \text{and} \quad u \in \mathbb{N}^5.$$

There are 19 different paths from this feasible solution to the optimal solution $(0, 3, 0, 3, 4)$ encoded by the monomial $b^3 d^3 e^4$. The generating function for these paths is:

$$z_1^2 + 3z_1 z_2^2 z_5 z_7 + 2z_1 z_2 z_5 z_7^2 z_{12} + 2z_1 z_2 z_5 z_8$$
$$+ 2z_1 z_2 z_{12} z_{13} + z_1 z_5 z_9 + z_2^3 z_4 z_5 z_7^2 + z_2^3 z_4 z_{13} + z_2^3 z_5 z_{11}$$
$$+ 2z_2 z_3 z_5 z_7 + z_3 z_5 z_7^2 z_{12} + z_3 z_5 z_8 + z_3 z_{12} z_{13}.$$

The difference of any two monomials in this generating function is a valid local equation for the toric Hilbert scheme of (1). For instance, the binomial $z_3 z_5 z_7^2 z_{12} - z_3 z_{12} z_{13}$ lies in J_M, and, conversely, J_M is generated by binomials obtained in this manner.

The scheme structure of J_M encodes obstructions to making certain reductions when solving our family of integer programs. For instance, the variable z_3 is a zero-divisor modulo J_M. If we factor it out from the binomial

$z_3 z_5 z_7^2 z_{12} - z_3 z_{12} z_{13} \in J_M$, we get $z_5 z_7^2 z_{12} - z_{12} z_{13}$, which does not lie in J_M. Thus there is no monomial $a^{i_1} b^{i_2} c^{i_3} d^{i_4} e^{i_5}$ for which both the paths $z_5 z_7^2 z_{12}$ and $z_{12} z_{13}$ are used to reach the optimum. It would be a worthwhile combinatorial project to study the path generating functions and their relation to the ideal J_M in more detail.

It is instructive to note that the binomials $z_{\mu_1} z_{\mu_2} \cdots z_{\mu_m} - z_{\nu_1} z_{\nu_2} \cdots z_{\nu_{m'}}$ in Theorem 3.3 do not form a vector space basis for the ideal J_M. We demonstrate this for the lexicographic Gröbner basis (with $a \succ b \succ c \succ d \succ e$) of the toric ideal defining the rational normal curve of degree 4. In this case, we can take $A = \begin{pmatrix} 1 & 1 & 1 & 1 & 1 \\ 0 & 1 & 2 & 3 & 4 \end{pmatrix}$ and the universal family in question is :

$$\{ac - z_1 b^2,\ ad - z_2 bc,\ ae - z_3 c^2,\ bd - z_4 c^2,\ be - z_5 cd,\ ce - z_6 d^2\}.$$

The corresponding ideal of local equations is $J_M = \langle z_3 - z_2 z_5, z_2 - z_1 z_4, z_5 - z_4 z_6 \rangle$, from which we see that M is a smooth point of $Hilb_A$. The binomial $z_1 z_5 - z_1 z_4 z_6$ lies in J_M but there is no monomial that has the reduction path $z_1 z_5$ or $z_5 z_1$ to optimality. Indeed, any monomial that admits the reductions $z_1 z_5$ or $z_5 z_1$ must be divisible by either ace or abe. The path generating functions for these two monomials are

$$abe \quad \to \quad (z_3 + z_1 z_4 z_5 + z_2 z_5) \cdot bc^2$$

$$ace \quad \to \quad (z_3 + z_1 z_4 z_5 + z_2 z_4 z_6) \cdot c^3.$$

Thus every reduction to optimality using z_1 and z_5 must also use z_4, and we conclude that $z_1 z_5 - z_1 z_4 z_6$ is not in the $\mathbb{C}$-span of the binomials listed in Theorem 3.3.

4 The Coherent Component of the Toric Hilbert Scheme

In this section we study the component of the toric Hilbert scheme $Hilb_A$ that contains the point corresponding to the toric ideal I_A. An A-graded ideal is coherent if and only if it is isomorphic to an initial ideal of I_A under the action of the torus $(\mathbb{C}^*)^n$. All coherent A-graded ideals lie on the same component of $Hilb_A$ as I_A. We will show that this component need not be normal, and we will describe how its local and global equations can be computed using *Macaulay 2*. Every term order for the toric ideal I_A can be realized by a weight vector that is an element in the lattice $N = Hom_{\mathbb{Z}}(ker_{\mathbb{Z}}(A), \mathbb{Z}) \simeq \mathbb{Z}^{n-d}$. Two weight vectors w and w' in N are considered *equivalent* if they define the same initial ideal $in_w(I_A) = in_{w'}(I_A)$. These equivalence classes are the relatively open cones of a projective fan Σ_A called the *Gröbner fan* of I_A [15], [23]. This fan lies in $\mathbb{R}^{n-d}$, the real vector space spanned by the lattice N.

Theorem 4.1. *The toric ideal I_A lies on a unique irreducible component of the toric Hilbert scheme $Hilb_A$, called the coherent component. The normalization of the coherent component is the projective toric variety defined by the Gröbner fan of I_A.*

Proof. The *divisor at infinity* on the toric Hilbert scheme $Hilb_A$ consists of all points at which at least one of the local coordinates (around some monomial A-graded ideal) is zero. This is a proper closed codimension one subscheme of $Hilb_A$, parametrizing all those A-graded ideals that contain at least one monomial. The complement of the divisor at infinity in $Hilb_A$ consists of precisely the orbit of I_A under the action of the torus $(\mathbb{C}^*)^n$. This is the content of [22, Lemma 10.12].

The closure of the $(\mathbb{C}^*)^n$-orbit of I_A is a reduced and irreducible component of $Hilb_A$. It is reduced because I_A is a smooth point on $Hilb_A$, as can be seen from the local equations, and it is irreducible since $(\mathbb{C}^*)^n$ is a connected group. It is a component of $Hilb_A$ because its complement lies in a divisor. We call this irreducible component the *coherent component* of $Hilb_A$.

Identifying $(\mathbb{C}^*)^n$ with $Hom_{\mathbb{Z}}(\mathbb{Z}^n, \mathbb{C}^*)$, we note that the stabilizer of I_A consists of those linear forms w that restrict to zero on the kernel of A. Therefore the coherent component is the closure in $Hilb_A$ of the orbit of the point I_A under the action of the torus $N \otimes \mathbb{C}^* = Hom_{\mathbb{Z}}(ker_{\mathbb{Z}}(A), \mathbb{C}^*)$. The $(N \otimes \mathbb{C}^*)$-fixed points on this component are precisely the coherent monomial A-graded ideals, and the same holds for the toric variety of the Gröbner fan.

Fix a maximal cone σ in the Gröbner fan Σ_A, and let $M = \langle x^{u_1}, \ldots, x^{u_r} \rangle$ be the corresponding (monomial) initial ideal of I_A. As before we write

$$\left\{ x^{u_1} - z_1 \cdot x^{v_1}, \ x^{u_2} - z_2 \cdot x^{v_2}, \ \ldots, \ x^{u_r} - z_r \cdot x^{v_r} \right\}$$

for the universal family arising from the corresponding reduced Gröbner basis of I_A. Let J_M be the ideal in $\mathbb{C}[z_1, z_2, \ldots, z_r]$ defining this family.

The restriction of the coherent component to the affine neighborhood of M on $Hilb_A$ is defined by $J_M : (z_1 z_2 \cdots z_r)^\infty$. It then follows from our combinatorial description of the ideal J_M that this ideal quotient is a binomial prime ideal. In fact, it is the ideal of algebraic relations among the Laurent monomials $x^{u_1 - v_1}, \ldots, x^{u_r - v_r}$. We conclude that the restriction of the coherent component to the affine neighborhood of M on $Hilb_A$ equals

$$\mathrm{Spec}\,\mathbb{C}\left[x^{u_1 - v_1}, x^{u_2 - v_2}, \ldots, x^{u_r - v_r}\right]. \tag{5}$$

The abelian group generated by the vectors $u_1 - v_1, \ldots, u_r - v_r$ equals $ker_{\mathbb{Z}}(A) = Hom_{\mathbb{Z}}(N, \mathbb{Z})$. This follows from [21, Lemma 12.2] because the binomials $x^{u_i} - x^{v_i}$ generate the toric ideal I_A. The cone generated by the vectors $u_1 - v_1, \ldots, u_r - v_r$ is precisely the polar dual $\sigma^\vee$ to the Gröbner cone σ. This follows from equation (2.6) in [21]. We conclude that the normalization of the affine variety (5) is the normal affine toric variety

$$\mathrm{Spec}\,\mathbb{C}\left[ker_{\mathbb{Z}}(A) \cap \sigma^\vee\right]. \tag{6}$$

The normalization morphism from (6) to (5) maps the identity point in the toric variety (6) to the point I_A in the affine chart (5) of the toric Hilbert scheme $Hilb_A$. Clearly, this normalization morphism is equivariant with respect to the action by the torus $N \otimes \mathbb{C}^*$. These two properties hold for every maximal cone σ of the Gröbner fan Σ_A. Hence there exists a unique $N \otimes \mathbb{C}^*$-equivariant morphism ϕ from the projective toric variety associated with Σ_A onto the coherent component of $Hilb_A$, such that ϕ maps the identity point to the point I_A on $Hilb_A$, and ϕ restricts to the normalization morphism (6) $\rightarrow$ (5) on each affine open chart. We conclude that ϕ is the desired normalization map from the projective toric variety associated with the Gröbner fan of I_A onto the coherent component of the toric Hilbert scheme $Hilb_A$. $\square$

We now present an example that shows that the coherent component of $Hilb_A$ need not be normal. This example is derived from the matrix that appears in Example 3.15 of [10]. This example is also mentioned in [17] without details. Let $d = 4$ and $n = 7$ and fix the matrix

$$A = \begin{pmatrix} 1 & 1 & 1 & 1 & 1 & 1 & 1 \\ 0 & 6 & 7 & 5 & 8 & 4 & 3 \\ 3 & 7 & 2 & 0 & 7 & 6 & 1 \\ 6 & 5 & 2 & 6 & 5 & 0 & 0 \end{pmatrix}. \tag{7}$$

The lattice $N = Hom_{\mathbb{Z}}(ker_{\mathbb{Z}}(A), \mathbb{Z})$ is three-dimensional. The toric ideal I_A is minimally generated by 30 binomials of total degree between 6 and 93.

```
i59 : A = {{1,1,1,1,1,1,1},{0,6,7,5,8,4,3},{3,7,2,0,7,6,1},
           {6,5,2,6,5,0,0}};

i60 : IA = toricIdeal A

             2 3        3 2   2      4 4     8 4    4 3 6    7 2 4      4  ...
o60 = ideal (a c e - b*d f , a c*d*e f  - b g , d e f  - b c g , a*b c  ...

o60 : Ideal of R
```

We fix the weight vector $w = (0, 0, 276, 220, 0, 0, 215)$ in N and compute the initial ideal $M = in_w(I_A)$. This initial ideal has 44 minimal generators.

```
i61 : Y = QQ[a..g, MonomialSize => 16,
                   Weights => {0,0,276,220,0,0,215},
                   Degrees =>transpose A];

i62 : IA = substitute(IA,Y);

o62 : Ideal of Y

i63 : M = ideal leadTerm IA

             2 3      8 4    7 2 4      4 7 3   5 4 3 5   2 6 5 4   3 3 1 ...
o63 = ideal (a c e, b g , b c g , a*b c f , b c d f , a b c g , a b c  ...

o63 : Ideal of Y
```

Proposition 4.2. *The three dimensional affine variety (5), for the initial ideal M with respect to $w = (0, 0, 276, 220, 0, 0, 215)$ of the toric ideal of A in (7), is not normal.*

Proof. The universal family for the toric Hilbert scheme $Hilb_A$ at M is:

$$\{ a^2 e^{15} g^{18} - z_1 b^3 c^6 d^{10} f^{16}, \; b^{13} d^{15} f^{16} - z_2 a^8 ce^{21} g^{14},$$
$$c^{59} d^{57} f^{110} - z_3 e^{92} g^{134}, ac^{14} d^{11} f^{23} - z_4 be^{19} g^{29},$$
$$b^7 c^2 g^4 - z_5 d^4 e^3 f^6, \; \ldots, \; bc^{34} d^{32} f^{62} - z_{44} e^{53} g^{76} \}.$$

The semigroup algebra in (5) is generated by 44 Laurent monomials gotten from this family. It turns out that the first four monomials suffice to generate the semigroup. In other words, for all $j \in \{5, 6, \ldots, 44\}$ there exist $i_1, i_2, i_3, i_4 \in \mathbb{N}$ such that $z_j - z_1^{i_1} z_2^{i_2} z_3^{i_3} z_4^{i_4} \in J_M : (z_1 \cdots z_{44})^\infty$. Hence the semigroup algebra in (5) is:

$$\mathbb{C}\left[\frac{a^2 e^{15} g^{18}}{b^3 c^6 d^{10} f^{16}}, \frac{b^{13} d^{15} f^{16}}{a^8 ce^{21} g^{14}}, \frac{c^{59} d^{57} f^{110}}{e^{92} g^{134}}, \frac{ac^{14} d^{11} f^{23}}{be^{19} g^{29}} \right] \; \simeq \; \frac{\mathbb{C}[z_1, z_2, z_3, z_4]}{\langle z_1^5 z_2 z_3 - z_4^2 \rangle}.$$

This algebra is not integrally closed, since a toric hypersurface is normal if and only if at least one of the two monomials in the defining equation is square-free. Its integral closure in $\mathbb{C}[ker_{\mathbb{Z}}(A)]$ is generated by the Laurent monomial

$$\frac{z_4}{z_1^2} \; = \; (z_1 z_2 z_3)^{\frac{1}{2}} \; = \; \frac{b^5 c^{26} d^{31} f^{55}}{a^3 e^{49} g^{65}}. \tag{8}$$

Hence the affine chart (6) of the toric variety of the Gröbner fan of I_A is the spectrum of the normal domain $\mathbb{C}[z_1, z_2, z_3, y]/\langle z_1 z_2 z_3 - y^2 \rangle$, where y maps to (8).   □

We now examine the local equations of $Hilb_A$ about M for this example.

```
i64 : JM = localCoherentEquations(IA)
```
```
                                                         . . .
o64 = ideal (z z  - z , z z  - z , z z  - z , z z  - z , z z  - z , z  . . .
              1 2    3   1 2    3   1 5    4   1 3    6   1 3    6   1 . . .
```
```
o64 : Ideal of B
```
```
i65 : G = removeRedundantVariables JM;
```
```
o65 : RingMap B <--- B
```
```
i66 : toString ideal gens gb(G JM)
```
```
o66 = ideal(z_32*z_42^2*z_44-z_37^2*z_42,z_32^3*z_35*z_37^2-z_42^2*z_4 · · ·
```

This ideal has six generators and decomposing it we see that there are five components through the monomial ideal M on this toric Hilbert scheme. They are defined by the ideals:

- $\langle z_{32} z_{42} z_{44} - z_{37}^2, z_{32}^4 z_{35} - z_{42}, z_{32}^3 z_{35} z_{37}^2 - z_{42}^2 z_{44}, z_{32}^2 z_{35} z_{37}^4 - z_{42}^3 z_{44}^2,$
  $z_{32} z_{35} z_{37}^6 - z_{42}^4 z_{44}^3, z_{35} z_{37}^8 - z_{42}^5 z_{44}^4 \rangle$
- $\langle z_{44}, z_{37} \rangle$
- $\langle z_{37}, z_{42}^2 \rangle$
- $\langle z_{42}, z_{35} \rangle$

$- \langle z_{42}, z_{32}^3 \rangle.$

All five components are three dimensional. The first component is an affine patch of the coherent component and two of the components are not reduced. Let K be the first of these ideals.

```
i67 : K = ideal(z_32*z_42*z_44-z_37^2,z_32^4*z_35-z_42,
            z_32^3*z_35*z_37^2-z_42^2*z_44,z_32^2*z_35*z_37^4-z_42^3*z_44^2,
            z_32*z_35*z_37^6-z_42^4*z_44^3,z_35*z_37^8-z_42^5*z_44^4);

o67 : Ideal of B
```

Applying `removeRedundantVariables` to K we see that the affine patch of the coherent component is, locally at M, a non-normal hypersurface singularity (agreeing with (8)). The labels on the variables depend on the order of elements in the initial ideal M computed by *Macaulay 2* in line i61.

```
i68 : GG = removeRedundantVariables K;

o68 : RingMap B <--- B

i69 : ideal gens gb (GG K)

              5      2
o69 = ideal(z  z  z   - z  )
             32 35 44    37

o69 : Ideal of B
```

There is a general algorithm due to de Jong [6] for computing the normalization of any affine variety. In the toric case, the problem of normalization amounts to computing the minimal *Hilbert basis* of a given convex rational polyhedral cone [20]. An efficient implementation can be found in the software package `Normaliz` by Bruns and Koch [4].

Our computational study of the toric Hilbert scheme in this chapter was based on local equations rather than global equations (arising from a projective embedding of $Hilb_A$), because the latter system of equations tends to be too large for most purposes. Nonetheless, they are interesting. In the remainder of this section, we present a canonical projective embedding of the coherent component of $Hilb_A$.

Let $G_1, G_2, G_3, \ldots, G_s$ denote all the *Graver fibers* of the matrix A. In Section 1 we showed how to compute them in *Macaulay 2*. Each set G_i consists of the monomials in $\mathbb{C}[x_1, \ldots, x_n]$ that have a fixed Graver degree. Consider the set $\mathbf{G} := G_1 G_2 G_3 \cdots G_s$ that consists of all monomials that are products of monomials, one from each of the distinct Graver fibers. Let t denote the cardinality of $\mathbf{G}$. We introduce an extra indeterminate z, and we consider the $\mathbb{N}$-graded semigroup algebra $\mathbb{C}[z\mathbf{G}]$, which is a subalgebra of $\mathbb{C}[x_1, \ldots, x_n, z]$. The grading of this algebra is $deg(z) = 1$ and $deg(x_i) = 0$. Labeling the elements of $\mathbf{G}$ with indeterminates y_i, we can write

$$\mathbb{C}[z\mathbf{G}] = \mathbb{C}[y_1, y_2, \ldots, y_t]/P_A,$$

where P_A is a homogeneous toric ideal associated with a configuration of t vectors in $\mathbb{Z}^{n+1}$. We note that the torus $(\mathbb{C}^*)^n$ acts naturally on $\mathbb{C}[z\mathbf{G}]$.

Example 4.3. Let $n = 4, d = 2$ and $A = \begin{pmatrix} 3 & 2 & 1 & 0 \\ 0 & 1 & 2 & 3 \end{pmatrix}$, so that I_A is the ideal of the twisted cubic curve. There are five Graver fibers:

```
i70 : A = {{1,1,1,1},{0,1,2,3}};

i71 : I = toricIdeal A;

o71 : Ideal of R

i72 : Graver = graver I;

             1      5
o72 : Matrix R  <--- R

i73 : fibers = graverFibers Graver;

i74 : peek fibers

o74 = HashTable{{2, 2} => | ac b2 |      }
                {2, 3} => | ad bc |
                {2, 4} => | bd c2 |
                {3, 3} => | a2d abc b3 |
                {3, 6} => | ad2 bcd c3 |
```

The set $\mathbf{G} = G_1 G_2 G_3 G_4 G_5$ consists of 22 monomials of degree 14.

```
i75 : G = trim product(values fibers, ideal)

             5     5    4 3 5    5 3 4    4 2 2 4    3 4    4    2 6 4    4  ···
o75 = ideal (a b*c*d , a b d , a c d , a b c d , a b c*d , a b d , a b ···

o75 : Ideal of R

i76 : numgens G

o76 = 22
```

We introduce a polynomial ring in 22 variables $y_1, y_2, \ldots, y_{22}$, and we compute the ideal P_A. It is generated by 180 binomial quadrics.

```
i77 : z = symbol z;

i78 : S = QQ[a,b,c,d,z];

i79 : zG = z ** substitute(gens G, S);

             1     22
o79 : Matrix S  <--- S

i80 : R = QQ[y_1 .. y_22];

i81 : F = map(S,R,zG)

             5     5    4 3 5    5 3 4    4 2 2 4    3 4    4    2 6 ···
o81 = map(S,R,{a b*c*d z, a b d z, a c d z, a b c d z, a b c*d z, a b  ···

o81 : RingMap S <--- R

i82 : PA = trim ker F

             2                                                       ···
o82 = ideal (y   - y y , y y  - y y , y y , y y  - y y , y y  -  ···
             21    20 22   19 21   18 22   18 21   17 22   17 21     ···

o82 : Ideal of R
```

These equations define a toric surface of degree 30 in projective 21-space.

```
i83 : codim PA

o83 = 19

i84 : degree PA

o84 = 30
```

The surface is smooth, but there are too many equations and the codimension is too large to use the Jacobian criterion for smoothness [7, §16.6] directly. Instead we check smoothness for each open set $y_i \neq 0$.

```
i85 : Aff = apply(1..22, v -> (
                     K = substitute(PA,y_v => 1);
                     FF = removeRedundantVariables K;
                     ideal gens gb (FF K)));

i86 : scan(Aff, i -> print toString i);
ideal()
ideal()
ideal()
ideal(y_1^4*y_5*y_21-1)
ideal(y_1^4*y_6^6*y_21-1)
ideal()
ideal(y_1^2*y_11^2*y_17-1)
ideal(y_1^3*y_9^2*y_21^2-1)
ideal(y_6^3*y_21-y_10,y_1*y_10^3-y_6^2,y_1*y_6*y_10^2*y_21-1)
ideal(y_6*y_15-1,y_2*y_15^2-y_6*y_14,y_6^2*y_14-y_2*y_15)
ideal()
ideal(y_11*y_13-1,y_1^2*y_21^3-y_13^2)
ideal(y_1^2*y_14^3*y_21^3-1)
ideal(y_10^2*y_21-1,y_1*y_15^4-y_10^3)
ideal()
ideal(y_11*y_20-1,y_3*y_20^2-y_11*y_17,y_11^2*y_17-y_3*y_20)
ideal(y_11*y_18*y_21-1,y_1*y_21^3-y_11*y_18^2,y_11^2*y_18^3-y_1*y_21^2)
ideal(y_1*y_19^4*y_21^4-1)
ideal(y_15*y_22-1)
ideal()
ideal(y_20*y_22-1)
ideal()
```

By examining these local equations, we see that $Hilb_A$ is smooth, and also that there are eight fixed points under the action of the 2-dimensional torus. They correspond to the variables $y_1, y_2, y_3, y_6, y_{11}, y_{15}, y_{20}$ and y_{22}. By setting any of these eight variables to 1 in the 180 quadrics above, we obtain an affine variety isomorphic to the affine plane.

Theorem 4.4. *The coherent component of the toric Hilbert scheme* $Hilb_A$ *is isomorphic to the projective spectrum* $Proj\, \mathbb{C}[z\mathbf{G}]$ *of the algebra* $\mathbb{C}[z\mathbf{G}]$.

Proof. The first step is to define a morphism from $Hilb_A$ to the $(t-1)$-dimensional projective space $\mathbb{P}(\mathbf{G}) = Proj\, \mathbb{C}[y_1, y_2, \ldots, y_t]$. Consider any point I on $Hilb_A$. We intersect the ideal I with the finite-dimensional vector space $\mathbb{C}G_i$, consisting of all homogeneous polynomials in $\mathbb{C}[x_1, \ldots, x_n]$ that lie in the i-th Graver degree. The definition of A-graded ideal implies that $I \cap \mathbb{C}G_i$ is a linear subspace of codimension 1 in $\mathbb{C}G_i$. We represent this subspace by an equation $g_i(I) = \sum_{u \in G_i} c_u x^u$, which is unique up to scaling. Taking the product of these polynomials for $i = 1, \ldots, t$, we get a unique (up to scaling) polynomial that is supported on $\mathbf{G} = G_1 G_2 \cdots G_t$. The map $I \mapsto g_1(I)g_2(I) \cdots g_t(I)$ defines a morphism from $Hilb_A$ to $\mathbb{P}(\mathbf{G})$. This morphism is equivariant with respect to the $(\mathbb{C}^*)^n$-action on both schemes.

Consider the restriction of this equivariant morphism to the coherent component of the toric Hilbert scheme. It maps the $(\mathbb{C}^*)^n$-orbit of the toric ideal I_A into the subvariety $Proj\, \mathbb{C}[z\mathbf{G}]$ of $\mathbb{P}(\mathbf{G})$. This inclusion is an isomorphism onto the dense torus, as the dimension of the Newton polytope of

$$g(I_A) = \prod_{i=1}^{t} \left(\sum_{u \in G_i} x^u \right)$$

equals the dimension of the kernel of A. Equivalently, the stabilizer of $g(I_A)$ in $(\mathbb{C}^*)^n$ consists only of those one-parameter subgroups w that restrict to zero on the kernel of A.

To show that our morphism is an isomorphism between the coherent component and $Proj\, \mathbb{C}[z\mathbf{G}]$, we consider the affine chart around an initial monomial ideal $M = in_w(I_A)$. The polynomial $g(M)$ is a monomial, namely, it is the product of all standard monomials whose degree is a Graver degree. Moreover, $g(M)$ is the leading monomial of $g(I_A)$ with respect to the weight vector w. The Newton polytope of $g(I_A)$ is the Minkowski sum of the Newton polytopes of the polynomials $g_1(I_A), \ldots, g_t(I_A)$, and it is a state polytope for I_A, by [22, Theorem 7.5].

Let $g(M) = x^q$, and let σ be the cone of the Gröbner fan Σ_A that has w in its interior. Then σ coincides with the normal cone at the vertex q of the state polytope described above [22, §3]. Consider the restriction of our morphism to the affine chart around M of the coherent component, as described in (5). This restriction defines an isomorphism onto the variety

$$Spec\, \mathbb{C}[\,x^{p-q} \,:\, x^p \in \mathbf{G}\,] \tag{9}$$

On the other hand, the semigroup algebra in (9) is isomorphic to that in (5) because each pair of vectors $\{u_i, v_i\}$ seen in the reduced Gröbner basis lies in one of the Graver fibers G_j. Hence our morphism restricts to an isomorphism from the affine chart around M of the coherent component onto (9). Finally, note that (9) is the principal affine open subset of $Proj\, \mathbb{C}[z\mathbf{G}]$ defined by the coordinate x^q. Hence we get an isomorphism between the coherent component of $Hilb_A$ and $Proj\, \mathbb{C}[z\mathbf{G}]$.   □

Appendix A. Fourier-Motzkin Elimination

We now give the *Macaulay 2* code for converting the generator/inequality representation of a rational convex polyhedron to the other. It is based on the Fourier-Motzkin elimination procedure for eliminating a variable from a system of inequalities [25]. This code was written by Greg Smith.

Given any cone $C \subset \mathbb{R}^d$, the polar cone of C is defined to be

$$C^\vee = \{x \in \mathbb{R}^d \mid x \cdot y \le 0, \text{for all } y \in C\}.$$

For a $d \times n$ matrix Z, define $cone(Z) = \{Zx \mid x \in \mathbb{R}^n_{\ge 0}\} \subset \mathbb{R}^d$, and $affine(Z) = \{Zx \mid x \in \mathbb{R}^n\} \subset \mathbb{R}^d$. For two integer matrices Z and H, both having d rows, polarCone(Z,H) returns a list of two integer matrices {A,E} such that

$$cone(Z) + affine(H) = \{x \in \mathbb{R}^d \mid A^t x \le 0, E^t x = 0\}.$$

Equivalently, $(cone(Z) + affine(H))^\vee = cone(A) + affine(E)$.

We now describe each routine in the package polarCone.m2. We have simplified the code for readability, sometimes at the cost of efficiency. We start with three simple subroutines: primitive, toZZ, and rotateMatrix.

The routine primitive takes a list of integers L, and divides each element of this list by their greatest common denominator.

```
i87 : code primitive

o87 = -- polarCone.m2:16-20
         primitive = (L) -> (
             n := #L-1;                      g  := L#n;
             while n > 0 do (n = n-1;        g = gcd(g, L#n);
                  if g === 1 then n = 0);
             if g === 1 then L else apply(L, i -> i // g));
```

The routine toZZ converts a list of rational numbers to a list of integers, by multiplying by their common denominator.

```
i88 : code toZZ

o88 = -- polarCone.m2:28-32
         toZZ = (L) -> (
             d := apply(L, e -> denominator e);
             R := ring d#0;                   l := 1_R;
             scan(d, i -> (l = (l*i // gcd(l,i))));
             apply(L, e -> (numerator(l*e))));
```

The routine rotateMatrix is a kind of transpose. Its input is a matrix, and its output is a matrix of the same shape as the transpose. It places the matrix in the form so that in the routine polarCone, computing a Gröbner basis will do the Gaussian elimination that is needed.

```
i89 : code rotateMatrix

o89 = -- polarCone.m2:41-43
         rotateMatrix = (M) -> (
             r := rank source M;           c := rank target M;
             matrix table(r, c, (i,j) -> M_(c-j-1, r-i-1)));
```

The procedure of Fourier-Motzkin elimination as presented by Ziegler in [25] is used, together with some heuristics that he presents as exercises. The following, which is a kind of S-pair criterion for inequalities, comes from Exercise 2.15(i) in [25].

The routine isRedundant determines if a row vector (inequality) is redundant. Its input argument V is the same input that is used in fourierMotzkin: it is a list of sets of integers. Each entry contains indices of the original rays that do *not* vanish at the corresponding row vector. vert is a set of integers; the original rays for the row vector in question. A boolean value is returned.

```
i90 : code isRedundant

o90 = -- polarCone.m2:57-65
      isRedundant = (V, vert) -> (
          -- the row vector is redundant iff 'vert' contains an
          -- entry in 'V'.
          x := 0;              k := 0;
          numRow := #V;         -- equals the number of inequalities
          while x < 1 and k < numRow do (
              if isSubset(V#k, vert) then x = x+1;
              k = k+1;);
          x === 1);
```

The main work horse of polarCone.m2 is the subroutine fourierMotzkin, which eliminates the first variable in the inequalities A using the double description version of Fourier-Motzkin elimination. The set A is a list of lists of integers, each entry corresponding to a row vector in the system of inequalities. The argument V is a list of sets of integers. Each entry contains the indices of the original rays that do *not* vanish at the corresponding row vector in A. Note that this set is the *complement* of the set V_i appearing in exercise 2.15 in [25]. The argument spot is the integer index of the variable being eliminated.

The routine returns a list {projA,projV} where projA is a list of lists of integers. Each entry corresponds to a row vector in the projected system of inequalities. The list projV is a list of sets of integers. Each entry contains indices of the original rays that do *not* vanish at the corresponding row vector in projA.

```
i91 : code fourierMotzkin

o91 = -- polarCone.m2:89-118
      fourierMotzkin = (A, V, spot) -> (
          -- initializing local variables
          numRow := #A;                    -- equal to the length of V
          numCol := #(A#0);                pos := {};
          neg := {};                        projA := {};
          projV := {};                      k := 0;
          -- divide the inequalities into three groups.
          while k < numRow do (
              if A#k#0 < 0 then neg = append(neg, k)
              else if A#k#0 > 0 then pos = append(pos, k)
              else (projA = append(projA, A#k);
                  projV = append(projV, V#k););
              k = k+1;);
```

```
            -- generate new irredundant inequalities.
            scan(pos, i -> scan(neg, j -> (vert := V#i + V#j;
                            if not isRedundant(projV, vert)
                            then (iRow := A#i;      jRow := A#j;
                                  iCoeff := - jRow#0;
                                  jCoeff := iRow#0;
                                  a := iCoeff*iRow + jCoeff*jRow;
                                  projA = append(projA, a);
                                  projV = append(projV, vert);););)));
            -- don't forget the implicit inequalities '-t <= 0'.
            scan(pos, i -> (vert := V#i + set{spot};
                    if not isRedundant(projV, vert) then (
                        projA = append(projA, A#i);
                        projV = append(projV, vert);););));
            -- remove the first column
            projA = apply(projA, e -> e_{1..(numCol-1)});
            {projA, projV});
```

As mentioned above, polarCone takes two matrices Z, H, both having d rows, and outputs a pair of matrices A, E such that $(\mathrm{cone}(Z)+\mathrm{affine}(H))^{\vee} = \mathrm{cone}(A) + \mathrm{affine}(E)$.

```
i92 : code(polarCone,Matrix,Matrix)

o92 = -- polarCone.m2:137-192
      polarCone(Matrix, Matrix) := (Z, H) -> (
          R := ring source Z;
          if R =!= ring source H then error ("polarCone: " |
              "expected matrices over the same ring");
          if rank target Z =!= rank target H then error (
              "polarCone: expected matrices to have the " |
              "same number of rows");
          if (R =!= ZZ) then error ("polarCone: expected " |
              "matrices over 'ZZ'");
          -- expressing 'cone(Y)+affine(B)' as '{x : Ax <= 0}'
          Y := substitute(Z, QQ);      B := substitute(H, QQ);
          if rank source B > 0 then Y = Y | B | -B;
          n := rank source Y;        d := rank target Y;
          A := Y | -id_(QQ^d);
          -- computing the row echelon form of 'A'
          A = gens gb rotateMatrix A;
          L := rotateMatrix leadTerm A;
          A = rotateMatrix A;
          -- find pivots
          numRow = rank target A;                  -- numRow <= d
          i := 0;                    pivotCol := {};
          while i < numRow do (j := 0;
                  while j < n+d and L_(i,j) =!= 1_QQ do j = j+1;
                  pivotCol = append(pivotCol, j);
                  i = i+1;);
          -- computing the row-reduced echelon form of 'A'
          A = ((submatrix(A, pivotCol))^(-1))  * A;
          -- converting 'A' into a list of integer row vectors
          A = entries A;
          A = apply(A, e -> primitive toZZ e);
          -- creating the vertex list 'V' for double description
          -- and listing the variables 'T' which remain to be
          -- eliminated
          V := {};                  T := toList(0..(n-1));
          scan(pivotCol, e -> (if e < n then (T = delete(e, T);
```

```
                        V = append(V, set{e});)));
        -- separating inequalities 'A' and equalities 'E'
        eqnRow := {};                 ineqnRow := {};
        scan(numRow, i -> (if pivotCol#i >= n then
                    eqnRow = append(eqnRow, i)
                    else ineqnRow = append(ineqnRow, i);));
        E := apply(eqnRow, i -> A#i);
        E = apply(E, e -> e_{n..(n+d-1)});
        A = apply(ineqnRow, i -> A#i);
        A = apply(A, e -> e_(T | toList(n..(n+d-1))));
        -- successive projections eliminate the variables 'T'.
        if A =!= {} then scan(T, t -> (
                    D := fourierMotzkin(A, V, t);
                    A = D#0;            V = D#1;));
        -- output formating
        A = apply(A, e -> primitive e);
        if A === {} then A = map(ZZ^d, ZZ^0, 0)
        else A = transpose matrix A;
        if E === {} then E = map(ZZ^d, ZZ^0, 0)
        else E = transpose matrix E;
        (A, E));
```

If the input matrix H has no columns, it can be omitted. A sequence of two matrices is returned, as above.

```
i93 : code(polarCone,Matrix)

o93 = -- polarCone.m2:199-200
      polarCone(Matrix) := (Z) -> (
          polarCone(Z, map(ZZ^(rank target Z), ZZ^0, 0)));
```

As a simple example, consider the permutahedron in $\mathbb{R}^3$ whose vertices are the following six points.

```
i94 : H = transpose matrix{
      {1,2,3},
      {1,3,2},
      {2,1,3},
      {2,3,1},
      {3,1,2},
      {3,2,1}};

              3        6
o94 : Matrix ZZ  <--- ZZ
```

The inequality representation of the permutahedron is obtained by calling polarCone on H: the facet normals of the polytope are the columns of the matrix in the first argument of the output. The second argument is trivial since our input is a polytope and hence there are is no non-trivial affine space contained in it. If we call polarCone on the output, we will get back H as expected.

```
i95 : P = polarCone H

o95 = (| 1   1   1   -1 -1 -5 |, 0)
        | -1  1  -5   1  -1  1 |
        | -1 -5   1  -1   1  1 |

o95 : Sequence

i96 : Q = polarCone P_0
```

```
o96 = (| 1 1 2 2 3 3 |, 0)
       | 2 3 1 3 1 2 |
       | 3 2 3 1 2 1 |

o96 : Sequence
```

Appendix B. Minimal Presentation of Rings

Throughout this chapter, we have used on several occasions the simple, yet useful subroutine `removeRedundantVariables`. In this appendix, we present *Macaulay 2* code for this routine, which is the main ingredient for finding minimal presentations of quotients of polynomial rings. Our code for this routine is a somewhat simplified, but less efficient version of a routine in the *Macaulay 2* package, `minPres.m2`, written by Amelia Taylor.

The routine `removeRedundantVariables` takes as input an ideal I in a polynomial ring A. It returns a ring map F from A to itself that sends redundant variables to polynomials in the non-redundant variables and sends non-redundant variables to themselves. For example:

```
i97 : A = QQ[a..e];

i98 : I = ideal(a-b^2-1, b-c^2, c-d^2, a^2-e^2)

               2           2         2       2   2
o98 = ideal (- b  + a - 1, - c  + b, - d  + c, a  - e )

o98 : Ideal of A

i99 : F = removeRedundantVariables I

              8       4   2
o99 = map(A,A,{d  + 1, d , d , d, e})

o99 : RingMap A <--- A
```

The non-redundant variables are d and e. The image of I under F gives the elements in this smaller set of variables. We take the ideal of a Gröbner basis of the image:

```
i100 : I1 = ideal gens gb(F I)

               16      8    2
o100 = ideal(d   + 2d  - e  + 1)

o100 : Ideal of A
```

The original ideal can be written in a cleaner way as

```
i101 : ideal compress (F.matrix - vars A) + I1

              8          4       2       16      8    2
o101 = ideal (d  - a + 1, d  - b, d  - c, d   + 2d  - e  + 1)

o101 : Ideal of A
```

Let us now describe the *Macaulay 2* code. The subroutine `findRedundant` takes a polynomial f, and finds a variable x_i in the ring of f such that $f = cx_i + g$ for a non-zero constant c and a polynomial g that does not involve the variable x_i. If there is no such variable, null is returned. Otherwise, if x_i is the first such variable , the list $\{i, c^{-1}g\}$ is returned.

```
i102 : code findRedundant

o102 = -- minPres.m2:1-12
       findRedundant=(f)->(
            A := ring(f);
            p := first entries contract(vars A,f);
            i := position(p, g -> g != 0 and first degree g === 0);
            if i === null then
                 null
            else (
                 v := A_i;
                 c := f_v;
                 {i,(-1)*(c^(-1)*(f-c*v))}
                 )
            )
```

The main function removeRedundantVariables requires an ideal in a polynomial ring (not a quotient ring) as input. The internal routine findnext finds the first entry of the (one row) matrix M that contains a redundancy. This redundancy is used to modify the list xmap, which contains the images of the redundant variables. The matrix M, and the list xmap are both updated, and then we continue to look for more redundancies.

```
i103 : code removeRedundantVariables

o103 = -- minPres.m2:14-39
       removeRedundantVariables = (I) -> (
            A := ring I;
            xmap := new MutableList from gens A;
            M := gens I;
            findnext := () -> (
                 p := null;
                 next := 0;
                 done := false;
                 ngens := numgens source M;
                 while next < ngens and not done do (
                   p = findRedundant(M_(0,next));
                   if p =!= null then
                        done = true
                   else next=next+1;
                 );
                 p);
            p := findnext();
            while p =!= null do (
                 xmap#(p#0) = p#1;
                 F1 := map(A,A,toList xmap);
                 F2 := map(A,A, F1 (F1.matrix));
                 xmap = new MutableList from first entries F2.matrix;
                 M = compress(F2 M);
                 p = findnext();
                 );
            map(A,A,toList xmap));
```

References

1. V.I. Arnold: A-graded algebras and continued fractions. *Communications in Pure and Applied Mathematics*, 42:993–1000, 1989.
2. A Bigatti, R. La Scala, and L. Robbiano: Computing toric ideals. *Journal of Symbolic Computation*, 27:351–365, 1999.
3. L. J. Billera, P. Filliman, and B. Sturmfels: Constructions and complexity of secondary polytopes. *Advances in Mathematics*, 83:155–179, 1990.
4. W. Bruns and R. Koch: Normaliz, a program to compute normalizations of semigroups. available by anonymous ftp from ftp.mathematik.Uni-Osnabrueck.DE/pub/osm/kommalg/software/.
5. D. Cox, J. Little, and D. O'Shea: *Ideals, Varieties, and Algorithms. An Introduction to Computational Algebraic Geometry and Commutative Algebra*. Springer-Verlag, New York, 1997.
6. T. de Jong: An algorithm for computing the integral closure. *Journal of Symbolic Computation*, 26:273–277, 1998.
7. D. Eisenbud: *Commutative Algebra with a View Toward Algebraic Geometry*. Springer-Verlag, New York, 1994.
8. I. M. Gel'fand, M. Kapranov, and A. Zelevinsky: *Multidimensional Determinants, Discriminants and Resultants*. Birkhäuser, Boston, 1994.
9. J.E. Graver: On the foundations of linear and integer programming. *Mathematical Programming*, 8:207–226, 1975.
10. S. Hoşten and D. Maclagan: The vertex ideal of a lattice. Preprint 2000.
11. S. Hoşten and J. Shapiro: Primary decomposition of lattice basis ideals. *Journal of Symbolic Computation*, 29:625–639, 2000.
12. B. Huber and R.R. Thomas: Computing Gröbner fans of toric ideals. *Experimental Mathematics*, 9:321–331, 2000. Software, TiGERS, available at http://www.math.washington.edu/~thomas/programs.html.
13. E. Korkina, G. Post, and M. Roelofs: Classification of generalized A-graded algebras with 3 generators. *Bulletin de Sciences Mathématiques*, 119:267–287, 1995.
14. D. Maclagan and R.R. Thomas: Combinatorics of the toric Hilbert scheme. *Discrete and Computational Geometry*. To appear.
15. T. Mora and L. Robbiano: The Gröbner fan of an ideal. *Journal of Symbolic Computation*, 6:183–208, 1998.
16. I. Peeva and M. Stillman: Local equations for the toric Hilbert scheme. *Advances in Applied Mathematics*. To appear.
17. I. Peeva and M. Stillman: Toric Hilbert schemes. Preprint 1999.
18. V. Reiner: The generalized Baues problem. In L. Billera, A. Björner, C. Greene, R. Simion, and R. Stanley, editors, *New Perspectives in Algebraic Combinatorics*. Cambridge University Press, 1999.
19. F. Santos: A point configuration whose space of triangulations is disconnected. *Journal of the American Math. Soc.*, 13:611–637, 2000.
20. A. Schrijver: *Theory of Linear and Integer Programming*. Wiley-Interscience, Chichester, 1986.
21. B. Sturmfels: The geometry of A-graded algebras. math.AG/9410032.
22. B. Sturmfels: *Gröbner Bases and Convex Polytopes*, volume 8. American Mathematical Society, University Lectures, 1996.
23. B. Sturmfels and R.R. Thomas: Variation of cost functions in integer programming. *Mathematical Programming*, 77:357–387, 1997.

24. R.R. Thomas: Applications to integer programming. In D.A. Cox and B. Sturm-fels, editors, *Applications of Computational Algebraic Geometry*. AMS Proceedings of Symposia in Applied Mathematics, 1997.
25. G. Ziegler: *Lectures on Polytopes*, volume 152. Springer-Verlag, New York, 1995.

Sheaf Algorithms Using the Exterior Algebra

Wolfram Decker and David Eisenbud

In this chapter we explain constructive methods for computing the cohomology of a sheaf on a projective variety. We also give a construction for the Beilinson monad, a tool for studying the sheaf from partial knowledge of its cohomology. Finally, we give some examples illustrating the use of the Beilinson monad.

1 Introduction

In this chapter V denotes a vector space of finite dimension $n+1$ over a field K with dual space $W = V^*$, and $S = \mathrm{Sym}_K(W)$ is the symmetric algebra of W, isomorphic to the polynomial ring on a basis for W. We write E for the exterior algebra on V. We grade S and E by taking elements of W to have degree 1, and elements of V to have degree -1. We denote the projective space of 1-quotients of W (or of lines in V) by $\mathbf{P}^n = \mathbf{P}(W)$.

Serre's sheafification functor $M \mapsto \tilde{M}$ allows one to consider a coherent sheaf on $\mathbf{P}(W)$ as an equivalence class of finitely generated graded S-modules, where we identify two such modules M and M' if, for some r, the truncated modules $M_{\geq r}$ and $M'_{\geq r}$ are isomorphic. A free resolution of M, sheafified, becomes a resolution of $\tilde{M}$ by sheaves that are direct sums of line bundles on $\mathbf{P}(W)$ – that is, a description of $\tilde{M}$ in terms of homogeneous matrices over S. Being able to compute syzygies over S one can compute the cohomology of $\tilde{M}$ starting from the minimal free resolution of M (see [16], [40] and Remark 3.2 below).

The Bernstein-Gel'fand-Gel'fand correspondence (BGG) is an isomorphism between the derived category of bounded complexes of finitely generated S-modules and the derived category of bounded complexes of finitely generated E-modules or of certain "Tate resolutions" of E-modules. In this chapter we show how to effectively compute the Tate resolution $\mathbf{T}(\mathcal{F})$ associated to a sheaf $\mathcal{F}$, and we use this construction to give relatively cheap computations of the cohomology of $\mathcal{F}$.

It turns out that by applying a simple functor to the Tate resolution $\mathbf{T}(\mathcal{F})$ one gets a finite complex of sheaves whose homology is the sheaf $\mathcal{F}$ itself. This complex is called a *Beilinson monad* for $\mathcal{F}$. The Beilinson monad provides a powerful method for getting information about a sheaf from partial knowledge of its cohomology. It is a representation of the sheaf in terms of direct sums of (suitably twisted) bundles of differentials and homomorphisms between these bundles, which are given by homogeneous matrices over E.

The following recipe for computing the cohomology of a sheaf is typical of our methods: Suppose that $\mathcal{F} = \tilde{M}$ is the coherent sheaf on $\mathbf{P}(W)$ associated to a finitely generated graded S-module $M = \oplus M_i$. To compute the cohomology of $\mathcal{F}$ we consider a sequence of free E-modules and maps

$$\mathbf{F}(M): \quad \cdots \longrightarrow F^{i-1} \xrightarrow{\phi_{i-1}} F^i \xrightarrow{\phi_i} F^{i+1} \longrightarrow \cdots .$$

Here we set $F^i = M_i \otimes_K E$ and define $\phi_i : F^i \longrightarrow F^{i+1}$ to be the map taking $m \otimes 1 \in M_i \otimes_K E$ to

$$\sum_j x_j m \otimes e_j \in M_{i+1} \otimes V \subset F^{i+1},$$

where $\{x_j\}$ and $\{e_j\}$ are dual bases of W and V respectively. It turns out that $\mathbf{F}(M)$ is a complex; that is, $\phi_i \phi_{i-1} = 0$ for every i (the reader may easily check this by direct computation; a proof without indices is given in [18]). If we regard M_i as a vector space concentrated in degree i, so that F^i is a direct sum of copies of $E(-i)$, then these maps are homogeneous of degree 0.

We shall see that if s is a sufficiently large integer then the truncation of the Tate resolution

$$F^s \xrightarrow{\phi_s} F^{s+1} \longrightarrow \cdots$$

is exact and is thus the minimal injective resolution of the finitely generated graded E-module $P_s = \ker \phi_{s+1}$. (In fact any value of s greater than the Castelnuovo-Mumford regularity of M will do.)

Because the number of monomials in E in any given degree is small compared to the number of monomials of that degree in the symmetric algebra, it is relatively cheap to compute a free resolution of P_s over E, and thus to compute the graded vector spaces $\mathrm{Tor}_t^E(P_s, K)$. Our algorithm exploits the fact, proved in [18], that the jth cohomology $\mathrm{H}^j \mathcal{F}$ of $\mathcal{F}$ in the Zariski topology is isomorphic to the degree $-n - 1$ part of $\mathrm{Tor}_{s-j}^E(P_s, K)$; that is,

$$\mathrm{H}^j \mathcal{F} \cong \mathrm{Tor}_{s-j}^E(P_s, K)_{-n-1}.$$

In addition, the linear parts of the matrices in the complex $\mathbf{T}(\mathcal{F})$ determine the graded S-modules

$$\mathrm{H}_*^j \mathcal{F} := \oplus_{i \in \mathbb{Z}} \mathrm{H}^j \mathcal{F}(i) .$$

In many cases this is the fastest known method for computing cohomology.

Section 2 of this paper is devoted to a sketch of the Eisenbud-Fløystad-Schreyer approach to the Bernstein-Gel'fand-Gel'fand correspondence, and the computation of cohomology, together with *Macaulay 2* programs that carry it out, is explained in Section 3.

The remainder of this paper is devoted to an explanation of the Beilinson monad, how to compute it in *Macaulay 2*, and what it is good for. This technique has played an important role in the construction and study of vector bundles and varieties. In the typical application one constructs or classifies monads in order to construct or classify sheaves.

The BGG correspondence and Beilinson's monad were originally formulated in the language of derived categories, and the proofs were rather complicated. The ideas of Eisenbud-Fløystad-Schreyer exposed above allow, for the first time, an explanation of these matters on a level that can be understood by an advanced undergraduate.

The Beilinson monad is similar in spirit to the technique of free resolutions. That theory essentially describes arbitrary sheaves by comparing them with direct sums of line bundles. In the Beilinson technique, one uses a different set of "elementary" sheaves, direct sums of exterior powers of the tautological sub-bundle. Beilinson's remarkable observation was that this comparison has a much more direct connection with cohomology than does the free resolution method.

Sections 4 and 5 are introductory in nature. In Section 4 we begin with a preparatory discussion of the necessary vector bundles on projective space and their cohomology. In Section 5 we define monads, a generalization of resolutions. We give a completely elementary account which constructs the Beilinson monad in a very special case, following ideas of Horrocks, and we use this to sketch part of one of the first striking applications of monads: the classification of stable rank 2 vector bundles on the projective plane by Barth, Hulek and Le Potier.

In Section 6 we give the construction of Eisenbud-Fløystad-Schreyer for the Beilinson monad in general. This is quite suitable for computation, and we give *Macaulay 2* code that does this job.

A natural question for the student at this point is: "Why should I bother learning Beilinson's theorem, what is it good for?" In section 7, we describe two more explicit applications of the theory developed. In the first, the classification of elliptic conic bundles in $\mathbf{P}^4$, computer algebra played a significant role, demonstrating that several published papers contained serious mistakes by constructing an example they had excluded! Using the routines developed earlier in the chapter we give a simpler account of the crucial computation.

In the second application, the construction of abelian surfaces in $\mathbf{P}^4$ and the related Horrocks-Mumford bundles, computer algebra allows one to greatly shorten some of the original arguments made. As the reader will see, everything follows easily with computation, once a certain 2×5 matrix of exterior monomials, given by Horrocks and Mumford, has been written down. One might compare the computations here with the original paper of Horrocks and Mumford [25] (for the cohomology) and the papers by Manolache [32] and Decker [13] (for the syzygies) of the Horrocks-Mumford bundle. A great deal of effort, using representation theory, was necessary to derive results that can be computed in seconds using the *Macaulay 2* programs here. Much more theoretical effort, however, is needed to derive classification results.

Another application of the construction of the Beilinson complex (in a slightly more general setting) is to compute Chow forms of varieties; see [19].

Perhaps the situation is similar to that in the beginning of the 1980's when it became clear that syzygies could be computed by a machine. Though syzygies had been used theoretically for many years it took quite a while until the practical computation of syzygies lead to applications, too, mostly through the greatly increased ability to study examples.

A good open problem of this sort is to extend and make more precise the very useful criterion given in 4.4: Can the reader find a necessary and sufficient condition to replace the necessary condition for surjectivity given there? How about a criterion for exactness?

2   Basics of the Bernstein-Gel'fand-Gel'fand Correspondence

In this section we describe the basic idea of the BGG correspondence, introduced in [8]. For a more complete treatment along the lines given here, see the first section of [18].

As a simple example of the construction given in Section 1, consider the case $M = S = \mathrm{Sym}_K(W)$. The associated complex, made from the homogeneous components $\mathrm{Sym}_i(W)$ of S, has the form

$$\mathbf{F}(S): \quad E \longrightarrow W \otimes E \longrightarrow \mathrm{Sym}_2(W) \otimes E \longrightarrow \cdots,$$

where we regard $\mathrm{Sym}_i W$ as concentrated in degree i. It is easy to see that the kernel of the first map, $E \longrightarrow W \otimes E$, is exactly the socle $\bigwedge^{n+1} V \subset E$, which is a 1-dimensional vector space concentrated in degree $-n - 1$. In fact $\mathbf{F}(S)$ is the minimal injective resolution of this vector space. If we tensor with the dual vector space $\bigwedge^{n+1} W$ (which is concentrated in degree $n + 1$), we obtain the minimal injective resolution of the vector space $\bigwedge^{n+1} W \otimes \bigwedge^{n+1} V$, which may be identified canonically with the residue field K of E. This resolution is called the *Cartan resolution* of K. To write it conveniently, we set $\omega_E = \bigwedge^{n+1} W \otimes E$. The socle of ω_E is K. Since E is injective (as well as projective) as an E-module, the same goes for ω_E, so ω_E is the injective envelope of the residue class field K and we have $\omega_E = \mathrm{Hom}_K(E, K)$. Thus we can write the injective resolution of the residue field as

$$\mathbf{R}(S): \quad \omega_E \longrightarrow W \otimes \omega_E \longrightarrow \mathrm{Sym}_2(W) \otimes \omega_E \longrightarrow \cdots,$$

or again as

$$\mathrm{Hom}_K(E, K) \longrightarrow \mathrm{Hom}_K(E, W) \longrightarrow \mathrm{Hom}_K(E, \mathrm{Sym}_2(W)) \longrightarrow \cdots.$$

Taking our cue from this situation, our primary object of study in the case of an arbitrary finitely generated graded S-module $M = \oplus M_i$ will be the complex

$$\mathbf{R}(M): \quad \cdots \longrightarrow M_i \otimes \omega_E \longrightarrow M_{i+1} \otimes \omega_E \longrightarrow \cdots,$$

which will have a more natural grading than $\mathbf{F}(M)$; in any case, it differs from $\mathbf{F}(M)$ only by tensoring over K with the one-dimensional K-vector space $\bigwedge^{n+1} W$, concentrated in degree $n+1$, and thus has the same basic properties. (Writing $\mathbf{R}(M)$ in terms of Hom as above suggests that the functor $\mathbf{R}$ might have a left adjoint, and indeed there is a left adjoint that produces linear free complexes over S from graded E-modules. $\mathbf{R}$ and its left adjoint are used to construct the isomorphisms of derived categories in the BGG correspondence; see [18] for a treatment in this spirit.)

An important fact for us is that the complex $\mathbf{R}(M)$ is eventually exact (and thus

$$ F^i \xrightarrow{\ \phi_i\ } F^{i+1} \longrightarrow \ \cdots $$

is the minimal injective resolution of $\ker \phi_i$ when $i \gg 0$). It turns out that the point at which exactness sets in is a well-known invariant, the Castelnuovo-Mumford regularity of M, whose definition we briefly recall:

If $M = \oplus M_i$ is a finitely generated graded S-module then for all large integers r the submodule $M_{\geq r} \subset M$ is generated in degree r and has a *linear free resolution*; that is, its first syzygies are generated in degree $r+1$, its second syzygies in degree $r+2$, etc. (see [17, chapter 20]). The *Castelnuovo-Mumford regularity* of M is the least integer r for which this occurs.

Theorem 2.1 ([18]). *Let M be a finitely generated graded S-module of Castelnuovo-Mumford regularity r. The complex $\mathbf{R}(M)$ is exact at $\mathrm{Hom}_K(E, M_i)$ for all $i \geq s$ if and only if $s > r$.* $\square$

More generally, it is shown in [18] that the components of the cohomology of the complex $\mathbf{R}(M)$ can be identified with the Koszul cohomology of M. An equivalent result was stated in [10].

For instance, it is not hard to show that if M is of finite length, then the regularity of M is the largest i such that $M_i \neq 0$. Let us verify Theorem 2.1 directly in a simple example:

Example 2.2. Let $S = K[x_0, x_1, x_2]$, and let $M = S/(x_0^2, x_1^2, x_2^2)$. The module $M_{\geq 3} = K \cdot x_0 x_1 x_2$ is a trivial S-module, and its resolution is the Koszul complex on x_0, x_1 and x_2, which is linear. Thus the Castelnuovo-Mumford regularity of M is ≤ 3. On the other hand $M_{\geq 2}$ is, up to twist, isomorphic to the dual of $S/(x_0, x_1, x_2)^2$, and it follows that the resolution of $M_{\geq 2}$ has the form

$$ 0 \longrightarrow S(-6) \longrightarrow 6S(-4) \longrightarrow 8S(-3) \longrightarrow 3S(-2), $$

which is not linear, so the Castelnuovo-Mumford regularity of M is exactly 3. Note that the regularity is larger than the degrees of the generators and relations of M—in general it can be much larger.

Over E the linear free complex corresponding to M has the form

$$ \cdots \to 0 \to M_0 \otimes \omega_E \to M_1 \otimes \omega_E \to M_2 \otimes \omega_E \to M_3 \otimes \omega_E \to 0 \to \cdots, $$

where all the terms not shown are 0. Using the isomorphism $\omega_E \cong E(-3)$ this can be written (non-canonically) as

$$0 \longrightarrow E(-3) \xrightarrow{\begin{pmatrix} e_0 \\ e_1 \\ e_2 \end{pmatrix}} 3E(-2) \xrightarrow{\begin{pmatrix} 0 & e_2 & e_1 \\ e_2 & 0 & e_0 \\ e_1 & e_0 & 0 \end{pmatrix}} 3E(-1) \xrightarrow{\begin{pmatrix} e_0 & e_1 & e_2 \end{pmatrix}} E \longrightarrow 0.$$

One checks easily that this complex is inexact at every non-zero term (despite its resemblance to a Koszul complex), verifying Theorem 2.1. □

Another case in which everything can be checked directly occurs when M is the homogeneous coordinate ring of a point:

Example 2.3. Take $M = S/I$ where I is generated by a codimension 1 space of linear forms in W, so that I is the homogeneous ideal of a point $p \in \mathbf{P}(W)$. The free resolution of M is the Koszul complex on n linear forms, so M is 0-regular. As M_i is 1-dimensional for every i the terms of the complex $\mathbf{R}(M)$ are all rank 1 free E-modules. One easily checks that $\mathbf{R}(M)$ takes the form

$$\mathbf{R}(M): \quad \omega_E \xrightarrow{a} \omega_E(-1) \xrightarrow{a} \omega_E(-2) \xrightarrow{a} \cdots,$$

where $a \in V = W^*$ is a linear functional that vanishes on all the linear forms in I; that is, a is a generator of the one-dimensional subspace of V corresponding to the point p. As for any linear form in E, the annihilator of a is generated by a, and it follows directly that the complex $\mathbf{R}(M)$ is acyclic in this case. □

We present two *Macaulay 2* functions, symExt and bgg, which compute a differential of the complex $\mathbf{R}(M)$ for a finitely generated graded module M defined over some polynomial ring $S = K[x_0, \ldots, x_n]$ with variables x_i of degree 1. Both functions expect as an additional input the name of an exterior algebra E with the same number $n + 1$ of generators, also supposed to be of degree 1 (and NOT -1). This convention, which makes the cohomology diagrams more naturally looking when printed in *Macaulay 2*, necessitates the adjustment of degrees in the second half of the programs.

The first of the functions, symExt, takes as input a matrix m with linear entries, which we think of as a presentation matrix for a positively graded S-module $M = \oplus_{i \geq 0} M_i$, and returns a matrix representing the map $M_0 \otimes \omega_E \to M_1 \otimes \omega_E$ which is the first differential of the complex $\mathbf{R}(M)$.

```
i1 : symExt = (m,E) ->(
          ev := map(E,ring m,vars E);
          mt := transpose jacobian m;
          jn := gens kernel mt;
          q  := vars(ring m)**id_(target m);
          ans:= transpose ev(q*jn);
          --now correct the degrees:
          map(E^{(rank target ans):1}, E^{(rank source ans):0},
              ans));
```

If M is a module whose presentation is not linear in the sense above, we can still apply symExt to a high truncation of M:

```
i2 : S=ZZ/32003[x_0..x_2];

i3 : E=ZZ/32003[e_0..e_2,SkewCommutative=>true];

i4 : M=coker matrix{{x_0^2, x_1^2}};

i5 : m=presentation truncate(regularity M,M);

             4       8
o5 : Matrix S  <--- S

i6 : symExt(m,E)

o6 = {-1} | e_2 e_1 e_0 0   |
     {-1} | 0   e_2 0   e_0 |
     {-1} | 0   0   e_2 e_1 |
     {-1} | 0   0   0   e_2 |

             4       4
o6 : Matrix E  <--- E
```

The function symExt is a quick-and-dirty tool which requires little computation. If it is called on two successive truncations of a module the maps it produces may NOT compose to zero because the choice of bases is not consistent. The second function, bgg, makes the computation in such a way that the bases are consistent, but does more computation to achieve this end. It takes as input an integer i and a finitely generated graded S-module M, and returns the ith map in $\mathbf{R}(M)$, which is an "adjoint" of the multiplication map between M_i and M_{i+1}.

```
i7 : bgg = (i,M,E) ->(
          S :=ring(M);
          numvarsE := rank source vars E;
          ev:=map(E,S,vars E);
          f0:=basis(i,M);
          f1:=basis(i+1,M);
          g :=((vars S)**f0)//f1;
          b:=(ev g)*((transpose vars E)**(ev source f0));
          --correct the degrees (which are otherwise
          --wrong in the transpose)
          map(E^{(rank target b):i+1},E^{(rank source b):i}, b));
```

For instance, in Example 2.2:

```
i8 : M=cokernel matrix{{x_0^2, x_1^2, x_2^2}};

i9 : bgg(1,M,E)

o9 = {-2} | e_1 e_0 0   |
     {-2} | e_2 0   e_0 |
     {-2} | 0   e_2 e_1 |

             3       3
o9 : Matrix E  <--- E
```

3   The Cohomology and the Tate Resolution of a Sheaf

Given a finitely generated graded S-module M we construct a (doubly infinite) E-free complex $\mathbf{T}(M)$ with vanishing homology, called the *Tate resolution* of M, as follows: Let r be the Castelnuovo-Mumford regularity of M. The truncation $\mathbf{T}^{>r}(M)$, the part of $\mathbf{T}(M)$ with cohomological degree $> r$, is $\mathbf{R}(M_{>r})$. We complete this to an exact complex by adjoining a minimal projective resolution of the kernel of $\mathrm{Hom}_K(E, M_{r+1}) \to \mathrm{Hom}_K(E, M_{r+2})$.

If, for example, M has finite length as in Example 2.2, the Tate resolution of M is the complex

$$\cdots \to 0 \to 0 \to 0 \to \cdots .$$

At the opposite extreme, take $M = S$, the free module of rank 1. Since S has regularity 0, it follows that $\mathbf{R}(S)$ is an injective resolution of the residue field K of E. Applying the exact functor $\mathrm{Hom}_K(-, K)$, and using the fact that it carries $\omega_E = \mathrm{Hom}_K(E, K)$ back to E, we see that the Tate resolution $\mathbf{T}(S)$ is the first row of the diagram

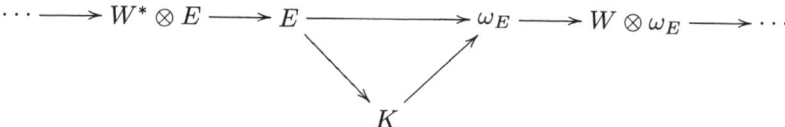

Another simple example occurs in the case where M is the homogeneous coordinate ring of a point $p \in \mathbf{P}(W)$. The complex $\mathbf{R}(M)$ constructed in Example 2.3 is periodic, so it may be simply continued to the left, giving

$$\mathbf{T}(M): \quad \cdots \xrightarrow{\ a\ } \omega_E(i) \xrightarrow{\ a\ } \omega_E(i-1) \xrightarrow{\ a\ } \cdots ,$$

where again $a \in V = W^*$ is a non-zero linear functional vanishing on the linear forms in the ideal of p.

For arbitrary M, by the results of the previous section, $\mathbf{R}(M_{>r})$ has no homology in cohomological degree $> r + 1$, so $\mathbf{T}(M)$ could be constructed by a similar recipe from any truncation $\mathbf{R}(M_{>s})$ with $s \geq r$. Thus the Tate resolution depends only on the sheaf $\tilde{M}$ on $\mathbf{P}(W)$ corresponding to M. We sometimes write $\mathbf{T}(M)$ as $\mathbf{T}(\tilde{M})$ to emphasize this point.

Using the *Macaulay 2* function symExt of the last section, one can compute any finite piece of the Tate resolution.

```
i10 : tateResolution = (m,E,loDeg,hiDeg)->(
          M := coker m;
          reg := regularity M;
          bnd := max(reg+1,hiDeg-1);
          mt  := presentation truncate(bnd,M);
          o   := symExt(mt,E);
          --adjust degrees, since symExt forgets them
          ofixed   := map(E^{(rank target o):bnd+1},
                     E^{(rank source o):bnd},
                     o);
          res(coker ofixed, LengthLimit=>max(1,bnd-loDeg+1)));
```

`tateResolution` takes as input a presentation matrix m of a finitely generated graded module M defined over some polynomial ring $S = K[x_0, \ldots, x_n]$ with variables x_i of degree 1, the name of an exterior algebra E with the same number $n + 1$ of generators, also supposed to be of degree 1, and two integers, say l and h. If r is the regularity of M, then `tateResolution(m,E,l,h)` computes the piece

$$\mathbf{T}^l(M) \to \cdots \to \mathbf{T}^{\max(r+2,h)}(M)$$

of $\mathbf{T}(M)$. For instance, for the homogeneous coordinate ring of a point in the projective plane:

```
i11 : m = matrix{{x_0,x_1}};

            1       2
o11 : Matrix S  <--- S

i12 : regularity coker m

o12 = 0

i13 : T = tateResolution(m,E,-2,4)

        1      1      1      1      1      1      1
o13 = E  <-- E  <-- E  <-- E  <-- E  <-- E  <-- E

        0      1      2      3      4      5      6

o13 : ChainComplex

i14 : betti T

o14 = total: 1 1 1 1 1 1 1
         -4: 1 1 1 1 1 1 1

i15 : T.dd_1

o15 = {-4} | e_2 |

            1       1
o15 : Matrix E  <--- E
```

For arbitrary M we have $M_i = \mathrm{H}^0 \tilde{M}(i)$ for large i, so the corresponding term of the complex $\mathbf{T}(\tilde{M})$ with cohomological degree i is $M_i \otimes \omega_E = \mathrm{H}^0(\tilde{M}(i)) \otimes \omega_E$. The following result generalizes this to a description of all the terms of the Tate resolution, and gives the formula for the cohomology described in the introduction.

Theorem 3.1 ([18]). *Let M be a finitely generated graded S-module. The term of the complex $\mathbf{T}(M) = \mathbf{T}(\tilde{M})$ with cohomological degree i is*

$$\oplus_j \mathrm{H}^j \tilde{M}(i - j) \otimes \omega_E ,$$

where $\mathrm{H}^j \tilde{M}(i - j)$ is regarded as a vector space concentrated in degree $i - j$, so that the summand $\mathrm{H}^j \tilde{M}(i - j) \otimes \omega_E$ is isomorphic to a direct sum of copies *of* $\omega_E(j-i)$. *Moreover the subquotient complex*

$$\cdots \to H^j \tilde{M}(i-j) \otimes \omega_E \to H^j \tilde{M}(i+1-j) \otimes \omega_E \to \cdots$$

is $\mathbf{R}(H^j_*(\tilde{M}(-j)))(j)$ *(up to twists and shifts it is* $\mathbf{R}(H^j_* \tilde{M})$. *)*   □

Thus each cohomology group of each twist of the sheaf $\tilde{M}$ occurs (exactly once) in a term of $\mathbf{T}(M)$. When we compute a part of $\mathbf{T}(M)$, we are computing the sheaf cohomology of various twists of the associated sheaf together with maps which describe the S-module structure of $H^j_* \tilde{M}$ in the sense that the linear maps in this complex are adjoints of the multiplication maps that determine the module structure (the multiplication maps themselves could be computed by a function similar to bgg). The higher degree maps in the complex $\mathbf{T}(M)$ determine certain higher cohomology operations, which we understand only in very special cases (see [19]).

If $M = \text{coker } m$, then betti tateResolution(m,E,l,h) prints the dimensions $h^j \tilde{M}(i-j) = \dim H^j \tilde{M}(i-j)$ for $\max(r+2, h) \geq i \geq l$, where r is the regularity of M. Truncating the Tate resolution if necessary allows one to restrict the size of the output.

```
i16 : sheafCohomology = (m,E,loDeg,hiDeg)->(
          T := tateResolution(m,E,loDeg,hiDeg);
          k := length T;
          d := k-hiDeg+loDeg;
          if d > 0 then
              chainComplex apply(d+1 .. k, i->T.dd_(i))
          else T);
```

The expression betti sheafCohomology(m,E,l,h) prints a cohomology table for $\tilde{M}$ of the form

$$
\begin{array}{ccc}
h^0 \tilde{M}(h) & \cdots & h^0 \tilde{M}(l) \\
h^1 \tilde{M}(h-1) & \cdots & h^1 \tilde{M}(l-1) \\
\vdots & & \vdots \\
h^n \tilde{M}(h-n) & \cdots & h^n \tilde{M}(l-n) \, .
\end{array}
$$

As a simple example we consider the cotangent bundle on projective 3-space (see the next section for the Koszul resolution of this bundle):

```
i17 : S=ZZ/32003[x_0..x_3];
```

```
i18 : E=ZZ/32003[e_0..e_3,SkewCommutative=>true];
```

The cotangent bundle is the cokernel of the third differential of the Koszul complex on the variables of S.

```
i19 : m=koszul(3,vars S);

                 6       4
o19 : Matrix S  <--- S
```

```
i20 : regularity coker m

o20 = 2

i21 : betti tateResolution(m,E,-6,2)

o21 = total: 45 20 6 1 4 15 36 70 120 189 280
         -4: 45 20 6 .  .  .  .  .   .   .   .
         -3:  .  . . 1  .  .  .  .   .   .   .
         -2:  .  . . .  .  .  .  .   .   .   .
         -1:  .  . . . 4 15 36 70 120 189 280

i22 : betti sheafCohomology(m,E,-6,2)

o22 = total: 6 1 4 15 36 70 120 189 280
         -2: 6 . .  .  .  .   .   .   .
         -1: . 1 .  .  .  .   .   .   .
          0: . . .  .  .  .   .   .   .
          1: . . 4 15 36 70 120 189 280
```

Of course these two results differ only in the precise point of truncation.

Remark 3.2. There is also a built-in sheaf cohomology function HH in *Macaulay 2* which is based on the algorithms in [16]. These algorithms are often much slower than sheafCohomology. To access it, first execute

    M=sheaf coker m;

and pick integers j and d. Then

    HH^j(M(>=d))

returns the truncated j^{th} cohomology module $\mathrm{H}^j_{i \geq d}\tilde{M}$. In the above example of the cotangent bundle $\mathcal{F}$ on projective 3-space we obtain the Koszul presentation of $H^1\mathcal{F} \cong K$ considered as an S-module sitting in degree 0:

```
i23 : M=sheaf coker m;

i24 : HH^1(M(>=0))

o24 = cokernel | x_3 x_2 x_1 x_0 |

                        1
o24 : S-module, quotient of S
```

□

The Tate resolutions of sheaves are, as the reader may easily check, precisely the doubly infinite, graded, exact complexes of finitely-generated free E-modules which are "eventually linear" on the right, in an obvious sense. What about other doubly exact graded free complexes? For example what if we take the dual of the Tate resolution of a sheaf? In general it will not be eventually linear. What is it?

To explain this we must generalize the construction of $\mathbf{R}(M)$: If

$$M^\bullet : \quad \cdots \longrightarrow M^{i+1} \longrightarrow M^i \longrightarrow M^{i-1} \longrightarrow \cdots$$

is a complex of S-modules, then applying the functor $\mathbf{R}$ gives a complex of free complexes over E. By changing some signs we get a double complex. In general the associated total complex is not minimal; but at least if $M^\bullet$ is a bounded complex then, just as one produces the unique minimal free resolution of a module from any free resolution, we can construct a unique minimal complex from it. We call this minimal complex $\mathbf{R}(M^\bullet)$. (See [18] for more information. This construction is a necessary part of interpreting the BGG correspondence as an equivalence of derived categories.)

Again if $M^\bullet$ is a bounded complex of finitely generated modules, then as before one shows that $\mathbf{R}(M^\bullet)$ is exact from a certain point on, and so we can form the Tate resolution $\mathbf{T}(M^\bullet)$ by adjoining a free resolution of a kernel. Once again, the Tate resolution depends only on the bounded complex of coherent sheaves $\mathcal{F}^\bullet$ associated to $M^\bullet$, and we write $\mathbf{T}(\mathcal{F}^\bullet) = \mathbf{T}(M^\bullet)$.

A variant of the theorem of Bernstein, Gel'fand and Gel'fand shows that every minimal graded doubly infinite exact sequence of finitely generated free E-modules is of the form $\mathbf{T}(\mathcal{F}^\bullet)$ for some complex of coherent sheaves $\mathcal{F}^\bullet$, unique up to quasi-isomorphism. The terms of the Tate resolution can be expressed using hypercohomology by a formula like that of Theorem 3.1.

One way that interesting complexes of sheaves arise is through duality. For simplicity, write $\mathcal{O}$ for the structure sheaf $\mathcal{O}_{\mathbf{P}(W)}$. If $\mathcal{F} = \tilde{M}$ is a sheaf on $\mathbf{P}(W)$ then the derived functor $RHom(\mathcal{F}, \mathcal{O})$ may be computed by applying the functor $Hom(-, \mathcal{O})$ to a sheafified free resolution of M; it's value is thus a complex of sheaves rather than an individual sheaf.

We can now identify the dual of the Tate resolution:

Theorem 3.3. $\mathrm{Hom}_K(\mathbf{T}(\mathcal{F}), K) \cong \mathbf{T}(RHom(\mathcal{F}, \mathcal{O}))[1]$. $\qquad\qquad \square$

Here the [1] denotes a shift by one in cohomological degree. For example, take $\mathcal{F} = \mathcal{O}$. We have $RHom(\mathcal{O}, \mathcal{O}) = \mathcal{O}$. The Tate resolution is given by

$$\mathbf{T}(\mathcal{O}): \quad \cdots \longrightarrow \underset{-1}{E} \longrightarrow \underset{0}{\omega_E} \longrightarrow \cdots$$

where the number under each term is its cohomological degree. Taking into account $\omega_E = \mathrm{Hom}_K(E, K)$, the dual of the Tate resolution is thus

$$\mathrm{Hom}_K(\mathbf{T}(\mathcal{O}), K): \quad \cdots \longleftarrow \underset{1}{\omega_E} \longleftarrow \underset{0}{E} \longleftarrow \cdots$$

which is the same as $\mathbf{T}(\mathcal{O})[1]$. A completely analogous computation gives the proof of Theorem 3.3 if $\mathcal{F} = \mathcal{O}(a)$ for some a, and the general case follows by taking free resolutions.

4   Cohomology and Vector Bundles

In this section we first recall how vector bundles, direct sums of line bundles, and bundles of differentials can be characterized among all coherent sheaves on $\mathbf{P}(W)$ in terms of cohomology (as usual we do not distinguish between vector bundles and locally free sheaves). Then we describe the homomorphisms between the suitably twisted bundles of differentials in terms of the exterior algebra E. This description plays an important role in the context of Beilinson monads.

Vector bundles on $\mathbf{P}(W)$ are characterized by a criterion of Serre [39] which can be formulated as follows: A coherent sheaf $\mathcal{F}$ on $\mathbf{P}(W)$ is locally free if and only if its module of sections $\mathrm{H}^0_*(\mathcal{F})$ is finitely generated and its *intermediate cohomology modules* $\mathrm{H}^j_*\mathcal{F}$, $1 \leq j \leq n-1$, are of finite length.

From a cohomological point of view, the simplest vector bundles are the direct sums of line bundles. Every vector bundle on the projective line splits into a direct sum of line bundles by Grothendieck's splitting theorem (see [37]). Induction yields Horrocks' splitting theorem (see [5]): A vector bundle on $\mathbf{P}(W)$ splits into a direct sum of line bundles if and only if its intermediate cohomology vanishes (originally, this theorem was proved as a corollary to a more general result, see [23] and [42]).

Just a little bit more complicated are the bundles of differentials. To fix our notation in this context we write $\mathcal{O} = \mathcal{O}_{\mathbf{P}(W)}$, $W \otimes \mathcal{O}$ for the trivial bundle on $\mathbf{P}(W)$ with fiber W, $U = \Omega_{\mathbf{P}(W)}(1)$ for the cotangent bundle twisted by 1, and

$$U^i = \bigwedge^i U = \bigwedge^i (\Omega_{\mathbf{P}(W)}(1)) = \Omega^i_{\mathbf{P}(W)}(i)$$

for the ith bundle of differentials twisted by i; in particular $U^0 = \mathcal{O}$, $U^n \cong \mathcal{O}(-1)$, and $U^i = 0$ if $i < 0$ or $i > n$.

Remark 4.1. For each $0 \leq i \leq n$ the pairing

$$U^i \otimes U^{n-i} \xrightarrow{\wedge} U^n \cong \mathcal{O}(-1)$$

induces an isomorphism

$$U^{n-i} \cong (U^i)^*(-1) . \qquad \square$$

The fiber of U at the point of $\mathbf{P}(W)$ corresponding to the line $\langle a \rangle \subset V$ is the subspace $(V/\langle a \rangle)^* \subset W$. Thus U fits into the short exact sequence

$$0 \to U \to W \otimes \mathcal{O} \to \mathcal{O}(1) \to 0 .$$

In fact, U is the *tautological subbundle* of $W \otimes \mathcal{O}$. Taking exterior powers, we get the short exact sequences

$$0 \to U^{i+1} \to \bigwedge^{i+1} W \otimes \mathcal{O} \to U^i \otimes \mathcal{O}(1) \to 0 .$$

Twisting the ith sequence by $-i - 1$, and gluing them together we get the exact sequence

$$0 \longrightarrow \bigwedge^{n+1} W \otimes \mathcal{O}(-n-1) \longrightarrow \cdots \longrightarrow \bigwedge^0 W \otimes \mathcal{O} \longrightarrow 0 .$$

This sequence is the sheafification of the Koszul complex, which is the free resolution of the "trivial" graded S-module $K = S/(W)$.

Remark 4.2. By taking cohomology in the short exact sequences above we find that

$$H_*^j U^i = \begin{cases} K(i) & j = i, \\ 0 & j \neq i, \end{cases} \quad 1 \leq i, j \leq n - 1,$$

where $K(i) = (S/(W))(i)$. Conversely, every vector bundle $\mathcal{F}$ on $\mathbf{P}(W)$ with this intermediate cohomology is *stably equivalent* to U^i; that is, there exists a direct sum $\mathcal{L}$ of line bundles such that $\mathcal{F} \cong U^i \oplus \mathcal{L}$. This follows by comparing the sheafified Koszul complex with the minimal free resolution of the dual bundle $\mathcal{F}^*$. $\qquad \square$

In what follows we describe the homomorphisms between the various U^i, $0 \leq i \leq n$. Note that since $U = U^1 \subset W \otimes \mathcal{O}$ each element of $V = \mathrm{Hom}_K(W, K)$ induces a homomorphism $U^1 \to U^0$ which is the composite

$$U^1 \subset W \otimes \mathcal{O} \to K \otimes \mathcal{O} = \mathcal{O} = U^0.$$

Similarly, using the diagonal map of the exterior algebra $U^i = \bigwedge^i U \to U \otimes U^{i-1}$, each element of V induces a homomorphism $U^i \to U^{i-1}$ which is the composite

$$U^i \to U \otimes U^{i-1} \to W \otimes U^{i-1} \to K \otimes U^{i-1} = U^{i-1}.$$

It is not hard to show that these maps induced by elements of V anticommute with each other (see for example [17, A2.4.1]). Thus we get maps $\bigwedge^j V \to \mathrm{Hom}(U^i, U^{i-j})$ which together give a graded ring homomorphism $\bigwedge V \to \mathrm{Hom}(\oplus_i U^i, \oplus_i U^i)$. In fact this construction gives all the homomorphisms between the U^i:

Lemma 4.3. *The maps*

$$\bigwedge^j V \to \mathrm{Hom}(U^i, U^{i-j}), \quad 0 \leq i, i - j \leq n,$$

described above are isomorphisms. Under these isomorphisms an element $e \in \bigwedge^j V$ acts by contraction on the fibers of the U^i:

$$\begin{array}{ccc} \bigwedge^i (V/\langle a \rangle)^* & \lhook\joinrel\longrightarrow & \bigwedge^i W \\ \downarrow & & \downarrow {\scriptstyle e} \\ \bigwedge^{i-j} (V/\langle a \rangle)^* & \lhook\joinrel\longrightarrow & \bigwedge^{i-j} W \ . \end{array}$$

Proof. Every homomorphism $U^i \to U^{i-j}$ lifts uniquely to a homomorphism between shifted Koszul complexes:

$$\begin{array}{ccccccccc} 0 \longrightarrow \bigwedge^{n+1} W \otimes \mathcal{O}(i - n - 1) \longrightarrow & \cdots & \longrightarrow \bigwedge^j W \otimes \mathcal{O}(i - j) \longrightarrow & \cdots \\ \downarrow & & \downarrow & \\ \cdots \longrightarrow \bigwedge^{n+1-j} W \otimes \mathcal{O}(i - n - 1) \longrightarrow & \cdots & \longrightarrow \mathcal{O}(i - j) \longrightarrow 0 \end{array}$$

Indeed, the corresponding obstructions vanish by Remarks 4.1 and 4.2. All results follow since the vertical arrows are necessarily given by contraction with an element in

$$\operatorname{Hom}(\textstyle\bigwedge^j W \otimes \mathcal{O}(i-j), \mathcal{O}(i-j)) \cong \textstyle\bigwedge^j V. \quad \square$$

In practical terms, these results say that a map $U^i \xrightarrow{\ e\ } U^{i-j}$ is represented as

$$
\begin{array}{ccc}
\bigwedge^{i+1} W \otimes \mathcal{O}(-1) & \longrightarrow & U^i \\
\downarrow{\scriptstyle e} & & \downarrow \\
\bigwedge^{i-j+1} W \otimes \mathcal{O}(-1) & \longrightarrow\!\!\!\rightarrow & U^{i-j}
\end{array}
$$

if $0 < i - j \le i \le n$, and as the composite

$$
\begin{array}{ccccc}
\bigwedge^{i+1} W \otimes \mathcal{O}(-1) & \longrightarrow\!\!\!\rightarrow & U^i & \lhook\joinrel\longrightarrow & \bigwedge^i W \otimes \mathcal{O} \\
& & \downarrow & & \downarrow{\scriptstyle e} \\
& & U^0 & = & \mathcal{O}
\end{array}
$$

if $0 = i - j < i \le n$.

A map from a sum of copies of various U^i to another such sum is given by a homogeneous matrix over the exterior algebra E. In general it is an interesting problem to relate properties of the matrix to properties of the map. Here is one relation which is easy. We will apply it later on in this chapter.

Proposition 4.4. *If*

$$r\,U^i \xrightarrow{\ B\ } s\,U^{i-1}$$

is a homomorphism, that is, if B is an $s \times r$-matrix with entries in V, then the following condition is necessary for B to be surjective: If $(b_1, \dots, b_r)$ is a non-trivial linear combination of the rows of B, then

$$\dim \operatorname{span}(b_1, \dots, b_r) \ge i + 1.$$

Proof. B is surjective if and only if its dual map is injective on fibers:

$$s \textstyle\bigwedge^{i-1}(V/\langle a\rangle) \xrightarrow{\ \wedge B^t\ } r \textstyle\bigwedge^i(V/\langle a\rangle)$$

is injective for any line $\langle a\rangle \subset V$. Consider a non-trivial linear combination $(b_1, \dots, b_r)^t$ of the columns of B^t, and write $d = \dim \operatorname{span}(b_1, \dots, b_r)$. If $d = i$, then B^t is not injective at any point of $\mathbf{P}(W)$ corresponding to a vector in $\operatorname{span}(b_1, \dots, b_r)$. If $d < i$, then B^t is not injective at any point of $\mathbf{P}(W)$. $\quad \square$

5 Cohomology and Monads

The technique of monads provides powerful tools for problems such as the construction and classification of coherent sheaves with prescribed invariants. This section is an introduction to monads. We demonstrate their usefulness, which is not obvious at first glance, by reviewing the classification of stable rank 2 vector bundles on the projective plane (see [4], [31], and [26]). Recall that stable bundles admit moduli (see [22], [33], and [34]).

The basic idea behind monads is to represent arbitrary coherent sheaves in terms of simpler sheaves such as line bundles or bundles of differentials, and in terms of homomorphisms between these simpler sheaves. If M is a finitely generated graded S-module, with associated sheaf $\mathcal{F} = \tilde{M}$, then the sheafification of the minimal free resolution of M is a monad for $\mathcal{F}$ which involves direct sums of line bundles and thus homogeneous matrices over S. The Beilinson monad for $\mathcal{F}$, which will be considered in the next section, involves direct sums of twisted bundles of differentials U^i, and thus homogeneous matrices over E.

Definition 5.1. A *monad* on $\mathbf{P}(W)$ is a bounded complex

$$\cdots \longrightarrow \mathcal{K}^{-1} \longrightarrow \mathcal{K}^0 \longrightarrow \mathcal{K}^1 \longrightarrow \cdots$$

of coherent sheaves on $\mathbf{P}(W)$ which is exact except at $\mathcal{K}^0$. The homology $\mathcal{F}$ at $\mathcal{K}^0$ is called the *homology of the monad*, and the monad is said to be a monad for $\mathcal{F}$. We say that the *type of a monad* is determined if the sheaves $\mathcal{K}^i$ are determined. □

There are different ways of representing a given sheaf as the homology of a monad, and the type of the monad depends on the way chosen.

When constructing or classifying sheaves in a given class via monads, one typically proceeds along the following lines.

Step 1. Compute cohomological information which determines the type of the corresponding monads.

Step 2. Construct or classify the differentials of the monads.

There are no general recipes for either step and some cases require sophisticated ideas and quite a bit of intuition (see Example 7.2 below). If one wants to classify, say, vector bundles, then a third step is needed:

Step 3. Determine which monads lead to isomorphic vector bundles.

One of the first successful applications of this approach was the classification of (Gieseker-)stable rank 2 vector bundles with even first Chern class $c_1 \in \mathbf{Z}$ on the complex projective plane by Barth [4], who detected geometric properties of the corresponding moduli spaces without giving an explicit description of the differentials in the second step. The same ideas apply in the case c_1 odd which we are going to survey in what follows (see [31], [26], and [37] for full details and proofs).

In general, rank 2 vector bundles enjoy properties which are not shared by all vector bundles.

Remark 5.2. Every rank 2 vector bundle $\mathcal{F}$ on $\mathbf{P}(W)$ is *self-dual*, that is, it admits a symplectic structure. Indeed, the map

$$\mathcal{F} \otimes \mathcal{F} \overset{\wedge}{\longrightarrow} \wedge^2 \mathcal{F} \cong \mathcal{O}_{\mathbf{P}(W)}(c_1)$$

induces an isomorphism $\varphi : \mathcal{F} \overset{\cong}{\to} \mathcal{F}^*(c_1)$ with $\varphi = -\varphi^*(c_1)$ (here c_1 is the first Chern class of $\mathcal{F}$). In particular there are isomorphisms

$$(\mathrm{H}^j \mathcal{F}(i))^* \cong \mathrm{H}^{n-j} \mathcal{F}(-i - n - 1 - c_1)$$

by Serre duality.   □

We will not give a general definition of stability here. For rank 2 vector bundles stability can be characterized as follows (see [37]).

Remark 5.3. If $\mathcal{F}$ is a rank 2 vector bundle on $\mathbf{P}(W)$, then the following hold:

(1) $\mathcal{F}$ is stable if and only if $\mathrm{Hom}(\mathcal{F}, \mathcal{F}) \cong K$. In this case the symplectic structure on $\mathcal{F}$ is uniquely determined up to scalars.

(2)   By tensoring with a line bundle we can *normalize* $\mathcal{F}$ so that its first Chern class is 0 or -1. In this case $\mathcal{F}$ is stable if and only if it has no global sections.   □

Example 5.4. By the results of the previous section the twisted cotangent bundle U on the projective plane is a stable rank 2 vector bundle with Chern classes $c_1 = -1$ and $c_2 = 1$.   □

Remark 5.5. The generalized theorem of Riemann-Roch yields a polynomial in $\mathbb{Q}[c_1, \ldots, c_r]$ which gives the Euler characteristic $\chi \mathcal{F} = \sum_j (-1)^j \mathrm{h}^j \mathcal{F}$ for every rank r vector bundle $\mathcal{F}$ on $\mathbf{P}(W)$ with Chern classes $c_1, \ldots, c_r$. This polynomial can be determined by interpreting the generalized theorem of Riemann-Roch or by computing the Euler characteristic for enough special bundles of rank r (like direct sums of line bundles). For a rank 2 vector bundle on the projective plane, for example, one obtains

$$\chi(\mathcal{F}) = (c_1^2 - 2c_2 + 3c_1 + 4)/2. \quad □$$

We now focus on stable rank 2 vector bundles on the complex projective plane $\mathbf{P}^2(\mathbf{C}) = \mathbf{P}(W)$ with first Chern class $c_1 = -1$. Let $\mathcal{F}$ be such a bundle.

Remark 5.6. Since $\mathcal{F}$ is stable and normalized its second Chern class c_2 must be ≥ 1. Indeed,

$$\mathrm{H}^2 \mathcal{F}(i - 2) = \mathrm{H}^0 \mathcal{F}(-i) = 0 \quad \text{for} \quad i \geq 0$$

by Remarks 5.2 and 5.3, and $\chi(\mathcal{F}(i)) = (i+1)^2 - c_2$ by Riemann-Roch. In particular the dimensions $h^j \mathcal{F}(i)$ in the range $-2 \leq i \leq 0$ are as in the following cohomology table (a zero is represented by an empty box):

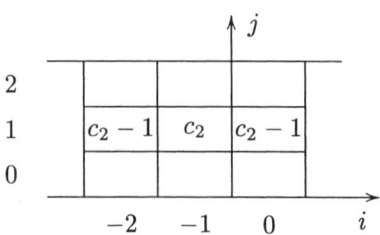

We abbreviate $\mathcal{O} = \mathcal{O}_{\mathbf{P}^2(\mathbf{C})}$ and go through the three steps above.

Step 1. In this step we show that $\mathcal{F}$ is the homology of a monad of type

$$0 \to \mathrm{H}^1\mathcal{F}(-2) \otimes U^2 \to \mathrm{H}^1\mathcal{F}(-1) \otimes U \to \mathrm{H}^1\mathcal{F} \otimes \mathcal{O} \to 0 ,$$

where the middle term occurs in cohomological degree 0. This actually follows from the general construction of Beilinson monads presented in the next chapter and the fact that $\mathrm{H}^2\mathcal{F}(i-2) = \mathrm{H}^0\mathcal{F}(-i) = 0$ for $2 \geq i \geq 0$ (see Remark 5.6). Here we derive the existence of the monad directly with Horrocks' technique of killing cohomology [24], which requires further cohomological information. Such information is typically obtained by restricting the given bundles to linear subspaces. In our case we consider the Koszul complex on the equations of a point $p \in \mathbf{P}^2(\mathbf{C})$:

$$0 \longrightarrow \mathcal{O}(-2) \xrightarrow{\binom{-x'}{x}} 2\mathcal{O}(-1) \xrightarrow{(x\ x')} \mathcal{O} \longrightarrow \mathcal{O}_p \longrightarrow 0 .$$

By tensoring with $\mathcal{F}(i+1)$ and taking cohomology we find that $\mathrm{H}^1\mathcal{F}$ generates $\mathrm{H}^1_{\geq 0}\mathcal{F}$. Indeed, the composite map

$$(x\ x') : 2\mathrm{H}^1\mathcal{F}(i) \longrightarrow \mathrm{H}^1(\mathcal{J}_p \otimes \mathcal{F}(i+1)) \longrightarrow \mathrm{H}^1\mathcal{F}(i+1)$$

is surjective if $i \geq -1$. In particular, if $c_2 = 1$, then $\mathrm{H}^1\mathcal{F}(i) = 0$ for $i \neq -1$ (apply Serre duality for the twists ≤ -2), so $\mathcal{F} \cong U$ is the twisted cotangent bundle by Remark 4.2 since both bundles have the same rank and intermediate cohomology.

If $c_2 \geq 2$ then $\mathrm{H}^1\mathcal{F} \neq 0$, and the identity in

$$\mathrm{Hom}(\mathrm{H}^1\mathcal{F}, \mathrm{H}^1\mathcal{F}) \cong \mathrm{Ext}^1(\mathrm{H}^1\mathcal{F} \otimes \mathcal{O}, \mathcal{F})$$

defines an extension

$$0 \to \mathcal{F} \to \mathcal{G} \to \mathrm{H}^1\mathcal{F} \otimes \mathcal{O} \to 0 ,$$

where $H^1_{\geq 0}\mathcal{G} = 0$, and where $\mathcal{G}$ is a vector bundle (apply Serre's criterion in Section 4). Similarly, by taking Serre duality into account, we obtain an extension

$$0 \to H^1\mathcal{F}(-2) \otimes U^2 \to \mathcal{H} \to \mathcal{F} \to 0 \ ,$$

where $\mathcal{H}$ is a vector bundle with $H^1_{\leq -2}\mathcal{H} = 0$. The two extensions fit into a commutative diagram with exact rows and and columns

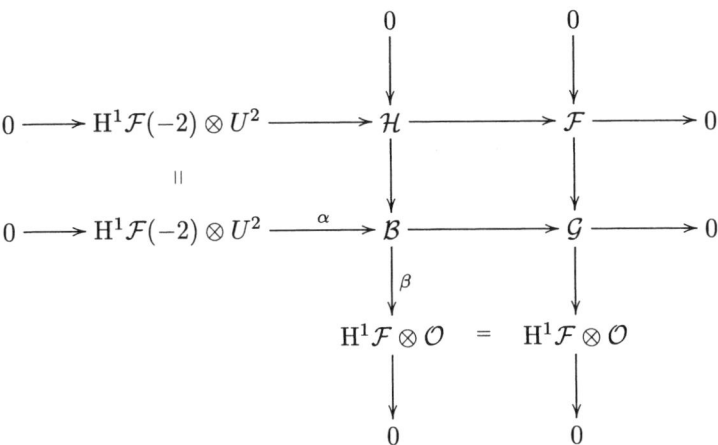

since, for example, the extension in the top row lifts uniquely to an extension as in the middle row (the obstructions in the corresponding Ext-sequence vanish). Then $\mathcal{B} \cong H^1\mathcal{F}(-1) \otimes U$ since by construction these bundles have the same rank and intermediate cohomology. What we have is the *display* of (the short exact sequences associated to) a monad

$$0 \longrightarrow H^1\mathcal{F}(-2) \otimes U^2 \xrightarrow{\alpha} H^1\mathcal{F}(-1) \otimes U \xrightarrow{\beta} H^1\mathcal{F} \otimes \mathcal{O} \longrightarrow 0$$

for $\mathcal{F}$.

Step 2. Our task in this step is to describe what maps α and β could be the differentials of a monad as above. In fact we give a description in terms of linear algebra for which it is enough to deal with one of the differentials, say α, since the self-duality of $\mathcal{F}$ and the vanishing of certain obstructions allows one to represent $\mathcal{F}$ as the homology of a "self-dual" monad. Let us abbreviate $A = H^1\mathcal{F}(-2)$, $B = H^1\mathcal{F}(-1)$ and $A^* \cong H^1\mathcal{F}$. By chasing the displays of a monad as above and its dual we see that the symplectic structure on $\mathcal{F}$ lifts to a unique isomorphism of monads

$$
\begin{array}{ccccccccc}
0 & \longrightarrow & A \otimes U^2 & \xrightarrow{\ \alpha\ } & B \otimes U & \xrightarrow{\ \beta\ } & A^* \otimes \mathcal{O} & \longrightarrow & 0 \\
& & \downarrow{\scriptstyle \Phi} & & \downarrow{\scriptstyle \Psi} & & \downarrow{\scriptstyle -\Phi^*(-1)} & & \\
0 & \longrightarrow & A \otimes \mathcal{O}(-1) & \xrightarrow{\beta^*(-1)} & B^* \otimes U^*(-1) & \xrightarrow{\alpha^*(-1)} & A^* \otimes (U^2)^*(-1) & \longrightarrow & 0
\end{array}
$$

with $\Psi = -\Psi^*(-1)$. Indeed, the corresponding obstructions vanish (see [5] and [37, II, 4.1] for a discussion of this argument in a general context). Ψ is the tensor product of an isomorphism $q : B \to B^*$ and a symplectic form $\iota \in \operatorname{Hom}(U, U^*(-1)) \cong \mathbf{C}$ on U. Note that q is symmetric since $-(q \otimes \iota) = (q \otimes \iota)^*(-1) = q^* \otimes \iota^*(-1) = -q^* \otimes \iota$. We may and will now assume that $\mathcal{F}$ is the homology of a *self-dual monad*, where self-dual means that $\beta = \alpha^d := \alpha^*(-1) \circ (q \otimes \iota)$. The monad conditions

(α_1) $\alpha^d \circ \alpha = 0$, and

(α_2) α is a vector bundle monomorphism (α^d is an epimorphism)

can be rewritten in terms of linear algebra as follows. The identifications in Lemma 4.3 allow one to view

$$\alpha \in \operatorname{Hom}(A \otimes U^2, B \otimes U) \cong V \otimes \operatorname{Hom}(A, B)$$

as a homomorphism $\alpha : W \to \operatorname{Hom}(A, B)$ operating by $\xi \otimes (x \wedge x') \to \alpha(x)(\xi) \otimes x' - \alpha(x')(\xi) \otimes x$ on the fibers of $A \otimes U^2$. Similarly we consider α^d as the homomorphism $\alpha^d : W \to \operatorname{Hom}(B, A^*), x \mapsto \alpha^*(x) \circ q$, operating by $\eta \otimes x \to \alpha^d(x)(\eta)$ on the fibers of $B \otimes U$. Then

(α_1') $\alpha^d(x) \circ \alpha(x') = \alpha^d(x') \circ \alpha(x)$ for all $x, x' \in W$, and

(α_2') for every $\xi \in A \setminus \{0\}$ the map $W \to B, x \to \alpha(x)(\xi)$ has rank ≥ 2.

Example 5.7. If $c_2 = 2$, then the monads can be written (non-canonically) as

$$0 \longrightarrow U^2 \xrightarrow{\binom{a}{b}} 2U \xrightarrow{(a\ b)} \mathcal{O} \longrightarrow 0 \,,$$

where a, b are two vectors in V. In this case (α_1) gives no extra condition and (α_2) means that a and b are linearly independent. If a and b are explicitly given, then we can compute the homology of the monad with the help of *Macaulay 2*:

```
i25 : S = ZZ/32003[x_0..x_2];
```

U is obtained from the Koszul complex resolving $S/(x_0, x_1, x_2)$ by tensoring the cokernel of the differential $\bigwedge^3 W \otimes S(-3) \to \bigwedge^2 W \otimes S(-2)$ with $S(1)$ (and sheafifying).

```
i26 : U = coker koszul(3,vars S) ** S^{1};
```

For representing α and α^d we also need the differential $\bigwedge^2 W \otimes S(-2) \to W \otimes S(-1)$ of the Koszul complex.

```
i27 : k2 = koszul(2,vars S)

o27 = {1} | -x_1 -x_2 0    |
      {1} | x_0  0    -x_2 |
      {1} | 0    x_0  x_1  |

              3        3
o27 : Matrix S <--- S
```

The expression `koszul(2,vars S)` computes a matrix representing the differential with respect to the monomial bases $x_0 \wedge x_1, x_0 \wedge x_2, x_1 \wedge x_2$ of $\bigwedge^2 W$ and x_0, x_1, x_2 of W. We pick $(a, b) = (e_1, e_2)$ and represent the corresponding maps α and α^d with respect to the monomial bases (see the discussion following Lemma 4.3).

```
i28 : alpha = map(U ++ U, S^{-1}, transpose{{0,-1,0,1,0,0}});

o28 : Matrix
```

```
i29 : alphad = map(S^1, U ++ U, matrix{{0,1,0,0,0,1}} * (k2 ++ k2));

o29 : Matrix
```

Prune computes a minimal presentation.

```
i30 : F = prune homology(alphad, alpha);

i31 : betti  F

o31 = relations : total: 3 1
                      1: 2 .
                      2: 1 1
```

In the next section we will present a more elegant way of computing the homology of Beilinson monads.   □

We go back to the general case and reverse our construction. Let A and B be $\mathbf{C}$-vector spaces of the appropriate dimensions, let q be a non-degenerate quadratic form on B, and let

$$\widetilde{\mathcal{M}} = \{\alpha \in \operatorname{Hom}(W, \operatorname{Hom}(A, B)) \mid \alpha \text{ satisfies } (\alpha_1') \text{ and } (\alpha_2')\} \ .$$

Then every $\alpha \in \widetilde{\mathcal{M}}$ defines a self-dual monad as above whose homology is a stable rank 2 vector bundle on $\mathbf{P}^2(\mathbf{C})$ with Chern classes $c_1 = -1$ and c_2. In this way we obtain a description of the differentials of the monads which is not as explicit as we might have hoped (with the exception of the case $c_2 = 2$). It is, however, enough for detecting geometric properties of the corresponding moduli spaces.

Step 3. Constructing the moduli spaces means to parametrize the isomorphism classes of our bundles in a convenient way. We very roughly outline how to do that. Let $O(B)$ be the orthogonal group of (B, q), and let $G := \operatorname{GL}(A) \times O(B)$. Then G acts on $\widetilde{\mathcal{M}}$ by $((\Phi, \Psi), \alpha) \mapsto \Psi\alpha\Phi^{-1}$, where $\Psi\alpha\Phi^{-1}(x) := \Psi\alpha(x)\Phi^{-1}$. We may consider an element $(\Phi, \Psi) \in G$ as an isomorphism between the monad defined by α and the monad defined by $\Psi\alpha\Phi^{-1}$. By going back and forth between isomorphisms of bundles and isomorphisms of monads one shows that the stabilizer of G in each point is $\{\pm 1\}$, and that our construction induces a bijection between the set of isomorphism classes of stable rank 2 vector bundles on $\mathbf{P}^2(\mathbf{C})$ with Chern classes $c_1 = -1$ and c_2 and $\mathcal{M} := \widetilde{\mathcal{M}}/G_0$, where $G_0 := G/\{\pm 1\}$. With the help of a universal monad over $\mathbf{P}^2(\mathbf{C}) \times \widetilde{\mathcal{M}}$ one proves that the analytic structure on $\widetilde{\mathcal{M}}$ descends to an analytic structure on $\mathcal{M}$ so that $\mathcal{M}$ is smooth of dimension $h^1 \mathcal{F}^* \otimes \mathcal{F} = 4c_2 - 4$ in each point (the obstructions for smoothness in the point corresponding to $\mathcal{F}$ lie in $H^2 \mathcal{F}^* \otimes \mathcal{F}$ which is zero). Moreover the homology of the universal monad tensored by a suitable line bundle descends to a universal family over $\mathcal{M}$ (here one needs $c_1 = -1$). In other words, $\mathcal{M}$ is what one calls a fine moduli space for our bundles. Further efforts show that $\mathcal{M}$ is irreducible and rational.

Remark 5.8. Horrocks' technique of killing cohomology always yields 3-term monads. In general, the bundle in the middle can be pretty complicated.

$\square$

6   The Beilinson Monad

We can use the Tate resolution associated to a sheaf to give a construction of a complex first described by Beilinson [6], which gives a powerful method for deriving information about a sheaf from information about a few of its cohomology groups. The general idea is the following:

Suppose that $\mathcal{A}$ is an additive category and consider a graded object $\oplus_{i=0}^{n+1} U^i$ in $\mathcal{A}$. Given a graded ring homomorphism $E \to \mathrm{End}_{\mathcal{A}}(\oplus_{i=0}^{n+1} U^i)$ we can make an additive functor from the category of free E-modules to $\mathcal{A}$: On objects we take

$$\omega_E(i) \mapsto \begin{cases} U^i & \text{for } 0 \leq i \leq n+1 \text{ and;} \\ 0 & \text{otherwise.} \end{cases}$$

To define the functor on maps, we use

$$\mathrm{Hom}_E(\omega_E(i), \omega_E(j)) = \mathrm{Hom}_E(E(i), E(j))$$
$$= E_{j-i} \longrightarrow \mathrm{End}(\oplus U^i)_{j-i} \longrightarrow \mathrm{Hom}(U^i, U^j) \,.$$

(Note that we could have taken any twist of E in place of $\omega_E \cong E(-n-1)$; the choice of ω_E is made to simplify the statement of Theorem 6.1, below.)

We shall be interested in the special case where $\mathcal{A}$ is the category of coherent sheaves on $\mathbf{P}(W)$ and where $U^i = \Omega^i_{\mathbf{P}(W)}(i)$ as in Section 4. Further examples may be obtained by taking U^i to be the ith exterior power of the tautological subbundle U_k on the Grassmannian of k-planes in W for any k; the case we have taken here is the case $k = n$. See [19] for more information on the general case and applications to the computation of resultants and more general Chow forms.

Applying the functor just defined to the Tate resolution $\mathbf{T}(\mathcal{F})$ of a coherent sheaf $\mathcal{F}$ on $\mathbf{P}(W)$, and using Theorem 3.1, we get a complex

$$\Omega(\mathcal{F}): \quad \cdots \longrightarrow \oplus_j H^j \mathcal{F}(i-j) \otimes U^{j-i} \longrightarrow \cdots,$$

where the term we have written down occurs in cohomological degree i. The resolution $\mathbf{T}(\mathcal{F})$ is well-defined up to homotopy, so the same is true of $\Omega(\mathcal{F})$. Since $U^k = 0$ unless $0 \le k \le n$ the only cohomology groups of $\mathcal{F}$ that are actually involved in $\Omega(\mathcal{F})$ are $\mathrm{H}^j\mathcal{F}(k)$ with $-n \le k \le 0$; $\Omega(\mathcal{F})$ is of type

$$0 \to \mathrm{H}^0\mathcal{F}(-n) \otimes U^n \to \cdots \to \oplus_{j=0}^n \mathrm{H}^j\mathcal{F}(-j) \otimes U^j \to \cdots \to \mathrm{H}^n\mathcal{F} \otimes U^0 \to 0$$

$$\|\qquad\qquad\qquad\qquad\qquad\| \qquad\qquad\qquad\qquad\qquad \|$$

$$0 \longrightarrow \Omega^{-n}(\mathcal{F}) \longrightarrow \cdots \longrightarrow \Omega^0(\mathcal{F}) \longrightarrow \cdots \longrightarrow \Omega^n(\mathcal{F}) \longrightarrow 0 \ .$$

For applications it is important to note that instead of working with $\Omega(\mathcal{F})$ one can also work with $\Omega(\mathcal{F}(i))$ for some twist i. This gives one some freedom in choosing the cohomology groups of $\mathcal{F}$ to be involved.

To see a simple example, consider again the structure sheaf $\mathcal{O}_p$ of the subvariety consisting of a point $p \in \mathbf{P}(W)$. Write I for the homogeneous ideal of p, and let $a \in V = W^*$ be a non-zero functional vanishing on the linear forms in I as before. The Tate resolution of the homogeneous coordinate ring S/I has already been computed, and we have seen that it depends only on the sheaf $\widetilde{S/I} = \mathcal{O}_p$. From the computation of $\mathbf{T}(S/I) = \mathbf{T}(\mathcal{O}_p)$ made in Section 3 we see that $\Omega(\mathcal{O}_p)$ takes the form

$$\Omega(\mathcal{O}_p): \quad 0 \to U^n \xrightarrow{\ a\ } U^{n-1} \xrightarrow{\ a\ } \cdots \xrightarrow{\ a\ } U^1 \xrightarrow{\ a\ } U^0 \longrightarrow 0 \ ,$$

with U^i in cohomological degree $-i$.

We have already noted that the map $a : U = U^1 \longrightarrow U^0 = \mathcal{O}_{\mathbf{P}(W)}$ is the composite of the tautological embedding $U \subset W \otimes \mathcal{O}_{\mathbf{P}(W)}$ with the map $a \otimes 1 : W \otimes \mathcal{O}_{\mathbf{P}(W)} \to \mathcal{O}_{\mathbf{P}(W)}$. Thus the image of $a : U^1 \to \mathcal{O}_{\mathbf{P}(W)}$ is the ideal sheaf of p, and we see that the homology of the complex $\Omega(\mathcal{O}_p)$ at U^0 is $\mathcal{O}_p$. One can check further that $\Omega(\mathcal{O}_p)$ is the Koszul complex associated with the map $a : U^1 \to \mathcal{O}_{\mathbf{P}(W)}$, and it follows that the homology of $\Omega(\mathcal{O}_p)$ at U^i is 0 for $i > 0$. The following result shows that this is typical.

Theorem 6.1 ([18]). *If $\mathcal{F}$ is a coherent sheaf on $\mathbf{P}(W)$, then the only non-vanishing homology of the complex $\Omega(\mathcal{F})$ is*

$$\mathrm{H}^0(\Omega(\mathcal{F})) = \mathcal{F}. \qquad \square$$

The existence of a complex satisfying the theorem and having the same terms as $\Omega(\mathcal{F})$ was first asserted by Beilinson in [6], and thus we will call $\Omega(\mathcal{F})$ a *Beilinson monad* for $\mathcal{F}$. Existence proofs via a somewhat less effective construction than the one given here may be found in [28] and [2].

The explicitness of the construction via Tate resolutions allows one to detect properties of the differentials of Beilinson monads. Let us write

$$d_{ij}^{(r)} \in \mathrm{Hom}(\mathrm{H}^j\mathcal{F}(i-j) \otimes U^{j-i}, \mathrm{H}^{j-r+1}\mathcal{F}(i-j+r) \otimes U^{j-i-r})$$

$$\cong \bigwedge{}^r V \otimes \mathrm{Hom}(\mathrm{H}^j\mathcal{F}(i-j), \mathrm{H}^{j-r+1}\mathcal{F}(i-j+r))$$

$$\cong \mathrm{Hom}(\bigwedge{}^r W \otimes \mathrm{H}^j\mathcal{F}(i-j), \mathrm{H}^{j-r+1}\mathcal{F}(i-j+r))$$

for the degree r maps actually occurring in $\Omega(\mathcal{F})$.

Remark 6.2. The constant maps $d_{ij}^{(0)}$ in $\Omega(\mathcal{F})$ are zero since $\mathbf{T}(\mathcal{F})$ is minimal.   □

Proposition 6.3 ([18]). *The linear maps $d_{ij}^{(1)}$ in $\Omega(\mathcal{F})$ correspond to the multiplication maps*

$$W \otimes \mathrm{H}^j \mathcal{F}(i - j) \to \mathrm{H}^j \mathcal{F}(i - j + 1) . \qquad □$$

This follows from the identification of the linear strands in $\mathbf{T}(\mathcal{F})$ (see the discussion following Theorem 3.1). The higher degree maps in $\mathbf{T}(\mathcal{F})$ and $\Omega(\mathcal{F})$, however, are not yet well-understood.

Since $(\mathbf{T}(\mathcal{F}))[1] = \mathbf{T}(\mathcal{F}(1))$ we can compare the differentials in $\Omega(\mathcal{F})$ with those in $\Omega(\mathcal{F}(1))$:

Proposition 6.4 ([18]). *If the maps $d_{ij}^{(r)}$ in $\Omega(\mathcal{F})$ and $d_{i-1,j}^{(r)}$ in $\Omega(\mathcal{F}(1))$ both actually occur, then they correspond to the same element in*

$$\bigwedge^r V \otimes \mathrm{Hom}(\mathrm{H}^j \mathcal{F}(i - j), \mathrm{H}^{j-r+1} \mathcal{F}(i - j + r)) . \qquad □$$

In what follows we present some *Macaulay 2* code for computing Beilinson monads. Our functions sortedBasis, beilinson1, U, and beilinson reflect what we did in Example 5.7 .

The expression sortedBasis(i,E) sorts the monomials of degree i in E to match the order of the columns of koszul(i,vars S), where our conventions with respect to S and E are as in Section 2, and where we suppose that the monomial order on E is reverse lexicographic, the *Macaulay 2* default order.

```
i32 : sortedBasis = (i,E) -> (
           m := basis(i,E);
           p := sortColumns(m,MonomialOrder=>Descending);
           m_p);
```

For example:

```
i33 : S=ZZ/32003[x_0..x_3];

i34 : E=ZZ/32003[e_0..e_3,SkewCommutative=>true];

i35 : koszul(2,vars S)

o35 = {1} | -x_1 -x_2 0    -x_3 0    0    |
      {1} | x_0  0    -x_2 0    -x_3 0    |
      {1} | 0    x_0  x_1  0    0    -x_3 |
      {1} | 0    0    0    x_0  x_1  x_2  |

                 4         6
o35 : Matrix S  <--- S
```

```
i36 : sortedBasis(2,E)

o36 = | e_0e_1 e_0e_2 e_1e_2 e_0e_3 e_1e_3 e_2e_3 |

                1        6
o36 : Matrix E   <--- E
```

If $e \in E$ is homogeneous of degree j, then beilinson1(e,j,i,S) computes the map $U^i \xrightarrow{e} U^{i-j}$ on $\mathbf{P}^n = \operatorname{Proj} S$. If $0 < i - j \le i \le n$, then the result is a matrix representing the map $\bigwedge^{i+1} W \otimes S(-1) \xrightarrow{e \otimes 1} \bigwedge^{i-j+1} W \otimes S(-1)$ defined by contraction with e. If $0 = i - j < i \le n$, then the result is a matrix representing the composite of the map $\bigwedge^i W \otimes S \xrightarrow{e \otimes 1} S$ with the Koszul differential $\bigwedge^{i+1} W \otimes S(-1) \to \bigwedge^i W \otimes S$. Note that the degrees of the result are not set correctly since the functions U and beilinson below are supposed to do that.

```
i37 : beilinson1=(e,dege,i,S)->(
           E := ring e;
           mi := if i < 0 or i >= numgens E then map(E^1, E^0, 0)
                   else if i === 0 then id_(E^1)
                   else sortedBasis(i+1,E);
           r := i - dege;
           mr := if r < 0 or r >= numgens E then map(E^1, E^0, 0)
                   else sortedBasis(r+1,E);
           s = numgens source mr;
           if i === 0 and r === 0 then
                   substitute(map(E^1,E^1,{{e}}),S)
           else if i>0 and r === i then substitute(e*id_(E^s),S)
           else if i > 0 and r === 0 then
                   (vars S) * substitute(contract(diff(e,mi),transpose mr),S)
           else substitute(contract(diff(e,mi), transpose mr),S));
```

For example:

```
i38 : beilinson1(e_1,1,3,S)

o38 = {-3} | 0 |
       {-3} | 0 |
       {-3} | 1 |
       {-3} | 0 |

                4       1
o38 : Matrix S   <--- S

i39 : beilinson1(e_1,1,2,S)

o39 = {-2} | 0   0   0 0 |
       {-2} | -1  0   0 0 |
       {-2} | 0   0   0 0 |
       {-2} | 0   -1  0 0 |
       {-2} | 0   0   0 0 |
       {-2} | 0   0   0 1 |

                6        4
o39 : Matrix S   <--- S

i40 : beilinson1(e_1,1,1,S)

o40 = | x_0 0 -x_2 0 -x_3 0 |
```

```
                    1      6
o40 : Matrix S  <--- S
```

The function U computes the bundles U^i on Proj S:

```
i41 : U = (i,S) -> (
              if i < 0 or i >= numgens S then S^0
              else if i === 0 then S^1
              else cokernel koszul(i+2,vars S) ** S^{i});
```

Finally, if $o : \oplus E(-a_i) \to \oplus E(-b_j)$ is a homogeneous matrix over E, then beilinson(o,S) computes the corresponding map $o : \oplus U^{a_i} \to \oplus U^{b_j}$ on Proj S by calling beilinson1 and U.

```
i42 : beilinson = (o,S) -> (
              coldegs := degrees source o;
              rowdegs := degrees target o;
              mats = table(numgens target o, numgens source o,
                      (r,c) -> (
                           rdeg = first rowdegs#r;
                           cdeg = first coldegs#c;
                           overS = beilinson1(o_(r,c),cdeg-rdeg,cdeg,S);
                           -- overS = substitute(overE,S);
                           map(U(rdeg,S),U(cdeg,S),overS)));
              if #mats === 0 then matrix(S,{{}})
              else matrix(mats));
```

With these functions the code in Example 5.7 can be rewritten as follows:

```
i43 : S=ZZ/32003[x_0..x_2];

i44 : E = ZZ/32003[e_0..e_2,SkewCommutative=>true];

i45 : alphad = map(E^1,E^{-1,-1},{{e_1,e_2}})

o45 = | e_1 e_2 |

                    1      2
o45 : Matrix E  <--- E

i46 : alpha = map(E^{-1,-1},E^{-2},{{e_1},{e_2}})

o46 = {1} | e_1 |
      {1} | e_2 |

                    2      1
o46 : Matrix E  <--- E

i47 : alphad=beilinson(alphad,S);

o47 : Matrix

i48 : alpha=beilinson(alpha,S);

o48 : Matrix

i49 : F = prune homology(alphad,alpha);

i50 : betti  F

o50 = relations : total: 3 1
                     1: 2 .
                     2: 1 1
```

7   Examples

In this section we give two examples of explicit constructions of Beilinson monads over $\mathbf{P}^4(\mathbf{C}) = \mathbf{P}(W)$ and of classification results based on these monads. As in Section 5 we proceed in three steps. Let us write $\mathcal{O} = \mathcal{O}_{\mathbf{P}^4(\mathbf{C})}$.

Example 7.1. Our first example is taken from the classification of *conic bundles* in $\mathbf{P}^4(\mathbf{C})$, that is, of smooth surfaces $X \subset \mathbf{P}^4(\mathbf{C})$ which are ruled in conics in the sense that there exists a surjective morphism $\pi : X \to C$ onto a smooth curve C such that the general fiber of π is a smooth conic in the given embedding of X. There are precisely three families of such surfaces (see [20] and [9]). Two families, the Del Pezzo surfaces of degree 4 and the Castelnuovo surfaces, are classical. The third family, consisting of *elliptic conic bundles* (conic bundles over an elliptic curve) of degree 8, had been falsely ruled out in two classification papers in the 1980's (see [36] and [27]). Only recently Abo, Decker, and Sasakura [1] constructed and classified such surfaces by considering the Beilinson monads for the suitably twisted ideal sheaves of the surfaces. Let us explain how this works.

Step 1. In this step we suppose that an elliptic conic bundle X as above exists, and we determine the type of the Beilinson monad for the suitably twisted ideal sheaf $\mathcal{J}_X$. We know from the classification of smooth surfaces in $\mathbf{P}^4(\mathbf{C})$ which are contained in a cubic hypersurface (see [38] and [3]) that $H^0 \mathcal{J}_X(i) = 0$ for $i \leq 3$. It follows from general results such as the theorem of Riemann-Roch that the dimensions $h^j \mathcal{J}_X(i)$ in range $-2 \leq i \leq 3$ are as follows (here, again, a zero is represented by an empty box):

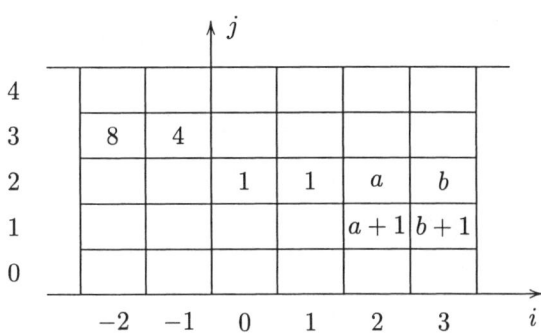

with $a := h^2 \mathcal{J}_X(2)$ and $b := h^2 \mathcal{J}_X(3)$ still to be determined. The Beilinson monad for $\mathcal{J}_X(2)$ is thus of type

$$0 \to 8\mathcal{O}(-1) \to 4U^3 \oplus U^2 \to U \oplus (a+1)\mathcal{O} \to a\mathcal{O} \to 0 ,$$

where $(a+1)\mathcal{O} \to a\mathcal{O}$ is the zero map (see Remark 6.2), and where consequently U is mapped surjectively onto $a\mathcal{O}$. By Proposition 4.4 this is only possible if $a = 0$. The same idea applied to $\mathcal{J}_X(3)$ shows that then also $b = 0$.

The cohomological information obtained so far determines the type of the Beilinson monad for $\mathcal{J}_X(2)$ and for $\mathcal{J}_X(3)$. We decide to concentrate on the monad for $\mathcal{J}_X(3)$ since its differentials are smaller in size than those of the monad for $\mathcal{J}_X(2)$. In order to ease our calculations further we kill the 4-dimensional space $H^3\mathcal{J}_X(-1)$. Let us write ω_X for the dualizing sheaf of X. Serre duality on $\mathbf{P}^4(\mathbf{C})$ respectively on X yields canonical isomorphisms

$$Z := \operatorname{Ext}^1(\mathcal{J}_X(-1), \mathcal{O}(-5))$$
$$\cong (H^3\mathcal{J}_X(-1))^* \cong (H^2\mathcal{O}_X(-1))^* \cong H^0(\omega_X(1)) .$$

The identity in

$$\operatorname{Hom}(Z, Z) \cong \operatorname{Ext}^1(\mathcal{J}_X(-1), Z^* \otimes \mathcal{O}(-5))$$

defines an extension which, twisted by 4, can be written as

$$0 \to 4\mathcal{O}(-1) \to \mathcal{G} \to \mathcal{J}_X(3) \to 0 .$$

Let us show that $\mathcal{G}$ is a vector bundle. We know from the classification of scrolls in $\mathbf{P}^4(\mathbf{C})$ (see [30] and [3]) that X is not a scroll. Hence adjunction theory implies that $\omega_X(1)$ is generated by the adjoint linear system $H^0(\omega_X(1))$ (see [7, Corollary 9.2.2]). It follows by Serre's criterion ([39], see also [35, Theorem 2.2]) that $\mathcal{G}$ is locally free. By construction $\mathcal{G}$ has a cohomology table as follows:

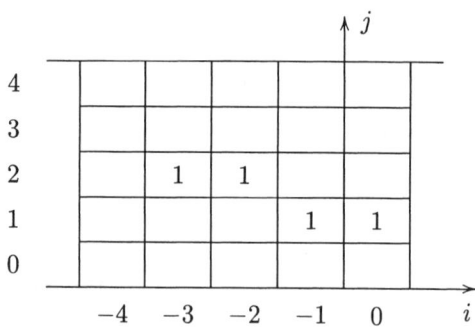

So the Beilinson monad of $\mathcal{G}$ is of type

$$0 \to U^3 \xrightarrow{\alpha} U^2 \oplus U \xrightarrow{\beta} \mathcal{O} \to 0 .$$

Step 2. Now we proceed the other way around. We show that a rank 5 bundle $\mathcal{G}$ as in the first step exists, and that the dependency locus of four general sections of $\mathcal{G}(1)$ is a surface of the desired type. Differentials which define a monad as above with a locally free homology can be easily found. By Lemma 4.3 α corresponds to a pair of vectors $\alpha = (\alpha_1, \alpha_2)^t \in V \oplus \bigwedge^2 V$. By dualizing (see Remark 4.1) we find that it is a vector bundle monomorphism if and only if $U^2 \oplus U^3 \xrightarrow{\alpha^t} U^1$ is an epimorphism. Equivalently, α_1 is non-zero and α_2 considered as a vector in $\bigwedge^2(V/\langle\alpha_1\rangle)$ is indecomposable (argue as in the proof of Proposition 4.4). Taking the other monad conditions into account we see that we may pick

$$\alpha = \begin{pmatrix} e_4 \\ e_0 \wedge e_2 + e_1 \wedge e_3 \end{pmatrix}$$

and

$$\beta = \begin{pmatrix} e_0 \wedge e_2 + e_1 \wedge e_3 \,, \; -e_4 \end{pmatrix},$$

where $e_0, \ldots, e_4$ is a basis of V, and that up to isomorphisms of monads and up to the choice of the basis this is the only possibility. We fix $\mathcal{G}$ as the homology of this monad and compute the syzygies of $\mathcal{G}$ with *Macaulay 2*.

```
i51 : S = ZZ/32003[x_0..x_4];

i52 : E = ZZ/32003[e_0..e_4,SkewCommutative=>true];

i53 : beta=map(E^1,E^{-2,-1},{{e_0*e_2+e_1*e_3,-e_4}})

o53 = | e_0e_2+e_1e_3 -e_4 |

                1      2
o53 : Matrix E  <--- E

i54 : alpha=map(E^{-2,-1},E^{-3},{{e_4},{e_0*e_2+e_1*e_3}})

o54 = {2} | e_4          |
      {1} | e_0e_2+e_1e_3 |

                2      1
o54 : Matrix E  <--- E

i55 : beta=beilinson(beta,S);

o55 : Matrix

i56 : alpha=beilinson(alpha,S);

o56 : Matrix

i57 : G = prune homology(beta,alpha);

i58 : betti res G

o58 = total: 10 9 5 1
          1: 10 4 1 .
          2:  . 5 4 1
```

We see in particular that $\mathcal{G}(1)$ is globally generated. Hence the dependency locus of four general sections of $\mathcal{G}(1)$ is indeed a smooth surface in $\mathbf{P}^4(\mathbf{C})$ by Kleiman's Bertini-type result [29]. The smoothness can also be checked with *Macaulay 2* in an example via the built-in Jacobian criterion (see [15] for a speedier method).

```
i59 : foursect = random(S^4, S^10) * presentation G;

                4    9
o59 : Matrix S   <--- S
```

The function `trim` computes a minimal presentation.

```
i60 : IX = trim minors(4,foursect);

o60 : Ideal of S

i61 : codim IX

o61 = 2

i62 : degree IX

o62 = 8

i63 : codim singularLocus IX

o63 = 5
```

By construction X has the correct invariants and is in fact an elliptic conic bundle as claimed: Since the adjoint linear system $H^0(\omega_X(1))$ is base point free and 4-dimensional by what has been said in the first step, the corresponding adjunction map $X \to \mathbf{P}^3$ is a morphism which exhibits, as is easy to see, X as a conic bundle over a smooth elliptic curve in $\mathbf{P}^3$ (see [1, Proposition 2.1]).

Step 3. Our discussion in the previous steps gives also a classification result. Up to projectivities the elliptic conic bundles of degree 8 in $\mathbf{P}^4(\mathbf{C})$ are precisely the smooth surfaces arising as the dependency locus of four sections of the bundle $\mathcal{G}(1)$ fixed in Step 2.   □

Example 7.2. This example is concerned with the construction and classification of abelian surfaces in $\mathbf{P}^4(\mathbf{C})$, and with the closely related Horrocks-Mumford bundles [25].

Step 1. Horrocks and Mumford found evidence for the existence of a family of abelian surfaces in $\mathbf{P}^4(\mathbf{C})$. Suppose that such a surface X exists. Then the dualizing sheaf of X is trivial, $\omega_X \cong \mathcal{O}_X$, and X has degree 10 (see [21, Example 3.2.15]). The same arguments as in Example 7.1 show that X arises as the zero scheme of a section of a rank 2 vector bundle: There is an extension

$$0 \to \mathcal{O} \to \mathcal{F}(3) \to \mathcal{J}_X(5) \to 0 ,$$

where $\mathcal{F}(3)$ is a rank 2 vector bundle with Chern classes $c_1 = 5$ and $c_2 = \deg X = 10$, and where $\mathcal{F}$ has a cohomology table as displayed in Figure 1. In particular $\mathcal{F}$, which has Chern classes $c_1 = -1$ and $c_2 = 4$, is stable by Remark 5.3. A discussion as in Section 5 shows that the Beilinson monad for $\mathcal{F}$ is of type

$$0 \longrightarrow A \otimes \mathcal{O}(-1) \xrightarrow{\ \alpha\ } B \otimes U^2 \xrightarrow{\ \alpha^d\ } A^* \otimes \mathcal{O} \longrightarrow 0 ,$$

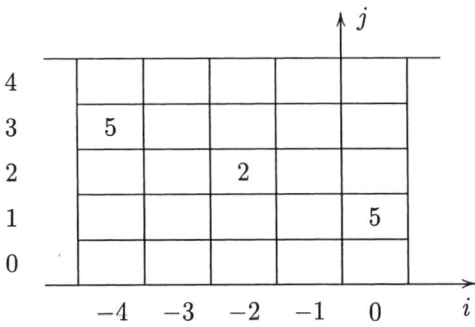

Fig. 1.

with **C**-vector spaces A and B of dimension 5 and 2 respectively, and with $\alpha^d = \alpha^*(-1) \circ (q \otimes \iota)$, where q is a symplectic form on B, and where $\iota : U^2 \xrightarrow{\cong} (U^2)^*(-1)$ is induced by the pairing $U^2 \otimes U^2 \xrightarrow{\wedge} U^4 \cong \mathcal{O}(-1)$. By choosing appropriate bases of A and B we may suppose that α is a 2×5 matrix with entries in $\bigwedge^2 V$ and that $\alpha^d = \alpha^t \cdot \begin{pmatrix} 0 & 1 \\ -1 & 0 \end{pmatrix}$.

Step 2. As in Example 7.1 we now proceed the other way around. But this time it is not obvious how to define α. Horrocks and Mumford remark that up to projectivities one may suppose that the abelian surfaces in $\mathbf{P}^4(\mathbf{C})$ are invariant under the action of the Heisenberg group H_5 in its Schrödinger representation, and they use the representation theory of H_5 and its normalizer N_5 in $\mathrm{SL}(5, \mathbf{C})$ to find

$$\alpha = \begin{pmatrix} e_2 \wedge e_3 & e_3 \wedge e_4 & e_4 \wedge e_0 & e_0 \wedge e_1 & e_1 \wedge e_2 \\ e_1 \wedge e_4 & e_2 \wedge e_0 & e_3 \wedge e_1 & e_4 \wedge e_2 & e_0 \wedge e_3 \end{pmatrix},$$

where $e_0, \ldots, e_4$ is a basis of V. A straightforward computation shows that with this α the desired monad conditions are indeed satisfied. The resulting Horrocks-Mumford bundle $\mathcal{F}_{\mathrm{HM}}$ on $\mathbf{P}^4(\mathbf{C})$ is essentially the only rank 2 vector bundle known on $\mathbf{P}^n(\mathbf{C})$, $n \geq 4$, which does not split as direct sum of two line bundles. Let us compute the syzygies of $\mathcal{F}_{\mathrm{HM}}$ with *Macaulay 2*.

```
i64 : alphad = matrix{{e_4*e_1, e_2*e_3},{e_0*e_2, e_3*e_4},
                      {e_1*e_3, e_4*e_0},{e_2*e_4, e_0*e_1},
                      {e_3*e_0, e_1*e_2}};

              5      2
o64 : Matrix E <--- E

i65 : alphad=map(E^5,E^{-2,-2},alphad)

o65 = | -e_1e_4  e_2e_3 |
      | e_0e_2   e_3e_4 |
      | e_1e_3  -e_0e_4 |
      | e_2e_4   e_0e_1 |
      | -e_0e_3  e_1e_2 |
```

```
                5         2
o65 : Matrix E    <--- E

i66 : alpha=syz alphad

o66 = {2} | e_2e_3 e_0e_4 e_1e_2 -e_3e_4 e_0e_1  |
      {2} | e_1e_4 e_1e_3 e_0e_3 e_0e_2  -e_2e_4 |

                2         5
o66 : Matrix E    <--- E

i67 : alphad=beilinson(alphad,S);

o67 : Matrix

i68 : alpha=beilinson(alpha,S);

o68 : Matrix

i69 : FHM = prune homology(alphad,alpha);

i70 : betti res FHM

o70 = total: 19 35 20 2
          3:  4  .  . .
          4: 15 35 20 .
          5:  .  .  . 2

i71 : regularity FHM

o71 = 5

i72 : betti sheafCohomology(presentation FHM,E,-6,6)

o72 = total: 210 100 37 14 10 5 2 5 10 14 37 100 210
         -6: 210 100 35  4  .  . . . .  .  .   .   .
         -5:   .   .  2 10 10  5 . . .  .  .   .   .
         -4:   .   .  .  .  .  2 . . .  .  .   .   .
         -3:   .   .  .  .  .  . . 5 10 10  2   .   .
         -2:   .   .  .  .  .  . . .  . 4 35 100 210
```

Since $H^0 \mathcal{F}_{HM}(i) = 0$ for $i < 3$ every non-zero section of $\mathcal{F}_{HM}(3)$ vanishes along a surface (with the desired invariants). Horrocks and Mumford need an extra argument to show that the general such surface is smooth (and thus abelian) since Kleiman's Bertini-type result does not apply ($\mathcal{F}_{HM}(3)$ is not globally generated). Our explicit construction allows one again to check the smoothness with *Macaulay 2* in an example.

```
i73 : sect =  map(S^1,S^15,0) | random(S^1, S^4);

              1         19
o73 : Matrix S    <--- S
```

We compute the equations of X via a mapping cone.

```
i74 : mapcone = sect || transpose presentation FHM;

              36        19
o74 : Matrix S    <--- S

i75 : fmapcone = res coker mapcone;
```

```
i76 : IX = trim ideal fmapcone.dd_2;

o76 : Ideal of S

i77 : codim IX

o77 = 2

i78 : degree IX

o78 = 10

i79 : codim singularLocus IX

o79 = 5
```

Step 3. Horrocks and Mumford showed that up to projectivities every abelian surface in $\mathbf{P}^4(\mathbf{C})$ arises as the zero scheme of a section of $\mathcal{F}_{HM}(3)$. In fact, one can show much more. By a careful analysis of possible Beilinson monads and their restrictions to various linear subspaces Decker [12] proved that every stable rank 2 vector bundle $\mathcal{F}$ on $\mathbf{P}^4(\mathbf{C})$ with Chern classes $c_1 = -1$ and $c_2 = 4$ is the homology of a monad of the type as in Step 1. From geometric properties of the "variety of unstable planes" of $\mathcal{F}$ Decker and Schreyer [14] deduced that up to isomorphisms and projectivities the differentials of the monad coincide with those of $\mathcal{F}_{HM}$. Together with results from [11] this implies that the moduli space of our bundles is isomorphic to the homogeneous space $SL(5, \mathbf{C})/N_5$. □

References

1. H. Abo, W. Decker, and N. Sasakura: An elliptic conic bundle on $\mathbf{P}^4$ arising from a stable rank-3 vector bundle. *Math. Z.*, 229:725–741, 1998.
2. V. Ancona and G. Ottaviani: An introduction to the derived categories and the theorem of Beilinson. *Atti Accad. Peloritana Pericolanti Cl. Sci. Fis. Mat. Natur.*, 67:99–110 (1991), 1989.
3. A. Aure: On surfaces in projective 4-space. PhD Thesis, Oslo, 1987.
4. W. Barth: Moduli of vector bundles on the projective plane. *Invent. Math.*, 42:63–91, 1977.
5. W. Barth and K. Hulek: Monads and moduli of vector bundles. *Manuscripta Math.*, 25:323–347, 1978.
6. A. A. Beĭlinson: Coherent sheaves on $\mathbf{P}^n$ and problems in linear algebra. *Functional Anal. Appl.*, 12:214–216, 1978.
7. M.C. Beltrametti and A. Sommese: *The adjunction theory of complex projective varieties.* de Gruyter, Berlin, 1995.
8. I. N. Bernšteĭn, I. M. Gel'fand, and S. I. Gel'fand: Algebraic vector bundles on $\mathbf{P}^n$ and problems of linear algebra. *Functional Anal. Appl.*, 12:212–214, 1978.
9. R. Braun and K. Ranestad: Conic bundles in projective fourspace. In P. Newstead, editor, *Algebraic geometry. Papers presented for the EUROPROJ conferences held in Catania, Italy, September 1993, and Barcelona, Spain, September 1994*, pages 331–339. Marcel Dekker, New York, 1994.

10. R.-O. Buchweitz: Appendix to Cohen-Macaulay modules on quadrics. In *Singularities, representation of algebras, and vector bundles (Lambrecht, 1985)*, pages 96–116. Springer, Berlin, 1987.

11. W. Decker: Das Horrocks-Mumford-Bündel und das Modul-Schema für stabile 2-Vektorbündel über P_4 mit $c_1=-1$, $c_2=4$. *Math. Z.*, 188:101–110, 1984.

12. W. Decker: Stable rank 2 vector bundles with Chern-classes $c_1=-1$, $c_2=4$. *Math. Ann.*, 275:481–500, 1986.

13. W. Decker: Monads and cohomoloy modules of rank 2 vector bundles. *Compositio Math.*, 76:7–17, 1990.

14. W. Decker and F.-O. Schreyer: On the uniqueness of the Horrocks-Mumford-bundle. *Math. Ann.*, 273:415–443, 1986.

15. W. Decker and F.-O. Schreyer: Non-general type surfaces in P^4: Some remarks on bounds and constructions. *J. Symbolic Computation*, 29:545–582, 2000.

16. D. Eisenbud: Computing cohomology. A chapter in [41].

17. D. Eisenbud: *Commutative algebra with a view toward algebraic geometry*. Springer-Verlag, New York, 1995.

18. D. Eisenbud, G. Fløystad, and F.-O. Schreyer: Sheaf cohomology and free resolutions over exterior algebras. Work in progress, 2001.

19. D. Eisenbud and F.-O. Schreyer: Resultants, chow forms, and free resolutions. In preparation, 2001.

20. P. Ellia and G. Sacchiero: Smooth surfaces of P^4 ruled in conics. In P. Newstead, editor, *Algebraic geometry. Papers presented for the EUROPROJ conferences held in Catania, Italy, September 1993, and Barcelona, Spain, September 1994*, pages 49–62. Marcel Dekker, New York, 1994.

21. W. Fulton: *Intersection theory*. Springer-Verlag, New York, 1984.

22. D. Gieseker: On the moduli of vector bundles on an algebraic surface. *Ann. of Math. (2)*, 106:45–60, 1977.

23. G. Horrocks: Vector bundles on the punctured spectrum of a local ring. *Proc. London Math. Soc. (3)*, 14:689–713, 1964.

24. G. Horrocks: Construction of bundles on P^n. In A. Douady and J.-L. Verdier, editors, *Les équations de Yang-Mills*, pages 197–203. Société Mathématique de France, Paris, 1980. Séminaire E. N. S., 1977-1978, Astérisqe 71-72.

25. G. Horrocks and D. Mumford: A rank 2 vector bundle on P^4 with 15,000 symmetries. *Topology*, 12:63–81, 1973.

26. Klaus Hulek: Stable rank-2 vector bundles on P_2 with c_1 odd. *Math. Ann.*, 242(3):241–266, 1979.

27. P. Ionescu: Embedded projective varieties of small invariants III. In A.J. Sommese, A. Biancofiore, and E. Livorni, editors, *Algebraic Geometry (L'Aquila 1988)*, pages 138–154. Springer, New York, 1990.

28. M. M. Kapranov: On the derived categories of coherent sheaves on some homogeneous spaces. *Invent. Math.*, 92(3):479–508, 1988.

29. S. Kleiman: Geometry on grassmanians and applications to splitting bundles and smoothing cycles. *Publ. Math. I.H.E.S.*, 36:281–297, 1969.

30. A. Lanteri: On the existence of scrolls in P^4. *Atti Accad. Naz. Lincei, VIII. Ser., Rend., Cl. Sci. Fis. Mat. Nat.*, 69:223–227, 1980.

31. J. Le Potier: Fibrés stables de rang 2 sur $P_2(C)$. *Math. Ann.*, 241:217–256, 1979.

32. N. Manolache: Syzygies of abelian surfaces embedded in $P^4(C)$. *J. reine angew. Math.*, 384:180–191, 1988.

33. M. Maruyama: Moduli of stable sheaves I. *J. Math. Kyoto Univ.*, 17:91–126, 1977.

34. M. Maruyama: Moduli of stable sheaves II. *J. Math. Kyoto Univ.*, 18:557–614, 1978.

35. Ch. Okonek: Reflexive Garben auf $\mathbf{P}^4$. *Math. Ann.*, 260:211–237, 1982.

36. Ch. Okonek: Flächen vom Grad 8 im $\mathbf{P}^4$. *Math. Z.*, 191:207–223, 1986.

37. Ch. Okonek, M. Schneider, and H. Spindler: *Vector bundles on complex projective spaces*. Birkhäuser Boston, Mass., 1980.

38. L. Roth: On the projective classification of surfaces. *Proc. of London Math. Soc.*, 42:142–170, 1937.

39. J.-P. Serre: Faisceaux algébriques cohérents. *Ann. of Math. (2)*, 61:197–278, 1955.

40. Gregory G. Smith: Computing global extension modules. *J. Symbolic Comput.*, 29(4-5):729–746, 2000. Symbolic computation in algebra, analysis, and geometry (Berkeley, CA, 1998).

41. Wolmer V. Vasconcelos: *Computational methods in commutative algebra and algebraic geometry*. Springer-Verlag, Berlin, 1998. With chapters by David Eisenbud, Daniel R. Grayson, Jürgen Herzog and Michael Stillman.

42. C. H. Walter: Pfaffian subschemes. *J. Algebraic Geom.*, 5:671–704, 1996.

Needles in a Haystack:
Special Varieties via Small Fields

Frank-Olaf Schreyer and Fabio Tonoli

In this article we illustrate how picking points over a finite field at random can help to investigate algebraic geometry questions. In the first part we develop a program that produces random curves of genus $g \leq 14$. In the second part we use the program to test Green's Conjecture on syzygies of canonical curves and compare it with the corresponding statement for Coble self-dual sets of points. In the third section we apply our techniques to produce Calabi-Yau 3-folds of degree 17 in $\mathbb{P}^6$.

Introduction

The advances in speed of modern computers and computer algebra systems gave life to the idea of solving equations by guessing a solution. Suppose $\mathbb{M} \subset \mathbb{G}$ is a subvariety of a rational variety of codimension c. Then we expect that the probability for a point $p \in \mathbb{G}(\mathbb{F}_q)$ to lie in $\mathbb{M}(\mathbb{F}_q)$ is about $1/q^c$. Here $\mathbb{F}_q$ denotes the field with q elements.

We will discuss this idea in the following setting: $\mathbb{M}$ will be a parameter space for objects in algebraic geometry, e.g., a Hilbert scheme, a moduli space, or a space dominating such spaces.

The most basic question we might have in this case is whether $\mathbb{M}$ is non-empty and whether an open part of $\mathbb{M}$ corresponds to smooth objects.

Typically in these cases we will not have explicit equations for $\mathbb{M} \subset \mathbb{G}$ but only an implicit algebraic description of $\mathbb{M}$, and our approach will be successful if the time required to check $p \notin \mathbb{M}(\mathbb{F}_q)$ is sufficiently small compared to q^c. The first author applied this method first in [32] to construct some rational surfaces in $\mathbb{P}^4$; see [15,11] for motivation.

In this first section we describe a program that picks curve of genus $g \leq 14$ at random. The moduli spaces $\mathfrak{M}_g$ are known to be unirational for $g \leq 13$; see [33,8].

Our approach based on this result can viewed as a computer aided proof of the unirationality. Many people might object that this not a proof because we cannot control every single step in the computation. We however think that such a proof is much more reliable than a proof based on man-made computations. A mistake in a computer aided approach most often leads to an output far away from our expectation, hence it is easy to spot. A substantial improvement of present computers and computer algebra systems would give us an explicit unirational parametrization of $\mathfrak{M}_g$ for $g \leq 13$.

In the second part we apply our "random curves" to probe the consequences of Green's conjecture on syzygies of canonical curves, and compare these results with the corresponding statements for "Coble self-dual" sets of $2g - 2$ points in $\mathbb{P}^{g-2}$.

In the last section we exploit our method to prove the existence of three components of the Hilbert scheme of Calabi-Yau 3-folds of degree 17 in $\mathbb{P}^6$ over the complex numbers. This is one of the main results of the second author's thesis [34, Chapter 4]. Calabi-Yau threefolds of lower degree in $\mathbb{P}^6$ are easy to construct, using the Pfaffian construction and a study of their Hartshorne-Rao modules. For degree 17 the Hartshorne-Rao module has to satisfy a subtle condition. Explicit examples of such Calabi-Yau 3-folds are first constructed over a finite field by our probabilistic method. Then a delicate semi-continuity argument gives us the existence of such Calabi-Yau 3-folds over some number field.

Acknowledgments. We thank Hans-Christian v. Bothmer and Dan Grayson for valuable discussions and remarks.

Notation. For a finitely generated graded module M over the polynomial ring $S = k[x_0, \ldots, x_r]$ we summarize the numerical information of a finite free resolution

$$0 \leftarrow M \leftarrow F_0 \leftarrow F_1 \leftarrow \ldots \leftarrow F_n \leftarrow 0$$

with $F_i = \oplus_j S(-j)^{\beta_{ij}}$ in a table of Betti numbers, whose ij^{th} entry is

$$\beta_{i,i+j} = \dim \mathrm{Tor}_i^S(M, k)_{i+j}.$$

As in the *Macaulay 2* command betti we suppress zeroes. For example the syzygies of the rational normal curve in $\mathbb{P}^3$ have the following Betti table.

$$\begin{bmatrix} 1 & - & - \\ - & 3 & 2 \end{bmatrix}$$

Note that the degrees of the entries of the matrices in the free resolution can be read off from the relative position of two numbers in consecutive columns. A pair of numbers in a line corresponds to linear entries. Quadratic entries correspond to two numbers of a square. Thus

$$\begin{bmatrix} 1 & - & - & - \\ - & 5 & 5 & - \\ - & - & - & 1 \end{bmatrix}$$

corresponds to a 4 term complex with a quadratic, a linear and another quadratic map. The Grassmannian $\mathbb{G}(2,5)$ in its Plücker embedding has such a free resolution.

1 How to Make Random Curves up to Genus 14

The moduli space of curves $\mathfrak{M}_g$ is known to be of general type for $g \geq 24$ and has non-negative Kodaira dimension for $g = 23$ by work of Harris, Mumford and Eisenbud [21,13]. For genus $g \leq 13$ unirationality is known [8,33]. In this section we present a *Macaulay 2* program that over a finite field $\mathbb{F}_q$ picks a point in $\mathfrak{M}_g(\mathbb{F}_q)$ for $g \leq 14$ at random.

By Brill-Noether theory [2] every curve of genus g has a linear system g_d^r of dimension r and degree d, provided that the Brill-Noether number ρ satisfies

$$\rho := \rho(g, d, r) := g - (r + 1)(g - d + r) \geq 0.$$

We utilize this to find appropriate (birational) models for general curves of genus g.

1.1 Plane Models, $g \leq 10$

This case was known to Severi; see [1]. Choose $d = g + 2 - \lfloor g/3 \rfloor$. Then $\rho(g, d, 2) \geq 0$ i.e., a general curve of genus g has a plane model C' of degree d. We expect that C' has

$$\delta = \binom{d - 1}{2} - g$$

double points. If the double points are in general position, then

$$s = h^0(\mathbb{P}^2, \mathcal{O}(d)) - 3\delta - 1$$

is the expected dimension of the linear system of curves of degree d with δ assigned double points. We have the following table:

g	1	2	3	4	5	6	7	8	9	10	11	12
ρ	1	2	0	1	2	0	1	2	0	1	2	0
d	3	4	4	5	6	6	7	8	8	9	10	10
δ	0	1	0	2	5	4	8	13	12	18	25	24
s	9	11	14	14	12	15	11	5	8	0	-10	-7

Thus for $g \leq 10$ we assume that these double points lie in general position. For $g > 10$ the double points cannot lie in general position because $s < 0$. Since it is difficult to describe the special locus $H_\delta(g) \subset \mathrm{Hilb}_\delta(\mathbb{P}^2)$ of double points of nodal genus g curves, the plane model approach collapses for $g > 10$.

Random Points. In our program, which picks plane models at random from an Zariski open subspace of $\mathfrak{M}_g$, we start by picking the nodes. However, over a small field $\mathbb{F}_q$ it is not a good idea to pick points individually, because there might be simply too few: $|\mathbb{P}^2(\mathbb{F}_q)| = 1 + q + q^2$. What we should do is to pick a collection Γ of δ points in $\mathbb{P}^2(\bar{\mathbb{F}}_q)$ that is defined over $\mathbb{F}_q$. General points in $\mathbb{P}^2$ satisfy the minimal resolution condition, that is, they have expected Betti numbers. This follows from the Hilbert-Burch theorem [12, Theorem 20.15]. If the ideal of such Γ has generators in minimal degree k, then $\binom{k+1}{2} \leq \delta < \binom{k+2}{2}$, which gives $\delta = \binom{k+1}{2} + \epsilon$ with $0 \leq \epsilon \leq k$. Thus $k = \lceil (-3 + \sqrt{9 + 8\delta})/2 \rceil$. The Betti table is one of the following two tables:

$2\epsilon \leq k:$

	0	1		
0	1	-		-
1	-	-		-
$\vdots$	$\vdots$	$\vdots$	$\vdots$	$\vdots$
$k-2$	-	-		-
$k-1$	-	$k+1-\epsilon$		$k-2\epsilon$
k	-	-		ϵ

$2\epsilon \geq k:$

	0	1		
0	1	-		-
1	-	-		-
$\vdots$	$\vdots$	$\vdots$	$\vdots$	$\vdots$
$k-2$	-	-		-
$k-1$	-	$k+1-\epsilon$		-
k	-	$2\epsilon-k$		ϵ

So we can specify a collection Γ of δ points by picking the Hilbert-Burch matrix of their resolution; see [12, Thm 20.15]. This is a matrix with linear and quadratic entries only, whose minors of size ϵ ($k - \epsilon$ if $2\epsilon \leq k$) generate the homogeneous ideal of Γ.

```
i1 : randomPlanePoints = (delta,R) -> (
         k:=ceiling((-3+sqrt(9.0+8*delta))/2);
         eps:=delta-binomial(k+1,2);
         if k-2*eps>=0
         then minors(k-eps,
                 random(R^(k+1-eps),R^{k-2*eps:-1,eps:-2}))
         else minors(eps,
                 random(R^{k+1-eps:0,2*eps-k:-1},R^{eps:-2})));
```

In unlucky cases these points might be infinitesimally near.

```
i2 : distinctPoints = (J) -> (
         singJ:=minors(2,jacobian J)+J;
         codim singJ == 3);
```

The procedure that returns the ideal of a random nodal curve is then straightforward:

```
i3 : randomNodalCurve = method();
```

```
i4 : randomNodalCurve (ZZ,ZZ,Ring) := (d,g,R) -> (
         delta:=binomial(d-1,2)-g;
         K:=coefficientRing R;
         if (delta==0)
         then (        --no double points
             ideal random(R^1,R^{-d}))
         else (        --delta double points
             Ip:=randomPlanePoints(delta,R);
             --choose the curve
             Ip2:=saturate Ip^2;
             ideal (gens Ip2 * random(source gens Ip2, R^{-d})))));
```

```
i5 : isNodalCurve = (I) -> (
         singI:=ideal jacobian I +I;delta:=degree singI;
         d:=degree I;g:=binomial(d-1,2)-delta;
         {distinctPoints(singI),delta,g});
```

We next ask if we indeed get in this way points in a parameter space that dominates $\mathfrak{M}_g$ for $g \leq 10$. Let $\mathrm{Hilb}_{(d,g)}(\mathbb{P}^2)$ denote the Hilbert scheme of nodal plane curves of degree d and geometric genus g. Our construction starts from a random element in $\mathrm{Hilb}_\delta(\mathbb{P}^2)$ and picks a random curve in the corresponding fiber of $\mathrm{Hilb}_{(d,g)}(\mathbb{P}^2) \to \mathrm{Hilb}_\delta(\mathbb{P}^2)$:

$$\mathrm{Hilb}_{(d,g)}(\mathbb{P}^2) \longrightarrow \mathfrak{M}_g \ .$$
$$\downarrow$$
$$\mathrm{Hilb}_\delta(\mathbb{P}^2)$$

So the question is whether $\mathrm{Hilb}_{(d,g)}(\mathbb{P}^2)$ dominates $\mathrm{Hilb}_\delta(\mathbb{P}^2)$. A naive dimension count suggests that this should be true: the dimension of our parameter space is given by $2\delta + s$, which is $3(g-1) + \rho + \dim \mathrm{PGL}(3)$, as it should be. To conclude this there is more to verify: it could be that the nodal models of general curves have double points in special position, while all curve constructed above lie over a subvariety of $\mathfrak{M}_g$. One way to exclude this is to prove that the variety $G(g,d,2)$ over $\mathfrak{M}_g$, whose fiber over a curve $\tilde{C} \in \mathfrak{M}_g$ is $G_d^2(\tilde{C}) = \{g_d^2\text{'s}\}$, is irreducible or, to put it differently, that the Severi Conjecture holds:

Theorem 1.1 (Harris [20]). *The space of nodal degree d genus g curves in $\mathbb{P}^2$ is irreducible.*

Another much easier proof for the few (d,g) we are interested in is to establish that our parameter space $\mathbb{M}$ of the construction is smooth of expected dimension at our random point $p \in \mathbb{M}$, as in [1]. Consider the following diagram:

$$\mathbb{H} = \mathrm{Hilb}_{(d,g)}(\mathbb{P}^2)/\mathrm{Aut}(\mathbb{P}^2) \xrightarrow{\pi} \mathfrak{M}_g.$$

For a given curve $\tilde{C} \in \mathfrak{M}_g$, the inverse image $\pi^{-1}(\tilde{C})$ consists of the variety $W_d^2(\tilde{C}) \subset \mathrm{Pic}^d(\tilde{C})$. Moreover the choice of a divisor $L \in W_d^2(\tilde{C})$ is equivalent to the choice of $p \in \mathbb{M}$, modulo $\mathrm{Aut}(\mathbb{P}^2)$: indeed p determines a morphism $\nu \colon \tilde{C} \longrightarrow C \subset \mathbb{P}^2$ and a line bundle $L = \nu^{-1}\mathcal{O}_{\mathbb{P}^2}(1)$, where $\tilde{C}$ is the normalization of the (nodal) curve C. Therefore $\mathbb{M}$ is smooth of expected dimension $3(g-1)+\rho+\dim \mathrm{PGL}(3)$ at $p \in \mathbb{M}$ if and only if $W_d^2(\tilde{C})$ is smooth of expected dimension ρ in L. This is well known to be equivalent to the injectivity of the multiplication map μ_L

$$H^0(L) \otimes H^0(K_{\tilde{C}} \otimes L^{-1}) \xrightarrow{\mu_L} H^0(K_{\tilde{C}}),$$

which can be easily checked in our cases, see [2, p. 189]. In our cases μ_L can be rewritten as

$$H^0(\mathcal{O}_{\mathbb{P}^2}(1)) \otimes H^0(I_\Gamma(d-4)) \xrightarrow{\mu_L} H^0(I_\Gamma(d-3)).$$

So we need two conditions:

(1) $H^0(I_\Gamma(d-5)) = 0$;
(2) there are no linear relations among the generators of $H^0(I_\Gamma(d-4))$ of degree $d-3$.

We proceed case by case. For genus $g \le 5$ this is clear, since $H^0(I_\Gamma(d-4)) = 0$ for $g = 2, 3$ and $\dim H^0(I_\Gamma(d-3)) = 1$ for $g = 4, 5$. For $g = 6$ we have $\dim H^0(I_\Gamma(d-3)) = \dim H^0(I_\Gamma(2)) = 2$ and the Betti numbers of Γ

$$
\begin{array}{|ccc}
1 & - & - \\
- & 2 & - \\
- & - & 1
\end{array}
$$

shows there are no relations with linear coefficients in $H^0(I_\Gamma(2))$. For $7 \le g \le 10$ the method is similar: everything is clear once the Betti table of resolution of the set of nodal points Γ is computed. As a further example we do here the case $g = 10$: we see that $\dim H^0(I_\Gamma(d-3)) = \dim H^0(I_\Gamma(5)) = 3$ and the Betti numbers of Γ are

$$
\begin{array}{|ccc}
1 & - & - \\
- & - & - \\
- & - & - \\
- & - & - \\
- & 3 & - \\
- & 1 & 3
\end{array}
$$

from which it is clear that there are no linear relations between the quintic generators of I_Γ.

1.2   Space Models and Hartshorne-Rao Modules

The Case of Genus $g = 11$. In this case we have $\rho(11, 12, 3) = 3$. Hence every general curve of genus 11 has a space model of degree 12. Moreover for a general curve the general space model of this degree is linearly normal, because $\rho(11, 13, 4) = -1$ takes a smaller value. If moreover such a curve $C \subset \mathbb{P}^3$ has maximal rank, i.e., for each $m \in \mathbb{Z}$ the map

$$
H^0(\mathbb{P}^3, \mathcal{O}(m)) \to H^0(C, \mathcal{O}_C(m))
$$

has maximal rank, then the Hartshorne-Rao module M, defined as $M = H_*^1(\mathcal{I}_C) = \oplus_m H^1(\mathbb{P}^3, \mathcal{I}_C(m))$, has Hilbert function with values $(0, 0, 4, 6, 3, 0, 0, \dots)$. For readers who want to know more about the Hartshorne-Rao module, we refer to the pleasant treatment in [24].

Since being of maximal rank is an open condition, we will try a construction of maximal rank curves. Consider the vector bundle $\mathcal{G}$ on $\mathbb{P}^3$ associated to the first syzygy module of I_C:

(1) $$ 0 \leftarrow \mathcal{I}_C \leftarrow \oplus_i \mathcal{O}(-a_i) \leftarrow \mathcal{G} \leftarrow 0 $$

In this set-up $H^2_*(\mathcal{G}) = H^1_*(\mathcal{I}_C)$. Thus $\mathcal{G}$ is, up to direct sum of line bundles, the sheafified second syzygy module of M; see e.g., [10, Prop. 1.5].

The expected Betti numbers of M are

$$
\begin{bmatrix}
4 & 10 & 3 & - & - \\
- & - & 8 & 2 & - \\
- & - & - & 6 & 3
\end{bmatrix}
$$

Thus the $\mathbb{F}$-dual $M^* = \mathrm{Hom}_{\mathbb{F}}(M, \mathbb{F})$ is presented as $\mathbb{F}[x_0, \ldots, x_3]$-module by a 3×8 matrix with linear and quadratic entries, and a general such matrix will give a general module (if the construction works, i.e., if the desired space of modules is non-empty), because all conditions we impose are semi-continuous and open. Thus M depends on

$$\dim \mathbb{G}(6, 3h^0\mathcal{O}(1)) + \dim \mathbb{G}(2, 3h^0\mathcal{O}(2) - 6h^0\mathcal{O}(1)) - \dim SL(3) = 36$$

parameters.

Assuming that C has minimal possible syzygies:

$$
\begin{bmatrix}
1 & - & - & - \\
- & - & - & - \\
- & - & - & - \\
- & - & - & - \\
- & 6 & 2 & - \\
- & - & 6 & 3
\end{bmatrix}
$$

we obtain, by dualizing the sequence (1), the following exact sequence

$$\mathcal{G}^* \leftarrow 6\mathcal{O}(5) \leftarrow \mathcal{O} \leftarrow 0.$$

If everything is as expected, i.e., the general curve is of maximal rank and its syzygies have minimal possible Betti numbers, then the entries of the right hand matrix are homogeneous polynomials that generate I_C. We will compute I_C by determining $\ker(\phi\colon 6\mathcal{O}(5) \to \mathcal{G}^*)$. Comparing with the syzygies of M we obtain the following isomorphism

$$\mathcal{G}^* \cong \ker(2\mathcal{O}(6) \oplus 6\mathcal{O}(7) \to 3\mathcal{O}(8)) \cong \mathrm{image}(3\mathcal{O}(4) \oplus 8\mathcal{O}(5) \to 2\mathcal{O}(6) \oplus 6\mathcal{O}(7)).$$

and $\mathcal{G}^* \leftarrow 6\mathcal{O}(5)$ factors over $\mathcal{G}^* \leftarrow 8\mathcal{O}(5) \oplus 3\mathcal{O}(4)$. A general $\phi \in \mathrm{Hom}(6\mathcal{O}(5), \mathcal{G}^*)$ gives a point in $\mathbb{G}(6, 8)$ and the Hilbert scheme of desired curves would have dimension $36 + 12 = 48 = 4 \cdot 12 = 30 + 3 + 15$ as expected, c.f. [19].

Therefore the computation for obtaining a random space curve of genus 11 is done as follows:

```
i6 : randomGenus11Curve = (R) -> (
        correctCodimAndDegree:=false;
        while not correctCodimAndDegree do (
            Mt=coker random(R^{3:8},R^{6:7,2:6});
            M=coker (transpose (res Mt).dd_4);
            Gt:=transpose (res M).dd_3;
            I:=ideal syz (Gt*random(source Gt,R^{6:5}));
            correctCodimAndDegree=(codim I==2 and degree I==12););
        I);
```

In general for these problems there is rarely an a priori reason why such a construction for general choices will give a smooth curve. Kleiman's global generation condition [23] is much too strong a hypothesis for many interesting examples. But it is easy to check an example over a finite field with a computer:

```
i7 : isSmoothSpaceCurve = (I) -> (
               --I generates the ideal sheaf of a pure codim 2 scheme in P3
               singI:=I+minors(2,jacobian I);
               codim singI==4);
```

Hence by semi-continuity this is true over $\mathbb{Q}$ and the desired unirationality of $G(11, 12, 3)/\mathfrak{M}_{11}$ holds for all fields, except possibly for those whose ground field has characteristic is in some finite set.

A calculation of an example over the integers would bound the number of exceptional characteristics, which then can be ruled out case by case, or by considering sufficiently many integer examples.

As in case of nodal curves, to prove unirationality of $\mathfrak{M}_{11}$ by computer aided computations we have to show the injectivity of

$$H^0(L) \otimes H^0(K_C \otimes L^{-1}) \xrightarrow{\mu_L} H^0(K_C),$$

where L is the restriction of $\mathcal{O}_{\mathbb{P}^3}(1)$ to the curve $C \subset \mathbb{P}^3$. The following few lines do the job:

```
i8 : K=ZZ/101;

i9 : R=K[x_0..x_3];

i10 : C=randomGenus11Curve(R);

o10 : Ideal of R

i11 : isSmoothSpaceCurve(C)

o11 = true

i12 : Omega=prune Ext^2(coker gens C,R^{-4});

i13 : betti Omega

o13 = relations : total: 5 10
                     -1: 2  .
                      0: 3 10
```

We see that there are no linear relations among the two generators of $H^0_*(\Omega_C)$ in degree -1.

Betti Numbers for Genus $g = 12, 13, 14, 15$. The approach in these cases is similar to $g = 11$. We choose here $d = g$, so $\rho(g, g, 3) = g - 12 \geq 0$. Under the maximal rank assumption the corresponding space curve has a Hartshorne-Rao module whose Hilbert function takes values $(0, 0, g - 9, 2g - 19, 3g - 34, 0, \ldots)$ in case $g = 12, 13$ and $(0, 0, g - 9, 2g - 19, 3g - 34, 4g - 55, 0, \ldots)$ in case $g = 14, 15$. Expected syzygies of M have Betti tables:

$$
g = 12 : \begin{bmatrix} 3 & 7 & - & - & - \\ - & - & 10 & 5 & - \\ - & - & - & 3 & 2 \end{bmatrix}
\qquad
g = 13 : \begin{bmatrix} 4 & 9 & 1 & - & - \\ - & - & 6 & - & - \\ - & - & 6 & 13 & 5 \end{bmatrix}
$$

$$
g = 14 : \begin{bmatrix} 5 & 11 & 2 & - & - \\ - & - & 3 & - & - \\ - & - & 13 & 17 & 4 \\ - & - & - & - & 1 \end{bmatrix}
\qquad
g = 15 : \begin{bmatrix} 6 & 13 & 3 & - & - \\ - & - & 3 & - & - \\ - & - & 8 & 3 & - \\ - & - & - & 9 & 5 \end{bmatrix}
$$

Comparing with the expected syzygies of C

$$
g = 12 : \begin{bmatrix} 1 & - & - & - \\ - & - & - & - \\ - & - & - & - \\ - & - & - & - \\ - & 7 & 5 & - \\ - & - & 3 & 2 \end{bmatrix}
\qquad
g = 13 : \begin{bmatrix} 1 & - & - & - \\ - & - & - & - \\ - & - & - & - \\ - & - & - & - \\ - & 3 & - & - \\ - & 6 & 13 & 5 \end{bmatrix}
$$

$$
g = 14 : \begin{bmatrix} 1 & - & - & - \\ - & - & - & - \\ - & - & - & - \\ - & - & - & - \\ - & - & - & - \\ - & 13 & 17 & 4 \\ - & - & - & 1 \end{bmatrix}
\qquad
g = 15 : \begin{bmatrix} 1 & - & - & - \\ - & - & - & - \\ - & - & - & - \\ - & - & - & - \\ - & - & - & - \\ - & 8 & 3 & - \\ - & - & 9 & 5 \end{bmatrix}
$$

we see that given M the choice of a curve corresponds to a point in $\mathbb{G}(7, 10)$ or $\mathbb{G}(3, 6)$ for $g = 12, 13$ respectively, while for $g = 14, 15$ everything is determined by the Hartshorne-Rao module. For $g = 12$ Kleiman's result guarantees smoothness for general choices, in contrast to the more difficult cases $g = 14, 15$. So the construction of M is the crucial step.

Construction of Hartshorne-Rao Modules. In case $g = 12$ the construction of M is straightforward. It is presented by a sufficiently general matrix of linear forms:

$$0 \leftarrow M \leftarrow 3S(-2) \leftarrow 7S(-3).$$

The procedure for obtaining a random genus 12 curve is:

```
i14 : randomGenus12Curve = (R) -> (
          correctCodimAndDegree:=false;
          while not correctCodimAndDegree do (
              M:=coker random(R^{3:-2},R^{7:-3});
              Gt:=transpose (res M).dd_3;
              I:=ideal syz (Gt*random(source Gt,R^{7:5}));
              correctCodimAndDegree=(codim I==2 and degree I==12););
          I);
```

In case $g = 13$ we have to make sure that M has a second linear syzygy. Consider the end of the Koszul complex:

$$6R(-2) \overset{\kappa}{\leftarrow} 4R(-3) \leftarrow R(-1) \leftarrow 0.$$

Any product of a general map $4R(-2) \overset{\alpha}{\leftarrow} 6R(-2)$ with the Koszul matrix κ yields $4R(-2) \leftarrow 4R(-3)$ with a linear syzygy, and concatenated with a general map $4R(-2) \overset{\beta}{\leftarrow} 5R(-3)$ gives a presentation matrix of a module M of desired type:

$$0 \leftarrow M \leftarrow 4R(-2) \leftarrow 4R(-3) \oplus 5R(-3).$$

The procedure for obtaining a random genus 13 curve is:

```
i15 : randomGenus13Curve = (R) -> (
          kappa:=koszul(3,vars R);
          correctCodimAndDegree:=false;
          while not correctCodimAndDegree do (
              test:=false;while test==false do (
                  alpha:=random(R^{4:-2},R^{6:-2});
                  beta:=random(R^{4:-2},R^{5:-3});
                  M:=coker(alpha*kappa|beta);
                  test=(codim M==4 and degree M==16););
              Gt:=transpose (res M).dd_3;
              --up to change of basis we can reduce phi to this form
              phi:=random(R^6,R^3)++id_(R^6);
              I:=ideal syz(Gt_{1..12}*phi);
              correctCodimAndDegree=(codim I==2 and degree I==13););
      I);
```

The case of genus $g = 14$ is about a magnitude more difficult. To start with, we can achieve two second linear syzygies by the same method as in the case $g = 13$. A general matrix $5R(-2) \overset{\alpha}{\leftarrow} 12R(-2)$ composed with $12R(-2) \overset{\kappa \oplus \kappa}{\leftarrow} 8R(-3)$ yields the first component of

$$5R(-2) \leftarrow (8+3)R(-3).$$

For a general choice of the second component $5R(-2) \overset{\beta}{\leftarrow} 3R(-3)$ the cokernel will be a module with Hilbert function $(0, 0, 5, 9, 8, 0, 0, \ldots)$ and syzygies

$$\begin{bmatrix} 5 & 11 & 2 & - & - \\ - & - & 2 & - & - \\ - & - & 17 & 23 & 8 \\ - & - & - & - & - \end{bmatrix}$$

What we want is to find α and β so that $\dim M_5 = 1$ and $\dim \operatorname{Tor}_2^R(M, \mathbb{F})_5 = 3$. Taking into account that we ensured $\dim \operatorname{Tor}_2^R(M, \mathbb{F})_4 = 2$ this amounts to asking that the 100×102 matrix $m(\alpha, \beta)$ obtained from

$$[0 \leftarrow 5R(-2)_5 \leftarrow 11R(-3)_5 \leftarrow 2R(-4)_5 \leftarrow 0] \cong [0 \leftarrow 100\mathbb{F} \overset{m(\alpha,\beta)}{\leftarrow} 102\mathbb{F} \leftarrow 0]$$

drops rank by 1. We do not know a systematic approach to produce such $m(\alpha, \beta)$'s. However, we can find such matrices in a probabilistic way. In the space of the matrices $m(\alpha, \beta)$, those which drop rank by 1 have expected codimension 3. Hence over a finite field $\mathbb{F} = \mathbb{F}_q$ we expect to find the desired modules M with a probability of $1/q^3$. The code to detect bad modules is rather fast.

```
i16 : testModulesForGenus14Curves = (N,p) ->(
            x := local x;
            R := ZZ/p[x_0..x_3];
            i:=0;j:=0;
            kappa:=koszul(3,vars R);
            kappakappa:=kappa++kappa;
            utime:=timing while (i<N) do (
                  test:=false;
                  alpha:=random(R^{5:-2},R^{12:-2});
                  beta:=random(R^{5:-2},R^{3:-3});
                  M:=coker (alpha*kappakappa|beta);
                  fM:=res (M,DegreeLimit =>3);
                  if (tally degrees fM_2)_{5}==3 then (
                        --further checks to pick up the right module
                        test=(tally degrees fM_2)_{4}==2 and
                        codim M==4 and degree M==23;);
                  i=i+1;if test==true then (j=j+1;););
            timeForNModules:=utime#0; numberOfGoodModules:=j;
            {timeForNModules,numberOfGoodModules});

i17 : testModulesForGenus14Curves(1000,5)

o17 = {41.02, 10}

o17 : List
```

For timing tests we used a Pentium2 400Mhz with 128Mb of memory running GNU Linux. On such a machine examples can be tested at a rate of 0.04 seconds per example. Hence an approximate estimation of the CPU-time required to find a good example is $q^3 \cdot 0.04$ seconds. Comparing this with the time to verify smoothness, which is about 12 seconds for an example of this degree, we see that up to $|\mathbb{F}_q| = q \le 13$ we can expect to obtain examples within few minutes. Actually the computations for $q = 2$ and $q = 3$ take longer than for $q = 5$ on average, because examples of "good modules" tend to give singular curves more often. Here is a table of statistics which summarizes the situation.

q	2	3	5	7	11	13
smooth curves	100	100	100	100	100	100
1-nodal curves	75	53	31	16	10	8
reduced more singular	1012	142	24	11	2	0
non reduced curves	295	7	0	0	0	0
total number of curves	1482	302	155	127	112	108
percentage of smooth curves	6.7%	33%	65%	79%	89%	93%
approx. time (in seconds)	7400	3100	2700	3400	6500	9500

The procedure for obtaining a random genus 14 curve is

```
i18 : randomGenus14Curve = (R) -> (
           kappa:=koszul(3,vars R);
           kappakappa:=kappa++kappa;
           correctCodimAndDegree:=false;
           count:=0;while not correctCodimAndDegree do (
               test:=false;
               t:=timing while test==false do (
                   alpha=random(R^{5:-2},R^{12:-2});
                   beta=random(R^{5:-2},R^{3:-3});
                   M:=coker (alpha*kappakappa|beta);
                   fM:=res (M,DegreeLimit =>3);
                   if (tally degrees fM_2)_{5}==3 then (
                       --further checks to pick up the right module
                       test=(tally degrees fM_2)_{4}==2 and
                           codim M==4 and degree M==23;);
                   count=count+1;);
               Gt:=transpose (res M).dd_3;
               I:=ideal syz (Gt_{5..17});
               correctCodimAndDegree=(codim I==2 and degree I==14););
           <<"    -- "<<t#0<<" seconds used for ";
           <<count<<" modules"<<endl;
           I);
```

For $g = 15$ we do not know a method along these lines that would give examples over small fields.

Counting Parameters. For genus $g = 12$ clearly the module M depends on $\dim \mathbb{G}(7, 3 \cdot h^0\mathcal{O}(1)) - \dim SL(3) = 7 \cdot 5 - 8 = 36$ parameters, and the family of curves has dimension $36 + \dim \mathbb{G}(7, 10) = 48 = 4 \cdot 12 = 33 + 0 + 15$, as expected.

For genus $g = 13$ and 14 the parameter count is more difficult. Let us make a careful parameter count for genus $g = 14$; the case $g = 13$ is similar and easier. The choice of α corresponds to a point in $\mathbb{G}(5, 12)$. Then β corresponds to a point $\mathbb{G}(3, B_\alpha)$ where $B_\alpha = U \otimes R_1/<\alpha>$ where U denotes the universal subbundle on $\mathbb{G}(5, 12)$ and $<\alpha>$ the subspace generated by the 8 columns of $\alpha \circ (\kappa \oplus \kappa)$. So $\dim B_\alpha = 20 - 8 = 12$ and $\mathbb{G}(3, B_\alpha) \to \mathbb{G}(5, 12)$ is a Grassmannian bundle with fiber dimension 27 and total dimension 62. In this space the scheme of good modules has codimension 3, so we get a 59 dimensional family. This is larger than the expected dimension $56 = 4 \cdot 14 = 39 + 2 + 15$ of the Hilbert scheme, c.f. [19]. Indeed the construction gives a curve together with a basis of $\mathrm{Tor}_2^R(M, \mathbb{F})_4$. Subtracting the dimension of the group of the projective coordinate changes we arrive at the desired dimension $59 - 3 = 56$.

The unirationality of $\mathfrak{M}_{12}$ and $\mathfrak{M}_{13}$ can be proved by computer as in case $\mathfrak{M}_{11}$, while in case $g = 14$ we don't know the unirationality of the parameter space of the modules M with $\dim M_5 = 1$ and $\dim \mathrm{Tor}_2^R(M, \mathbb{F})_5 = 3$.

2 Comparing Green's Conjecture for Curves and Points

2.1 Syzygies of Canonical Curves

One of the most outstanding conjectures about free resolutions is Green's prediction for the syzygies of canonical curves.

A canonical curve $C \subset \mathbb{P}^{g-1}$, i.e., a linearly normal curve with $\mathcal{O}_C(1) \equiv \omega_C$, the canonical line bundle, is projectively normal by a result of Max Noether, and hence has a Gorenstein homogeneous coordinate ring and is 3-regular.

Therefore the Betti numbers of the free resolution of a canonical curve are symmetric, that is, $\beta_{j,j+1} = \beta_{g-2-j,g-j}$, and essentially only two rows of Betti numbers occur. The situation is summarized in the following table.

1	-	-	-	-	-	-	-	-	-	-	-	-
-	$\beta_{1,2}$	$\beta_{2,3}$	$\cdots$	$\beta_{p,p+1}$	$\cdots$	$\bullet$	$\cdots$	$\beta_{p,p+2}$	-	-	-	-
-	-	-	-	$\beta_{p,p+2}$	$\cdots$		$\cdots$	$\beta_{p,p+1}$	$\cdots$	$\beta_{2,3}$	$\beta_{1,2}$	-
-	-	-	-	-	-	-	-	-	-	-	-	1

The first p such that $\beta_{p,p+2} \neq 0$ is conjecturally precisely the Clifford index of the curve.

Conjecture 2.1 (Green [16]). Let C be a smooth canonical curve over $\mathbb{C}$. Then $\beta_{p,p+2} \neq 0$ if and only if $\exists L \in \operatorname{Pic}^d(C)$ with $h^0(C, L), h^1(C, L) \geq 2$ and $\operatorname{cliff}(L) := d - 2(h^0(C, L)) - 1) \leq p$. In particular, $\beta_{j,j+2} = 0$ for $j \leq \lfloor \frac{g-3}{2} \rfloor$ for a general curve of genus g.

The "if" part is proved by Green and Lazarsfeld in [18] and holds for arbitrary ground fields. For some partial results see [35,31,30,4,5,22,25]. The conjecture is known to be false for some (algebraically closed) fields of finite characteristic, e.g., genus $g = 7$ and characteristic char $\mathbb{F} = 2$; see [29].

2.2 Coble Self-Dual Sets of Points

The free resolution of a hyperplane section of a Cohen-Macaulay ring has the same Betti numbers. Thus we may ask for a geometric interpretation of the syzygies of $2g - 2$ points in $\mathbb{P}^{g-2}$ (hyperplane section of a canonical curve), or syzygies of a graded Artinian Gorenstein algebra with Hilbert function $(1, g - 2, g - 2, 1, 0, \ldots)$ (twice a hyperplane section). Any collection of $2g - 2$ points obtained as a hyperplane section of a canonical curve is special in the sense that it imposes only $2g - 3$ conditions on quadrics. An equivalent condition for points in linearly uniform position is that they are Coble (or Gale) self-dual; see [14]. Thus if we distribute the $2g - 2$ points into two collections each of $g - 1$ points, with, say, the first consisting of the coordinate points and the second corresponding to the rows of a $(g - 1) \times (g - 1)$ matrix

$A = (a_{ij})$, then A can be chosen to be an orthogonal matrix, i.e., $A^t A = 1$; see [14].

To see what the analogue of Green's Conjecture for the general curve means for orthogonal matrices we recall a result of [28].

Set $n = g - 2$. We identify the homogeneous coordinate ring of $\mathbb{P}^n$ with the ring $S = \mathbb{F}[\partial_0, \ldots, \partial_n]$ of differential operators with constant coefficients, $\partial_i = \frac{\partial}{\partial x_i}$. The ring S acts on $\mathbb{F}[x_0, \ldots, x_n]$ by differentiation. The annihilator of $q = x_0^2 + \ldots + x_n^2$ is a homogeneous ideal $J \subset S$ such that S/J is a graded Artinian Gorenstein ring with Hilbert function $(1, n+1, 1)$ and socle induced by q, see [27], [12, Section 21.2 and related exercise 21.7]. The syzygy numbers of S/J are

$$
\begin{vmatrix}
1 & - & - & - & - & - & - \\
- & \frac{n}{n+2}\binom{n+3}{2} & \cdots & \frac{p(n+1-p)}{n+2}\binom{n+3}{p+1} & \cdots & \frac{n}{n+2}\binom{n+3}{n+1} & - \\
- & - & - & - & - & - & 1
\end{vmatrix}
$$

A collection $H_0, \ldots, H_n$ of hyperplanes in $\mathbb{P}^n$ is said to form a polar simplex to q if and only if the collection $\Gamma = \{p_0, \ldots, p_n\} \subset \check{\mathbb{P}}^n$ of the corresponding points in the dual space has its homogeneous ideal $I_\Gamma \subset S$ contained in J.

In particular the set Λ consisting of the coordinate points correspond to a polar simplex, because $\partial_i \partial_j$ annihilates q for $i \neq j$.

For any polar collection of points Γ the free resolution $\mathbf{S}_\Gamma$ is a subcomplex of the resolution $\mathbf{S}_{S/J}$. Green's conjecture for the generic curve of genus $g = n + 2$ would imply:

Conjecture 2.2. For a general Γ and the given Λ the corresponding Tor-groups

$$
\operatorname{Tor}_k^S(S/I_\Gamma, \mathbb{F})_{k+1} \cap \operatorname{Tor}_k^S(S/I_\Lambda, \mathbb{F})_{k+1} \subset \operatorname{Tor}_k^S(S/J, \mathbb{F})_{k+1}
$$

intersect transversally.

Proof. A zero-dimensional non-degenerate scheme $\Gamma \subset \mathbb{P}^n$ of degree $n+1$ has syzygies

$$
\begin{vmatrix}
1 & - & - & - & - & - \\
- & \binom{n+1}{2} & \cdots & k\binom{n+1}{k+1} & \cdots & n
\end{vmatrix}
$$

Since both Tor groups are contained in $\operatorname{Tor}_k^S(S/J, \mathbb{F})_{k+1}$, the claim is equivalent to saying that for a general polar simplex Γ the expected dimension of their intersection is $\dim \operatorname{Tor}_k^S(S/\Gamma, \mathbb{F})_{k+1} + \dim \operatorname{Tor}_k^S(S/\Gamma, \mathbb{F})_{k+1} - \dim \operatorname{Tor}_k^S(S/J, \mathbb{F})_{k+1}$, which is

$$
2k\binom{g-1}{k+1} - \frac{k(g-1-k)}{g}\binom{g+1}{k+1}.
$$

On the other hand, $I_{\Gamma \cup \Lambda} = I_\Gamma \cap I_\Lambda$, hence

$$\mathrm{Tor}_k^S(S/I_\Gamma, \mathbb{F})_{k+1} \cap \mathrm{Tor}_k^S(S/I_\Lambda, \mathbb{F})_{k+1} = \mathrm{Tor}_k^S(S/I_{\Gamma \cup \Lambda}, \mathbb{F})_{k+1},$$

and Green's conjecture would imply

$$\dim \mathrm{Tor}_k^S(S/I_{\Gamma \cup \Lambda}, \mathbb{F})_{k+1} = \beta_{k,k+1}(\Gamma \cup \Lambda) = k\binom{g-2}{k+1} - (g-1-k)\binom{g-2}{k-2}.$$

Now a calculation shows that the two dimensions above are equal.

The family of all polar simplices V is dominated by the family defined by the ideal of 2×2 minors of the matrix

$$\begin{pmatrix} \partial_0 & \cdots & \partial_i & \cdots & \partial_n \\ \sum_j b_{0j} \partial_j & \cdots & \sum_j b_{ij} \partial_j & \cdots & \sum_j b_{nj} \partial_j \end{pmatrix}$$

depending on a symmetric matrix $B = (b_{ij})$, i.e., $b_{ij} = b_{ji}$ as parameters. For B a general diagonal matrix we get Λ together with a specific element in $\mathrm{Tor}_n^S(S/I_\Lambda, \mathbb{F})_{n+1}$.

2.3 Comparison and Probes

One of the peculiar consequences of Green's conjecture for odd genus $g = 2k+1$ is that, if $\beta_{k,k+1} = \beta_{k-1,k+1} \neq 0$, then the curve C lies in the closure of the locus of $k+1$-gonal curve. Any $k+1$-gonal curve lies on a rational normal scroll X of codimension k that satisfies $\beta_{k,k+1}(X) = k$. Hence

$$\beta_{k,k+1}(C) \neq 0 \Rightarrow \beta_{k,k+1}(C) \geq k$$

We may ask whether a result like this is true for the union of two polar simplices $\Lambda \cup \Gamma \subset \mathbb{P}^{2k-1}$. Define

$$\tilde{D} = \{\Gamma \in V | \Gamma \cup \Lambda \text{ is syzygy special}\}$$

where, as above, V denotes the variety of polar simplices and Λ denotes the coordinate simplex. If $\tilde{D}$ is a proper subvariety, then it is a divisor, because $\beta_{k,k+1}(\Gamma \cup \Lambda) = \beta_{k-1,k+1}(\Gamma \cup \Lambda)$.

Conjecture 2.3. The subscheme $\tilde{D} \subset V$ is an irreducible divisor, for $g = n+2 = 2k+1 \in \{7, 9, 11\}$. The value of $\beta_{k,k+1}$ on a general point of D is $3, 2, 1$ respectively.

We can prove this for $g = 7$ by computer algebra. For $g = 9$ and $g = 11$ a proof is computationally out of reach with our methods, but we can get some evidence from examples over finite fields.

Evidence. Since $\tilde{D}$ is a divisor, we expect that if we pick symmetric matrices B over $\mathbb{F}_q$ at random, we will hit points on every component of $\tilde{D}$ at a probability of $1/q$. For a general point on $\tilde{D}$ the corresponding Coble self-dual set of points will have the generic number of extra syzygies of that component. Points with even more syzygies will occur in higher codimension, hence only with a probability of $1/q^2$. Some evidence for irreducibility can be obtained from the Weil formulas: for sufficiently large q we should see points on $\tilde{D}$ with probability $Cq^{-1}+O(q^{-\frac{3}{2}})$, where C is the number of components.

The following tables give for small fields $\mathbb{F}_q$ the number s_i of examples with i extra syzygies in a test of 1000 examples for $g = 9$ and 100 examples for $g = 11$. The number $s_{\mathrm{tot}} = \sum_{i>0} s_i$ is the total number of examples with extra syzygies.

Genus $g = 9$:

q	$1000/q$	s_{tot}	s_1	s_2	s_3	s_4
2	500	925	0	130	0	63
3	333	782	0	273	0	33
4	250	521	0	279	0	99
5	200	350	0	217	0	74
7	143	197	0	144	0	36
8	125	199	0	147	0	43
9	111	218	0	98	0	0
11	91	118	0	102	0	15
13	77	90	0	79	0	10
16	62	72	0	68	0	4
17	59	76	0	69	0	6

Genus $g = 11$:

q	$100/q$	s_{tot}	s_1	s_2	s_3	s_4
7	14.3	16	14	0	0	0
17	5.9	7	7	0	0	0

In view of these numbers, it is more likely that the set $\tilde{D}$ of syzygy special Coble points is irreducible than that it is reducible. For a more precise statement we refer to [6].

A test of Green's Conjecture for Curves. In view of 2.3 it seems plausible that for a general curve of odd genus $g \geq 11$ with $\beta_{k,k+1}(C) \neq 0$ the value might be $\beta_{k,k+1} = 1$ contradicting Green's conjecture. It is clear that the syzygy exceptional locus has codimension 1 in $\mathfrak{M}_g$ for odd genus, if it is proper, i.e., if Green's conjecture holds for the general curve of that genus. So picking points at random we might be able to find such curve over a finite field $\mathbb{F}_q$ with probability $1/q$, roughly.

Writing code that does this is straightforward. One makes a loop that picks up randomly a curve, computes its canonical image, and resolves its ideal, counting the possible values $\beta_{k,k+1}$ until a certain amount of special curves is reached. The result for 10 special curves in $\mathbb{F}_7$ is as predicted:

		total	special	possible values of $\beta_{k,k+1}$					
g	seconds	curves	curves	≤ 2	3	4	5	6	≥ 7
7	148	75	10	0	10	0	0	0	0
9	253	58	10	0	0	9	0	0	1
11	25640	60	10	0	0	0	9	0	1

(The test for genus 9 and 11 used about 70 and 120 megabytes of memory, respectively.)

So Green's conjecture passed the test for $g = 9, 11$. Shortly after the first author tried this test for the first time, a paper of Hirschowitz and Ramanan appeared proving this in general:

Theorem 2.4 ([22]). *If the general curve of odd genus $g = 2k + 1$ satisfies Green's conjecture then the syzygy special curves lie on the divisor $D = \{C \in \mathfrak{M}_g | W^1_{k+1}(C) \neq \emptyset\}$*

The theorem gives strong evidence for the full Green's conjecture in view of our study of Coble self-dual sets of points.

Our findings suggest that the variety of points arising as hyperplane sections of smooth canonical curves has the strange property that it intersects the divisor of syzygy special sets of points $\tilde{D}$ only in its singular locus.

The conjecture for general curves is known to us up to $g \leq 17$, which is as far as a computer allows us to do a ribbon example; see [4].

3   Pfaffian Calabi-Yau Threefolds in $\mathbb{P}^6$

Calabi-Yau 3-folds caught the attention of physicists because they can serve as the compact factor of the Kaluza-Klein model of spacetime in superstring theory. One of the remarkable things that grows out of the work in physics is the discovery of mirror symmetry, which associates to a family of Calabi-Yau 3-folds (M_λ), another family (W_μ) whose Hodge diamond is the mirror of the Hodge diamond of the original family.

Although there is an enormous amount of evidence at present, the existence of a mirror is still a hypothesis for general Calabi-Yau 3-folds. The thousands of cases where this was established all are close to toric geometry, where through the work of Batyrev and others [3,9] rigorous mirror constructions were given and parts of their conjectured properties proved.

From a commutative algebra point of view the examples studied so far are rather trivial, because nearly all are hypersurfaces or complete intersections on toric varieties, or zero loci of sections in homogeneous bundles on homogeneous spaces.

Of course only a few families of Calabi-Yau 3-folds should be of this kind. Perhaps the easiest examples beyond the toric/homogeneous range are Calabi-Yau 3-folds in $\mathbb{P}^6$. Here examples can be obtained by the Pfaffian construction of Buchsbaum-Eisenbud [7] with vector bundles; see section 3.1 below. Indeed a recent theorem of Walter [36] says that any smooth Calabi-Yau in $\mathbb{P}^6$ can be obtained in this way. In this section we report on our construction of such examples.

As is quite usual in this kind of problem, there is a range where the construction is still quite easy, e.g., for surfaces in $\mathbb{P}^4$ the work in [10,26] shows that the construction of nearly all the 50 known families of smooth non-general type surfaces is straight forward and their Hilbert scheme component unirational. Only in very few known examples is the construction more difficult and the unirationality of the Hilbert scheme component an open problem.

The second author did the first "non-trivial" case of a construction of Calabi-Yau 3-folds in $\mathbb{P}^6$. Although in the end the families turned out to be unirational, the approach utilized small finite field constructions as a research tool.

3.1   The Pfaffian Complex

Let $\mathcal{F}$ be a vector bundle of odd rank $\operatorname{rk} \mathcal{F} = 2r + 1$ on a projective manifold M, and let $\mathcal{L}$ be a line bundle. Let $\varphi \in H^0(M, \Lambda^2 \mathcal{F} \otimes \mathcal{L})$ be a section. We can think of φ as a skew-symmetric twisted homomorphism

$$\mathcal{F}^* \xrightarrow{\varphi} \mathcal{F} \otimes \mathcal{L}.$$

The rth divided power of φ is a section $\varphi^{(r)} = \frac{1}{r!}(\varphi \wedge \cdots \wedge \varphi) \in H^0(M, \Lambda^{2r} \mathcal{F} \otimes \mathcal{L}^r)$. Wedge product with $\varphi^{(r)}$ defines a morphism

$$\mathcal{F} \otimes \mathcal{L} \xrightarrow{\psi} \Lambda^{2r+1} \mathcal{F} \otimes \mathcal{L}^{r+1} = \det(\mathcal{F}) \otimes \mathcal{L}^{r+1}.$$

The twisted image $\mathcal{I} = \operatorname{image}(\psi) \otimes \det(\mathcal{F}^*) \otimes \mathcal{L}^{-r-1} \subset \mathcal{O}_M$ is called the *Pfaffian ideal* of φ, because working locally with frames, it is given by the ideal generated by the $2r \times 2r$ principle Pfaffians of the matrix describing φ. Let $\mathcal{D}$ denote the determinant line bundle $\det(\mathcal{F}^*)$.

Theorem 3.1 (Buchsbaum-Eisenbud [7]). *With this notation*

$$0 \to \mathcal{D}^2 \otimes \mathcal{L}^{-2r-1} \xrightarrow{\psi^t} \mathcal{D} \otimes \mathcal{L}^{-r-1} \otimes \mathcal{F}^* \xrightarrow{\varphi} \mathcal{F} \otimes \mathcal{D} \otimes \mathcal{L}^{-r} \xrightarrow{\psi} \mathcal{O}_M$$

is a complex. $X = V(\mathcal{I}) \subset M$ has codimension ≤ 3 at every point, and in case equality holds (everywhere along X) then this complex is exact and resolves the structure sheaf $\mathcal{O}_X = \mathcal{O}_M/\mathcal{I}$ of the locally Gorenstein subscheme X.

We will apply this to construct Calabi-Yau 3-folds in $\mathbb{P}^6$. In that case we want X to be smooth and $\det(\mathcal{F})^{-2} \otimes \mathcal{L}^{-2r-1} \cong \omega_\mathbb{P} \cong \mathcal{O}(-7)$, so we may conclude that $\omega_X \cong \mathcal{O}_X$. A result of Walter [36] for $\mathbb{P}^n$ guarantees the existence of a Pfaffian presentation in $\mathbb{P}^6$ for every subcanonical embedded 3-fold. Moreover Walter's choice of $\mathcal{F} \otimes \mathcal{D} \otimes \mathcal{L}^{-r}$ for Calabi-Yau 3-folds $X \subset \mathbb{P}^6$ is the sheafified first syzygy module $H^1_*(\mathcal{I}_X)$ plus possibly a direct sum of line bundles (indeed $H^2_*(\mathcal{I}_X) = 0$ because of the Kodaira vanishing theorem). Under the maximal rank assumption for

$$H^0(\mathbb{P}^6, \mathcal{O}(m)) \to H^0(X, \mathcal{O}_X(m))$$

the Hartshorne-Rao module is zero for $d = \deg X \in \{12, 13, 14\}$ and an arithmetically Cohen-Macaulay X is readily found. For $d \in \{15, 16, 17, 18\}$ the Hartshorne-Rao modules M have Hilbert functions with values $(0, 0, 1, 0, \ldots)$, $(0, 0, 2, 1, 0, \ldots)$, $(0, 0, 3, 5, 0, \ldots)$ and $(0, 0, 4, 9, 0, \ldots)$ respectively.

We do not discuss the cases $d \leq 16$ further. The construction in those cases is obvious; see [34].

3.2   Analysis of the Hartshorne-Rao Module for Degree 17

Denote with $\mathcal{F}_1$ the sheaf $\mathcal{F} \otimes \mathcal{D} \otimes \mathcal{L}^{-r}$. We try to construct $\mathcal{F}_1$ as the sheafified first syzygy module of M. The construction of a module with the desired Hilbert function is straightforward. The cokernel of $3S(-2) \xleftarrow{b} 16S(-3)$ for a general matrix of linear forms has this property. However, for a general b and $\mathcal{F}_1 = \ker(16\mathcal{O}(-3) \xrightarrow{b} 3\mathcal{O}(-2))$ the space of skew-symmetric maps $\mathrm{Hom}_{\mathrm{skew}}(\mathcal{F}_1^*(-7), \mathcal{F}_1)$ is zero: M has syzygies

$$
\begin{array}{|ccccccc}
3 & 16 & 28 & - & - & - & - \\
- & - & - & 70 & 112 & 84 & 32 & 5
\end{array}
$$

Any map $\varphi \colon \mathcal{F}_1^*(-7) \to \mathcal{F}_1$ induces a map on the free resolutions:

$$
\begin{array}{ccccccc}
0 \longleftarrow & \mathcal{F}_1 & \longleftarrow 28\mathcal{O}(-4) & \longleftarrow 70\mathcal{O}(-6) & \longleftarrow 112\mathcal{O}(-7) \\
& \uparrow{\scriptstyle \varphi} & \uparrow{\scriptstyle \varphi_0} & \uparrow{\scriptstyle \varphi_1} & \\
0 \longleftarrow & \mathcal{F}_1^*(-7) & \longleftarrow 16\mathcal{O}(-4) & \longleftarrow 3\mathcal{O}(-5) & \longleftarrow 0
\end{array}
$$

Since $\varphi_1 = 0$ for degree reasons, $\varphi = 0$ as well, and $\mathrm{Hom}(\mathcal{F}_1^*(-7), \mathcal{F}_1) = 0$ for a general module M.

What we need are special modules M that have extra syzygies

$$
\begin{array}{|ccccccc}
3 & 16 & 28 & k & - & - & - \\
- & - & k & 70 & 112 & 84 & 32 & 5
\end{array}
$$

with k at least 3.

In a neighborhood of $o \in \operatorname{Spec} B$, where denotes B the base space of a semi-universal deformation of M, the resolution above would lift to a complex over $B[x_0, \ldots, x_6]$ and in the lifted complex there is a $k \times k$ matrix Δ with entries in the maximal ideal $o \subset B$. By the principal ideal theorem we see that Betti numbers stay constant in a subvariety of codimension at most k^2. To check for second linear syzygies on a randomly chosen M is a computationally rather easy task. The following procedure tests the computer speed of this task.

```
i19 : testModulesForDeg17CY = (N,k,p) -> (
        x:=symbol x;R:=(ZZ/p)[x_0..x_6];
        numberOfGoodModules:=0;i:=0;
        usedTime:=timing while (i<N) do (
            b:=random(R^3,R^{16:-1});
            --we put SyzygyLimit=>60 because we expect
            --k<16 syzygies, so 16+28+k<=60
            fb:=res(coker b,
                DegreeLimit =>0,SyzygyLimit=>60,LengthLimit =>3);
            if rank fb_3>=k and (dim coker b)==0 then (
                fb=res(coker b, DegreeLimit =>0,LengthLimit =>4);
                if rank fb_4==0
                then numberOfGoodModules=numberOfGoodModules+1;);
            i=i+1;);
        collectGarbage();
        timeForNModules:=usedTime#0;
        {timeForNModules,numberOfGoodModules});
```

Running this procedure we see that it takes not more than 0.64 seconds per example. Hence we can hope to find examples with $k = 3$ within a reasonable time for a very small field, say $\mathbb{F}_3$.

The first surprise is that examples with k extra syzygies are found much more often, as can be seen by looking at the second value output by the function testModulesForDeg17CY().

This is not only a "statistical" remark, in the sense that the result is confirmed by computing the semi-universal deformations of these modules. Indeed define $\mathbb{M}_k = \{M \mid \operatorname{Tor}_3^S(M, \mathbb{F})_5 \geq k\}$ and consider a module $M \in \mathbb{M}_k$: "generically" we obtain $\operatorname{codim}(\mathbb{M}_k)_M = k$ instead of k^2 (and in fact one can diagonalize the matrix Δ over the algebraic closure $\bar{\mathbb{F}}$).

The procedure is straightforward but a bit long. First we pick up an example with k-extra syzygies.

```
i20 : randomModuleForDeg17CY = (k,R) -> (
        isGoodModule:=false;i:=0;
        while not isGoodModule do (
            b:=random(R^3,R^{16:-1});
            --we put SyzygyLimit=>60 because we expect
            --k<16 syzygies, so 16+28+k<=60
            fb:=res(coker b,
                DegreeLimit =>0,SyzygyLimit=>60,LengthLimit =>3);
            if rank fb_3>=k and (dim coker b)==0 then (
                fb=res(coker b, DegreeLimit =>0,LengthLimit =>4);
                if rank fb_4==0 then isGoodModule=true;);
            i=i+1;);
        <<"     -- Trial n. " << i <<", k="<< rank fb_3 <<endl;
        b);
```

Notice that the previous function returns a presentation matrix b of M, and not M.

Next we compute the tangent codimension of $\mathbb{M}_k$ in the given example $M = \operatorname{Coker} b$ by computing the codimension of the space of the infinitesimal deformations of M that still give an element in $\mathbb{M}_k$. Denote with b_i the maps in the linear strand of a minimal free resolution of M, and with b'_2 the quadratic part in the second map of this resolution. Over $B = \mathbb{F}[\epsilon]/\epsilon^2$ let $b_1 + \epsilon f_1$ be an infinitesimal deformation of b_1. Then f_1 lifts to a linear map $f_2 \colon 28S(-4) \to 16S(-3)$ determined by $(b_1 + \epsilon f_1) \circ (b_2 + \epsilon f_2) = 0$, and f_2 to a map $f_3 \oplus \Delta \colon kS(-5) \to 28S(-4) \oplus kS(-5)$ determined by $(b_2 + b'_2) \circ \epsilon (f_3 \oplus \Delta) = 0$. Therefore we can determine Δ as:

```
i21 : getDeltaForDeg17CY = (b,f1) -> (
        fb:=res(coker b, LengthLimit =>3);
        k:=numgens target fb.dd_3-28; --# of linear syzygies
        b1:=fb.dd_1;b2:=fb.dd_2_{0..27};b2':=fb.dd_2_{28..28+k-1};
        b3:=fb.dd_3_{0..k-1}^{0..27};
        --the equation for f2 is b1*f2+f1*b2=0,
        --so f2 is a lift of (-f1*b2) through b1
        f2:=-(f1*b2)//b1;
        --the equation for A=(f3||Delta) is -f2*b3 = (b2|b2') * A
        A:=(-f2*b3)//(b2|b2');
        Delta:=A^{28..28+k-1});
```

Now we just parametrize all possible maps $f_1 \colon 16S(-3) \to 3S(-2)$, compute their respective maps Δ, and find the codimension of the condition that Δ is the zero map:

```
i22 : codimInfDefModuleForDeg17CY = (b) -> (
        --we create a parameter ring for the matrices f1's
        R:=ring b;K:=coefficientRing R;
        u:=symbol u;U:=K[u_0..u_(3*16*7-1)];
        i:=0;while i<3 do (
            <<endl<< " " << i+1 <<":" <<endl;
            j:=0;while j<16 do(
                << "    " << j+1 <<". "<<endl;
                k:=0;while k<7 do (
                    l=16*7*i+7*j+k; --index parametrizing the f1's
                    f1:=matrix(R,apply(3,m->apply(16,n->
                        if m==i and n==j then x_k else 0)));
                    Delta:=substitute(getDeltaForDeg17CY(b,f1),U);
                    if l==0 then (equations=u_l*Delta;) else (
                        equations=equations+u_l*Delta;);
                    k=k+1;);
                collectGarbage(); --frees up memory in the stack
                j=j+1;);
            i=i+1;);
        codim ideal equations);
```

The second surprise is that for $\mathcal{F}_1 = syz_1(M)$ we find

$$\dim \operatorname{Hom}_{skew}(\mathcal{F}_1^*(-7), \mathcal{F}_1) = k = \dim \operatorname{Tor}_3^S(M, \mathbb{F})_5.$$

$\operatorname{Hom}_{skew}(\mathcal{F}_1^*(-7), \mathcal{F}_1)$ is the vector space of skew-symmetric linear matrices φ such that $b \circ \varphi = 0$. The following procedure gives a matrix of size $\binom{16}{2} \times$

$\dim \mathrm{Hom}_{\mathrm{skew}}(\mathcal{F}_1^*(-7), \mathcal{F}_1)$ whose i-th column gives the entries of a 16×16 skew-symmetric matrix inducing the i-th basis element of the vector space $\mathrm{Hom}_{\mathrm{skew}}(\mathcal{F}_1^*(-7), \mathcal{F}_1)$.

```
i23 : skewSymMorphismsForDeg17CY = (b) -> (
          --we create a parameter ring for the morphisms:
          K:=coefficientRing ring b;
          u:=symbol u;U:=K[u_0..u_(binomial(16,2)-1)];
          --now we compute the equations for the u_i's:
          UU:=U**ring b;
          equationsInUU:=flatten (substitute(b,UU)*
              substitute(genericSkewMatrix(U,u_0,16),UU));
          uu:=substitute(vars U,UU);
          equations:=substitute(
              diff(uu,transpose equationsInUU),ring b);
          syz(equations,DegreeLimit =>0));
```

A morphism parametrized by a column skewSymMorphism is then recovered by the following code.

```
i24 : getMorphismForDeg17CY = (SkewSymMorphism) -> (
          u:=symbol u;U:=K[u_0..u_(binomial(16,2)-1)];
          f=map(ring SkewSymMorphism,U,transpose SkewSymMorphism);
          f genericSkewMatrix(U,u_0,16));
```

Rank 1 Linear Syzygies of M. To understand this phenomenon we consider the multiplication tensor of M:

$$\mu: M_2 \otimes V \to M_3$$

where $V = H^0(\mathbb{P}^6, \mathcal{O}(1))$.

Definition 3.2. A decomposable element of $M_2 \otimes V$ in the kernel of μ is called a rank 1 linear syzygy of M. The (projective) space of rank 1 syzygies is

$$Y = (\mathbb{P}^2 \times \mathbb{P}^6) \cap \mathbb{P}^{15} \subset \mathbb{P}^{20}$$

where $\mathbb{P}^2 = \mathbb{P}(M_2^*), \mathbb{P}^6 = \mathbb{P}(V^*)$ and $\mathbb{P}^{15} = \mathbb{P}(\ker(\mu)^*)$ inside the Segre space $\mathbb{P}((M_2 \otimes V)^*) \cong \mathbb{P}^{20}$.

Proposition 1.5 of [17] says that, for $\dim M_2 \leq j$, the existence of a jth linear syzygy implies $\dim Y \geq j - 1$. This is automatically satisfied for $j = 3$ in our case: $\dim Y \geq 3$ with equality expected.

The projection $Y \to \mathbb{P}^2$ has linear fibers, and the general fiber is a $\mathbb{P}^1$. However, special fibers might have higher dimension. In terms of the presentation matrix b a special 2-dimensional fiber (defined over $\mathbb{F}$) corresponds to a block

$$b = \begin{pmatrix} 0 & 0 & 0 & * & \dots \\ 0 & 0 & 0 & * & \dots \\ l_1 & l_2 & l_3 & * & \dots \end{pmatrix},$$

where l_1, l_2, l_3 are linear forms, in the 3×16 presentation matrix of M. Such a block gives a

$$\begin{array}{|cccccccc}
1 & 3 & 3 & 1 & - & - & - & - \\
- & - & - & - & - & - & - & -
\end{array}$$

subcomplex in the free resolution of M and an element $s \in H^0(\mathbb{P}^6, \Lambda^2 \mathcal{F}_1 \otimes \mathcal{O}(7))$ since the syzygy matrix

$$\begin{pmatrix} 0 & -l_3 & l_2 \\ l_3 & 0 & -l_1 \\ -l_2 & l_1 & 0 \end{pmatrix}$$

is skew.

This answers the questions posed by both surprises: we want a module M with at least $k \geq 3$ special fibers and these satisfy $h^0(\mathbb{P}^6, \Lambda^2 \mathcal{F}_1 \otimes \mathcal{O}(7)) \geq k$, if the k sections are linearly independent. The condition for k special fibers is of expected codimension k in the parameter space $\mathbb{G}(16, 3h^0(\mathbb{P}^6, \mathcal{O}(1))$ of the presentation matrices. In a given point M the actual codimension can be readily computed by a first order deformations and that $H^0(\mathbb{P}^6, \Lambda^2 \mathcal{F}_1 \otimes \mathcal{O}(7))$ is k-dimensional, and spanned by the k sections corresponding to the k special fibers can be checked as well.

First we check that M has k distinct points in $\mathbb{P}(M_2^*)$ where the multiplication map drops rank. (Note that this condition is likely to fail over small fields. However, the check is computationally easy).

```
i25 : checkBasePtsForDeg17CY = b -> (
          --firstly the number of linear syzygies
          fb:=res(coker b, DegreeLimit=>0, LengthLimit =>4);
          k:=#select(degrees source fb.dd_3,i->i=={3});
          --then the check
          a=symbol a;A=K[a_0..a_2];
          mult:=(id_(A^7)**vars A)*substitute(
              syz transpose jacobian b,A);
          basePts=ideal mingens minors(5,mult);
          codim basePts==2 and degree basePts==k and distinctPoints(
              basePts));
```

Next we check that $H^0(\mathbb{P}^6, \Lambda^2 \mathcal{F}_1 \otimes \mathcal{O}(7))$ is k-dimensional, by looking at the numbers of columns of skewSymMorphismsForDeg17CY(b). Finally we do the computationally hard part of the check, which is to verify that the k special sections corresponding to the k special fibers of $Y \to \mathbb{P}^2$ span $H^0(\mathbb{P}^6, \Lambda^2 \mathcal{F}_1 \otimes \mathcal{O}(7))$.

```
i26 : checkMorphismsForDeg17CY = (b,skewSymMorphisms) -> (
          --first the number of linear syzygies
          fb:=res(coker b, DegreeLimit=>0, LengthLimit =>4);
          k:=#select(degrees source fb.dd_3,i->i=={3});
          if (numgens source skewSymMorphisms)!=k then (
              error "the number of skew-sym morphisms is wrong";);
          --we parametrize the morphisms:
          R:=ring b;K:=coefficientRing R;
          w:=symbol w;W:=K[w_0..w_(k-1)];
          WW:=R**W;ww:=substitute(vars W,WW);
```

```
genericMorphism:=getMorphismForDeg17CY(
    substitute(skewSymMorphisms,WW)*transpose ww);
--we compute the scheme of the 3x3 morphisms:
equations:=mingens pfaffians(4,genericMorphism);
equations=diff(
    substitute(symmetricPower(2,vars R),WW),equations);
equations=saturate ideal flatten substitute(equations,W);
CorrectDimensionAndDegree:=(
    dim equations==1 and degree equations==k);
isNonDegenerate:=#select(
    (flatten degrees source gens equations),i->i==1)==0;
collectGarbage();
isOK:=CorrectDimensionAndDegree and isNonDegenerate;
if isOK then (
    --in this case we also look for a skew-morphism f
    --which is a linear combination of the special
    --morphisms with all coefficients nonzero.
    isGoodMorphism:=false;while isGoodMorphism==false do (
        evRandomMorphism:=random(K^1,K^k);
        itsIdeal:=ideal(
            vars W*substitute(syz evRandomMorphism,W));
        isGoodMorphism=isGorenstein(
            intersect(itsIdeal,equations));
        collectGarbage());
    f=map(R,WW,vars R|substitute(evRandomMorphism,R));
    randomMorphism:=f(genericMorphism);
    {isOK,randomMorphism}) else {isOK});
```

The code above is structured as follows. First we parametrize the skew-symmetric morphisms with new variables. The ideal of 4×4 Pfaffians is generated by forms of bidegree $(2,2)$ over $\mathbb{P}^6 \times \mathbb{P}^{k-1}$. We are interested in points $p \in \mathbb{P}^{k-1}$ such that the whole fiber $\mathbb{P}^6 \times \{p\}$ is contained in the zero locus of the Pfaffian ideal. The next two lines produce the ideal of these points on $\mathbb{P}^{k-1}$. Since we already know of k distinct points by the previous check, it suffices to establish that the set consists of collection of k spanning points. Finally, if this is the case, a further point, i.e., a further skew morphism, is a linear combination with all coefficients non-zero, if and only if the union with this point is a Gorenstein set of $k + 1$ points in $\mathbb{P}^{k-1}$.

```
i27 : isGorenstein = (I) -> (
        codim I==length res I and rank (res I)_(length res I)==1);
```

It is clear that all 16 relations should take part in the desired skew homomorphism $\mathcal{F}_1^*(-7) \xrightarrow{\varphi} \mathcal{F}_1$. Thus we need $k \geq 6$ to have a chance for a Calabi-Yau. Since $3 \cdot 5 < 16$ it easy to guarantee 5 special fibers by suitable choice of the presentation matrix. So the condition $k \geq 6$ is only of codimension $k - 5$ on this subspace, and we have a good chance to find a module of the desired type.

```
i28 : randomModule2ForDeg17CY = (k,R) -> (
        isGoodModule:=false;i:=0;
        while not isGoodModule do (
            b:=(random(R^1,R^{3:-1})++
                random(R^1,R^{3:-1})++
                random(R^1,R^{3:-1})|
                matrix(R,{{1},{1},{1}})**random(R^1,R^{3:-1})|
                random(R^3,R^1)**random(R^1,R^{3:-1})|
```

```
        random(R^3,R^{1:-1}));
    --we put SyzygyLimit=>60 because we expect
    --k<16 syzygies, so 16+28+k<=60
    fb:=res(coker b,
        DegreeLimit =>0,SyzygyLimit=>60,LengthLimit =>3);
    if rank fb_3>=k and dim coker b==0 then (
        fb=res(coker b, DegreeLimit =>0,LengthLimit =>4);
        if rank fb_4==0 then isGoodModule=true;);
    i=i+1;);
<<"    -- Trial n. " << i <<", k="<< rank fb_3 <<endl;
b);
```

Some modules M with $k = 8, 9, 11$ lead to smooth examples of Calabi-Yau 3-folds X of degree 17. To check the smoothness via the Jacobian criterion is computationally too heavy for a common computer today. For a way to speed up this computation considerably and to reduce the required amount of memory to a reasonable value (128MB), we refer to [34].

Since $h^0(\mathbb{P}^6, \Lambda^2 \mathcal{F}_1 \otimes \mathcal{O}(7)) = k$ and $\mathrm{codim}\{M \mid \mathrm{Tor}_3^S(M, \mathbb{F})_5 \geq k\} = k$ all three families have the same dimension. In particular no family lies in the closure of another.

A deformation computation verifies $h^1(X, \mathcal{T}) = h^1(X, \Omega^2) = 23$. Hence a computation of the Hodge numbers $h^q(X, \Omega^p)$ gives the diamond

$$
\begin{array}{ccccccc}
 & & & 1 & & & \\
 & & 0 & & 0 & & \\
 & 0 & & 1 & & 0 & \\
1 & & 23 & & 23 & & 1 \\
 & 0 & & 1 & & 0 & \\
 & & 0 & & 0 & & \\
 & & & 1 & & & \\
\end{array}
$$

Example 3.3. The following commands give an example of a Calabi-Yau 3-fold in $\mathbb{P}^6$:

```
i29 : K=ZZ/13;

i30 : R=K[x_0..x_6];

i31 : time b=randomModule2ForDeg17CY(8,R);
      -- Trial n. 1757, k=8
      -- used 764.06 seconds

                3       16
o31 : Matrix R  <--- R

i32 : betti res coker b

o32 = total: 3 16 36 78 112 84 32 5
          0: 3 16 28  8  .  .  .  .
          1: .  .  8 70 112 84 32 5

i33 : betti (skewSymMorphisms=skewSymMorphismsForDeg17CY b)

o33 = total: 120 8
         -1: 120 8
```

We check whether the base points in M_0 are all distinct.

```
i34 : checkBasePtsForDeg17CY b
```

```
o34 = true
```

Now we check whether the k sections span the morphisms. If we get `true` then this is a good module.

```
i35 : finalTest=checkMorphismsForDeg17CY(b,skewSymMorphisms);
```

```
i36 : finalTest#0
```

```
o36 = true
```

We pick up a random morphism involving all k sections.

```
i37 : n=finalTest#1;
```

```
               16         16
o37 : Matrix R     <--- R
```

If all the tests are okay, there should be a high degree syzygy.

```
i38 : betti (nn=syz n)
```

```
o38 = total: 16 4
         1: 16 3
         2:  . .
         3:  . 1
```

```
i39 : n2t=transpose submatrix(nn,{0..15},{3});
```

```
              1         16
o39 : Matrix R     <--- R
```

```
i40 : b2:=syz b;
```

```
              16         36
o40 : Matrix R     <--- R
```

Finally, compute the ideal of the Calabi-Yau 3-fold in $\mathbb{P}^6$.

```
i41 : j:=ideal mingens ideal flatten(n2t*b2);
```

```
o41 : Ideal of R
```

```
i42 : degree j
```

```
o42 = 17
```

```
i43 : codim j
```

```
o43 = 3
```

```
i44 : betti res j
```

```
o44 = total: 1 20 75 113 84 32 5
         0: 1  .   .    .   .  . .
         1: .  .   .    .   .  . .
         2: .  .   .    .   .  . .
         3: . 12   5    .   .  . .
         4: .  8  70  113  84 32 5
```

3.3   Lift to Characteristic Zero

At this point we have constructed Calabi-Yau 3-folds $X \subset \mathbb{P}^6$ over the finite field $\mathbb{F}_5$ or $\mathbb{F}_7$. However, our main interest is the field of complex numbers $\mathbb{C}$. The existence of a lift to characteristic zero follows by the following argument.

The set $\mathbb{M}_k = \{M \mid \mathrm{Tor}_3^S(M,\mathbb{F})_5 \geq k\}$ has codimension at most k. A deformation calculation shows that at our special point $M^{\mathrm{special}} \in \mathbb{M}(\mathbb{F}_p)$ the codimension is achieved and that $\mathbb{M}_k$ is smooth at this point. Thus taking a transversal slice defined over $\mathbb{Z}$ through this point we find a number field K and a prime $\mathfrak{p}$ in its ring of integers O_K with $O_K/\mathfrak{p} \cong \mathbb{F}_p$ such that M^{special} is the specialization of an $O_{K,\mathfrak{p}}$-valued point of $\mathbb{M}_k$. Over the generic point of $\mathrm{Spec}\, O_{K,\mathfrak{p}}$ we obtain a K-valued point. From our computations with checkBasePtsForDeg17CY() and checkMorphismsForDeg17CY(), which explained why $h^0(\mathbb{P}^6, \Lambda^2 \mathcal{F}_1^{\mathrm{special}} \otimes \mathcal{O}(7)) = k$, it follows that

$$H^0(\mathbb{P}_{\mathbb{Z}}^6 \times \mathrm{Spec}\, O_{K,\mathfrak{p}}, \Lambda^2 \mathcal{F}_1 \otimes \mathcal{O}(7))$$

is free of rank k over $O_{K,\mathfrak{p}}$. Hence $\varphi^{\mathrm{special}}$ extends to $O_{K,\mathfrak{p}}$ as well, and by semi-continuity we obtain a smooth Calabi-Yau 3-fold defined over $K \subset \mathbb{C}$.

Theorem 3.4 ([34]). *The Hilbert scheme of smooth Calabi-Yau 3-folds of degree 17 in $\mathbb{P}^6$ has at least 3 components. These three components are reduced and have dimension $23 + 48$. The corresponding Calabi-Yau 3-folds differ in the number of quintic generators of their homogeneous ideals, which are 8, 9 and 11 respectively.*

See [34] for more details.

Note that we do not give a bound on the degree $[K : \mathbb{Q}]$ of the number field, and certainly we are far away from a bound of its discriminant.

This leaves the question open whether these parameter spaces of Calabi-Yau 3-folds are unirational. Actually they are, as the geometric construction of modules $M \in \mathbb{M}_k$ in [34] shows.

A construction of one or several mirror families of these Calabi-Yau 3-folds is an open problem.

References

1. E. Arbarello and M. Cornalba: A few remarks about the variety of irreducible plane curve of given degree and genus. *Ann. Sci. École Norm. Sup(4)*, 16:467–488, 1983.
2. E. Arbarello, M. Cornalba, P. Griffith, and J. Harris: Geometry of algebraic curves, vol I. *Springer Grundlehren*, 267:xvi+386, 1985.
3. V. Batyrev: Dual polyhedra and mirror symmetry for Calabi-Yau hypersurfaces in algebraic tori. *J. Alg. Geom.*, 3:493–535, 1994.
4. D. Bayer and D. Eisenbud: Ribbons and their canonical embeddings. *Trans. Am. Math. Soc.*, 347,:719–756, 1995.

5. H.-C. v. Bothmer: Geometrische Syzygien kanonischer Kurven. *Thesis Bayreuth*, pages 1–123, 2000.

6. H.-C. v. Bothmer and F.-O. Schreyer: A quick and dirty irreduciblity test. *manuscript*, to appear.

7. D. Buchsbaum and D. Eisenbud: Algebra structure on finite free resolutions and some structure theorem for ideals of codimension 3. *Am. J. Math.*, 99:447–485, 1977.

8. M.-C. Chang and Z. Ran: Unirationality of the moduli space of curves genus 11, 13 (and 12). *Invent. Math.*, 76:41–54, 1984.

9. D. Cox and S. Katz: Mirror symmetry and algebraic geometry. *AMS, Mathematical Surveys and Monographs*, 68:xxii+469, 1999.

10. W. Decker, L. Ein, and F.-O. Schreyer: Construction of surfaces in $\mathbb{P}^4$. *J. of Alg. Geom.*, 2:185–237, 1993.

11. W. Decker and F.-O. Schreyer: Non-general type surfaces in $\mathbb{P}^4$ - Remarks on bounds and constructions. *J. Symbolic Comp.*, 29:545–585, 2000.

12. D. Eisenbud: Commutative Algebra. With a view towards algebraic geometry. *Springer Graduate Texts in Mathematics*, 150:xvi+785, 1995.

13. D. Eisenbud and J. Harris: The Kodaira dimension of the moduli space of curves of genus $g \geq 23$. *Invent. Math.*, 87:495–515, 1987.

14. D. Eisenbud and S. Popescu: Gale duality and free resolutions of ideals of points.. *Invent. Math.*, 136:419–449, 1999.

15. G. Ellingsrud and C. Peskine: Sur les surfaces de $\mathbb{P}^4$. *Invent. Math.*, 95:1–11, 1989.

16. M. Green: Koszul cohomology and the geometry of projective varieties. *J. Diff. Geom.*, 19:125–171, 1984.

17. M. Green: The Eisenbud-Koh-Stillman conjecture on linear syzygies. *Invent. Math*, 163:411–418, 1999.

18. M. Green and R. Lazarsfeld: Appendix to Koszul cohomology and the geometry of projective varieties. *J. Diff. Geom.*, 19:168–171, 1984.

19. J. Harris: Curves in projective space. With collaboration of David Eisenbud. *Séminaire de Mathématiques Supérieures*, 85:Montreal, 138 pp., 1982.

20. J. Harris: On the Severi problem. *Invent. Math.*, 84:445–461, 1986.

21. J. Harris and D. Mumford: On the Kodaira dimension of the moduli space of curves. With an appendix by William Fulton. *Invent. Math.*, 67:23–88, 1982.

22. A. Hirschowitz and S. Ramanan: New evidence for Green's conjecture on syzgyies of canonical curves. *Ann. Sci. École Norm. Sup. (4)*, 31:145–152, 1998.

23. S. Kleiman: Geometry on Grassmannians and application to splitting bundles and smoothing cycles. *Inst. Hautes Études Sci. Publ. Math. No.*, 36:281–287, 1969.

24. M. Martin-Deschamps and D. Perrin: Sur la classification des courbes gauches. *Asterisque*, 184-185:1–208, 1990.

25. S. Mukai: Curves and symmetric spaces, I.. *Amer. J. Math.*, 117:1627–1644, 1995.

26. S. Popescu: Some examples of smooth non general type surfaces in $\mathbb{P}^4$. *Thesis Saarbrücken*, pages ii+123, 1994.

27. K. Ranestad and F.-O. Schreyer: Varieties of sums of powers. *J. Reine Angew. Math.*, 525:147–182, 2000.

28. K. Ranestad and F.-O. Schreyer: On the variety of polar simplices. *manuscript*, to be completed.

29. F.-O. Schreyer: Syzygies of canonical curves and special linear series. *Math. Ann.*, 275:105–137, 1986.

30. F.-O. Schreyer: Green's conjecture for p-gonal curves of large genus. Algebraic curves and projective geometry (Trento 1988). *Lecture Notes in Math.*, 1396:254–260, 1989.

31. F.-O. Schreyer: A standard basis approach to syzygies of canonical curves. *J. Reine Angew. Math.*, 421:83–123, 1991.

32. F.-O. Schreyer: Small fields in constructive algebraic geometry. In *Moduli of vector bundles (Sanda, 1994; Kyoto, 1994)*, pages 221–228. Dekker, New York, 1996.

33. E. Sernesi: L' unirazionalità della varieta dei moduli delle curve di genere dodici. *Ann. Scuola Norm. Sup. Pisa Cl. Sci. (4)*, 8:405–439, 1981.

34. F. Tonoli: Canonical surfaces in $\mathbb{P}^5$ and Calabi-Yau threefolds in $\mathbb{P}^6$. *Thesis Padova*, pages 1–86, 2000.

35. C. Voisin: Courbes tétragonales et cohomologie de Koszul. *J. Reine Angew. Math.*, 387:111–121, 1988.

36. C. Walter: Pfaffian subschemes. *J. Alg. Geom.*, 5:671–704, 1996.

D-modules and Cohomology of Varieties

Uli Walther

In this chapter we introduce the reader to some ideas from the world of differential operators. We show how to use these concepts in conjunction with *Macaulay 2* to obtain new information about polynomials and their algebraic varieties.

Gröbner bases over polynomial rings have been used for many years in computational algebra, and the other chapters in this book bear witness to this fact. In the mid-eighties some important steps were made in the theory of Gröbner bases in non-commutative rings, notably in rings of differential operators. This chapter is about some of the applications of this theory to problems in commutative algebra and algebraic geometry.

Our interest in rings of differential operators and *D*-modules stems from the fact that some very interesting objects in algebraic geometry and commutative algebra have a *finite* module structure over an appropriate ring of differential operators. The prime example is the ring of regular functions on the complement of an affine hypersurface. A more general object is the Čech complex associated to a set of polynomials, and its cohomology, the local cohomology modules of the variety defined by the vanishing of the polynomials. More advanced topics are restriction functors and de Rham cohomology.

With these goals in mind, we shall study applications of Gröbner bases theory in the simplest ring of differential operators, the Weyl algebra, and develop algorithms that compute various invariants associated to a polynomial f. These include the Bernstein-Sato polynomial $b_f(s)$, the set of differential operators $J(f^s)$ which annihilate the germ of the function f^s (where s is a new variable), and the ring of regular functions on the complement of the variety of f.

For a family $f_1, \ldots, f_r$ of polynomials we study the associated Čech complex as a complex in the category of modules over the Weyl algebra. The algorithms are illustrated with examples. We also give an indication what other invariants associated to polynomials or varieties are known to be computable at this point and list some open problems in the area.

Acknowledgments. It is with great pleasure that I acknowledge the help of A. Leykin, M. Stillman and H. Tsai while writing this chapter. The *D*-module routines used or mentioned here have all been written by them and I would like to thank them for this marvelous job. I also would like to thank D. Grayson for help on *Macaulay 2* and D. Eisenbud and B. Sturmfels for inviting me to contribute to this volume.

1   Introduction

1.1   Local Cohomology – Definitions

Let R be a commutative Noetherian ring (always associative, with identity) and M an R-module. For $f \in R$ one defines a *Čech complex* of R-modules

$$\check{C}^\bullet(f) = (0 \to \underbrace{R}_{\text{degree } 0} \hookrightarrow \underbrace{R[f^{-1}]}_{\text{degree } 1} \to 0) \tag{1.1}$$

where the injection is the natural map sending $g \in R$ to $g/1 \in R[f^{-1}]$ and "degree" refers to cohomological degree. For a family $f_1, \ldots, f_r \in R$ one defines

$$\check{C}^\bullet(f_1, \ldots, f_r) = \bigotimes_{i=1}^r \check{C}^\bullet(f_i), \tag{1.2}$$

and for an R-module M one sets

$$\check{C}^\bullet(M; f_1, \ldots, f_r) = M \otimes_R \check{C}^\bullet(f_1, \ldots, f_r). \tag{1.3}$$

The i-th (algebraic) *local cohomology functor* with respect to $f_1, \ldots, f_r$ is the i-th cohomology functor of $\check{C}^\bullet(-; f_1, \ldots, f_r)$. If $I = R \cdot (f_1, \ldots, f_r)$ then this functor agrees with the i-th right derived functor of the functor $H^0_I(-)$ which sends M to the I-torsion $\bigcup_{k=1}^\infty (0 :_M I^k)$ of M and is denoted by $H^i_I(-)$. This means in particular, that $H^\bullet_I(-)$ depends only on the (radical of the) ideal generated by the f_i. Local cohomology was introduced by A. Grothendieck [13] as an algebraic analog of (classical) relative cohomology. For instance, if X is a scheme, Y is a closed subscheme and $U = X \setminus Y$ then there is a long exact sequence

$$\cdots \to H^i(X, \mathfrak{F}) \to H^i(U, \mathfrak{F}) \to H^{i+1}_Y(X, \mathfrak{F}) \to \cdots$$

for all quasi-coherent sheaves $\mathfrak{F}$ on X. (To make sense of this one has to generalize the definition of local cohomology to be the right derived functor of $H^0_Y(-) : \mathfrak{F} \to (U \to \{f \in \mathfrak{F}(U) : \operatorname{supp}(f) \subseteq Y \cap U\})$.) An introduction to algebraic local cohomology theory may be found in [8].

The *cohomological dimension of I in R*, denoted by $\operatorname{cd}(R, I)$, is the smallest integer c such that $H^i_I(M) = 0$ for all $i > c$ and all R-modules M. If R is the coordinate ring of an affine variety X and $I \subseteq R$ is the defining ideal of the Zariski closed subset $Y \subseteq X$ then the *local cohomological dimension of Y in X* is defined as $\operatorname{cd}(R, I)$. It is not hard to show that if X is smooth, then the integer $\dim(X) - \operatorname{cd}(R, I)$ depends only on Y but neither on X nor on the embedding $Y \hookrightarrow X$.

1.2   Motivation

As one sees from the definition of local cohomology, the modules $H_I^i(R)$ carry information about the sections of the structure sheaf on Zariski open sets, and hence about the topology of these open sets. This is illustrated by the following examples. Let $I \subseteq R$ and $c = \mathrm{cd}(R, I)$. Then I cannot be generated by fewer than c elements – in other words, $\mathrm{Spec}(R) \setminus \mathrm{Var}(I)$ cannot be covered by fewer than c affine open subsets (i.e., $\mathrm{Var}(I)$ cannot be cut out by fewer than c hypersurfaces). In fact, no ideal J with the same radical as I will be generated by fewer than c elements, [8].

Let $H_{\mathrm{Sing}}^i(-; \mathbb{C})$ stand for the i-th singular cohomology functor with complex coefficients. The classical Lefschetz Theorem [12] states that if $X \subseteq \mathbb{P}_{\mathbb{C}}^n$ is a variety in projective n-space and Y a hyperplane section of X such that $X \setminus Y$ is smooth, then $H_{\mathrm{Sing}}^i(X; \mathbb{C}) \to H_{\mathrm{Sing}}^i(Y; \mathbb{C})$ is an isomorphism for $i < \dim(X) - 1$ and injective for $i = \dim(X) - 1$. The Lefschetz Theorem has generalizations in terms of local cohomology, called Theorems of Barth Type. For example, let $Y \subseteq \mathbb{P}_{\mathbb{C}}^n$ be Zariski closed and $I \subseteq R = \mathbb{C}[x_0, \ldots, x_n]$ the defining ideal of Y. Then $H_{\mathrm{Sing}}^i(\mathbb{P}_{\mathbb{C}}^n; \mathbb{C}) \to H_{\mathrm{Sing}}^i(Y; \mathbb{C})$ is an isomorphism for $i < n - \mathrm{cd}(R, I)$ and injective if $i = n - \mathrm{cd}(R, I)$ ([16], Theorem III.7.1).

A consequence of the work of Ogus and Hartshorne ([38], 2.2, 2.3 and [16], Theorem IV.3.1) is the following. If $I \subseteq R = \mathbb{C}[x_0, \ldots, x_n]$ is the defining ideal of a complex smooth variety $Y \subseteq \mathbb{P}_{\mathbb{C}}^n$ then, for $i < n - \mathrm{codim}(Y)$,

$$\dim_{\mathbb{C}} \mathrm{soc}_R(H_{\mathfrak{m}}^0(H_I^{n-i}(R))) = \dim_{\mathbb{C}} H_x^i(\tilde{Y}; \mathbb{C})$$

where $H_x^i(\tilde{Y}; \mathbb{C})$ stands for the i-th singular cohomology group of the affine cone $\tilde{Y}$ over Y with support in the vertex x of $\tilde{Y}$ and with coefficients in $\mathbb{C}$ (and $\mathrm{soc}_R(M)$ denotes the socle $(0 :_M (x_0, \ldots, x_n)) \subseteq M$ for any R-module M), [25]. These iterated local cohomology modules have a special structure (cf. Subsection 4.3).

Local cohomology relates to the connectedness of the underlying spaces as is shown by the following facts. If Y is a complete intersection of positive dimension in $\mathbb{P}_{\mathbb{C}}^n$, then Y cannot be disconnected by the removal of closed subsets of codimension 2 in Y or higher, [7]. This is a consequence of the so-called Hartshorne-Lichtenbaum vanishing theorem, see [8].

In a similar spirit one can show that if $(A, \mathfrak{m})$ is a complete local domain of dimension n and $f_1, \ldots, f_r$ are elements of the maximal ideal with $r + 2 \leq n$, then $\mathrm{Var}(f_1, \ldots, f_r) \setminus \{\mathfrak{m}\}$ is connected, [7].

In fact, as we will discuss to some extent in Section 5, over the complex numbers the complex $\check{C}^\bullet(R; f_1, \ldots, f_r)$ for $R = \mathbb{C}[x_1, \ldots, x_n]$ determines the Betti numbers $\dim_{\mathbb{C}}(H_{\mathrm{Sing}}^i(\mathbb{C}^n \setminus \mathrm{Var}(f_1, \ldots, f_r); \mathbb{C}))$.

1.3   The Master Plan

The cohomological dimension has been studied by many authors. For an extensive list of references and some open questions we recommend to consult the very nice survey article [17].

It turns out that for the determination of $\text{cd}(R, I)$ it is in fact enough to find a test to decide whether or not the local cohomology module $H_I^i(R) = 0$ for given i, R, I. This is because $H_I^i(R) = 0$ for all $i > c$ implies $\text{cd}(R, I) \leq c$ (see [14], Section 1).

Unfortunately, calculations are complicated by the fact that $H_I^i(M)$ is rarely finitely generated as R-module, even for very nice R and M. In this chapter we show how in an important class of examples one may still carry out explicit computations, by enlarging R.

We shall assume that $I \subseteq R_n = K[x_1, \ldots, x_n]$ where K is a computable field containing the rational numbers. (By a *computable field* we mean a subfield K of $\mathbb{C}$ such that K is described by a finite set of data and for which addition, subtraction, multiplication and division as well as the test whether the result of any of these operations is zero in the field can be executed by the Turing machine. For example, K could be $\mathbb{Q}[\sqrt{2}]$ stored as a 2-dimensional vector space over $\mathbb{Q}$ with an appropriate multiplication table.)

The ring of K-linear differential operators $D(R, K)$ of the commutative K-algebra R is defined inductively: one sets $D_0(R, K) = R$, and for $i > 0$ defines

$$D_i(R, K) = \{P \in \text{Hom}_K(R, R) : Pr - rP \in D_{i-1}(R, K) \text{ for all } r \in R\}.$$

Here, $r \in R$ is interpreted as the endomorphism of R that multiplies by r.

The local cohomology modules $H_I^i(R_n)$ have a natural structure of finitely generated left $D(R_n, K)$-modules (see for example [20,25]). The basic reason for this finiteness is that in this case $R_n[f^{-1}]$ is a cyclic $D(R_n, K)$-module, generated by f^a for $\mathbb{Z} \ni a \ll 0$ (compare [5]):

$$R_n[f^{-1}] = D(R_n, K) \bullet f^a. \tag{1.4}$$

Using this finiteness we employ the theory of Gröbner bases in $D(R_n, K)$ to develop algorithms that give a presentation of $H_I^i(R_n)$ and $H_\mathfrak{m}^i(H_I^j(R_n))$ for all triples $i, j \in \mathbb{N}$, $I \subseteq R_n$ in terms of generators and relations over $D(R_n, K)$ (where $\mathfrak{m} = R_n \cdot (x_1, \ldots, x_n)$), see Section 4. This also leads to an algorithm for the computation of the invariants

$$\lambda_{i,j}(R_n/I) = \dim_K \text{soc}_{R_n}(H_\mathfrak{m}^i(H_I^{n-j}(R_n)))$$

introduced in [25].

At the basis for the computation of local cohomology are algorithms that compute the localization of a $D(R_n, K)$-module at a hypersurface $f \in R_n$. That means, if the left module $M = D(R_n, K)^d/L$ is given by means of a finite number of generators for the left module $L \subseteq D(R_n, K)^d$ then we want to compute a finite number of generators for the left module $L' \subseteq D(R_n, K)^{d'}$ which satisfies

$$D(R_n, K)^{d'}/L' \cong (D(R_n, K)^d/L) \otimes_{R_n} R_n[f^{-1}],$$

which we do in Section 3.

Let L be a left ideal of $D(R_n, K)$. The computation of the localization of $M = D(R_n, K)/L$ at $f \in R_n$ is closely related to the $D(R_n, K)[s]$-module $\mathcal{M}_f$ generated by

$$\overline{1} \otimes 1 \otimes f^s \in M \otimes_{R_n} R_n[f^{-1}, s] \otimes f^s \tag{1.5}$$

and the minimal polynomial $b_f(s)$ of s on the quotient of $\mathcal{M}_f$ by its submodule $\mathcal{M}_f \cdot f$ generated over $D(R_n, K)[s]$ by $\overline{1} \otimes f \otimes f^s$, cf. Section 3. Algorithms for the computation of these objects have been established by T. Oaku in a sequence of papers [31–33].

Astonishingly, the roots of $b_f(s)$ prescribe the exponents a that can be used in the isomorphism (1.4) between $R_n[f^{-1}]$ and the $D(R_n, K)$-module generated by f^a. Moreover, any good exponent a can be used to transform $\mathcal{M}_f$ into $M \otimes R_n[f^{-1}]$ by a suitable "plugging in" procedure.

Thus the strategy for the computation of local cohomology will be to compute $\mathcal{M}_f$ and a good a for each $f \in \{f_1, \ldots, f_r\}$, and then assemble the Čech complex.

1.4 Outline of the Chapter

The next section is devoted to a short introduction of results on the Weyl algebra $D(R_n, K)$ and D-modules as they apply to our work. We start with some remarks on the theory of Gröbner bases in the Weyl algebra.

In Section 3 we investigate Bernstein-Sato polynomials, localizations and the Čech complex. The purpose of that section is to find a presentation of $M \otimes R_n[f^{-1}]$ as a cyclic $D(R_n, K)$-module if $M = D(R_n, K)/L$ is a given holonomic D-module (for a definition and some properties of holonomic modules, see Subsection 2.3 below).

In Section 4 we describe algorithms that for arbitrary i, j, k, I determine the structure of $H_I^k(R), H_{\mathfrak{m}}^i(H_I^j(R))$ and find $\lambda_{i,j}(R/I)$. The final section is devoted to comments on implementations, efficiency, discussions of other topics, and open problems.

2 The Weyl Algebra and Gröbner Bases

D-modules, that is, rings or sheaves of differential operators and modules over these, have been around for several decades and played prominent roles in representation theory, some parts of analysis and in algebraic geometry. The founding fathers of the theory are M. Sato, M. Kashiwara, T. Kawai, J. Bernstein, and A. Beilinson. The area has also benefited much from the work of P. Deligne, J.-E. Björk, J.-E. Roos, B. Malgrange and Z. Mebkhout. The more computational aspects of the theory have been initiated by T. Oaku and N. Takayama.

The simplest example of a ring of differential operators is given by the Weyl algebra, the ring of K-linear differential operators on R_n. In characteristic zero, this is a finitely generated K-algebra that resembles the ring of polynomials in $2n$ variables but fails to be commutative.

2.1   Notation

Throughout we shall use the following notation: K will denote a computable field of characteristic zero and $R_n = K[x_1, \ldots, x_n]$ the ring of polynomials over K in n variables. The K-linear differential operators on R_n are then the elements of

$$D_n = K\langle x_1, \partial_1, \ldots, x_n, \partial_n \rangle,$$

the n-*th Weyl algebra* over K, where the symbol x_i denotes the operator "multiply by x_i" and ∂_i denotes the operator "take partial derivative with respect to x_i". We therefore have in D_n the relations

$$\begin{aligned}
x_i x_j &= x_j x_i && \text{for all } 1 \le i, j \le n, \\
\partial_i \partial_j &= \partial_j \partial_i && \text{for all } 1 \le i, j \le n, \\
x_i \partial_j &= \partial_j x_i && \text{for all } 1 \le i \ne j \le n, \\
\text{and } x_i \partial_i + 1 &= \partial_i x_i && \text{for all } 1 \le i \le n.
\end{aligned}$$

The last relation is nothing but the *product* (or *Leibniz*) *rule*, $xf' + f = (xf)'$. We shall use multi-index notation: $x^\alpha \partial^\beta$ denotes the monomial

$$x_1{}^{\alpha_1} \cdots x_n{}^{\alpha_n} \cdot \partial_1{}^{\beta_1} \cdots \partial_n{}^{\beta_n}$$

and $|\alpha| = \alpha_1 + \cdots + \alpha_n$.

In order to keep the product $\partial_i x_i \in D_n$ and the application of $\partial_i \in D_n$ to $x_i \in R_n$ apart, we shall write $\partial_i \bullet (g)$ to mean the result of the action of ∂_i on $g \in R_n$. So for example, $\partial_i x_i = x_i \partial_i + 1 \in D_n$ but $\partial_i \bullet x_i = 1 \in R_n$. The action of D_n on R_n takes precedence over the multiplication in R_n (and is of course compatible with the multiplication in D_n), so for example $\partial_2 \bullet (x_1) x_2 = 0 \cdot x_2 = 0 \in R_n$.

The symbol $\mathfrak{m}$ will stand for the maximal ideal $R_n \cdot (x_1, \ldots, x_n)$ of R_n, Δ will denote the maximal left ideal $D_n \cdot (\partial_1, \ldots, \partial_n)$ of D_n and I will stand for the ideal $R_n \cdot (f_1, \ldots, f_r)$ in R_n. Every D_n-module becomes an R_n-module via the embedding $R_n \hookrightarrow D_n$ as $D_0(R_n, K)$.

All tensor products in this chapter will be over R_n and all D_n-modules (resp. ideals) will be left modules (resp. left ideals) unless specified otherwise.

2.2   Gröbner Bases in D_n

This subsection is a severely shortened version of Chapter 1 in [40] (and we strongly recommend that the reader take a look at this book). The purpose is to see how Gröbner basis theory applies to the Weyl algebra.

The elements in D_n allow a *normally ordered expression*. Namely, if $P \in D_n$ then we can write it as

$$P = \sum_{(\alpha,\beta) \in E} c_{\alpha,\beta} x^\alpha \partial^\beta$$

where E is a finite subset of $\mathbb{N}^{2n}$. Thus, as K-vector spaces there is an isomorphism

$$\Psi : K[x, \xi] \to D_n$$

(with $\xi = \xi_1, \ldots, \xi_n$) sending $x^\alpha \xi^\beta$ to $x^\alpha \partial^\beta$. We will assume that every $P \in D_n$ is normally ordered.

We shall say that $(u, v) \in \mathbb{R}^{2n}$ is a *weight vector* for D_n if $u + v \geq 0$, that is $u_i + v_i \geq 0$ for all $1 \leq i \leq n$. We set the *weight* of the monomial $x^\alpha \partial^\beta$ under (u, v) to be $u \cdot \alpha + v \cdot \beta$ (scalar product). The weight of an operator is then the maximum of the weights of the nonzero monomials appearing in the normally ordered expression of P. If (u, v) is a weight vector for D_n, there is an associated graded ring $\mathrm{gr}_{(u,v)}(D_n)$ with

$$\mathrm{gr}^r_{(u,v)}(D_n) = \frac{\{P \in D_n : w(P) \leq r\}}{\{P \in D_n : w(P) < r\}}.$$

So $\mathrm{gr}_{(u,v)}(D_n)$ is the K-algebra on the symbols $\{x_i : 1 \leq i \leq n\} \cup \{\partial_i : u_i + v_i = 0\} \cup \{\xi_i : u_i + v_i > 0\}$. Here all variables commute with each other except ∂_i and x_i for which the Leibniz rule holds.

Each $P \in D_n$ has an *initial form* or *symbol* $\mathrm{in}_{(u,v)}(P)$ in $\mathrm{gr}_{(u,v)}(D_n)$ defined by taking all monomials in the normally ordered expression for P that have maximal weight, and replacing all ∂_i with $u_i + v_i > 0$ by the corresponding ξ_i.

The inequality $u_i + v_i \geq 0$ is needed to assure that the product of the initial forms of two operators equals the initial form of their product: one would not want to have $\mathrm{in}(\partial_i \cdot x_i) = \mathrm{in}(x_i \cdot \partial_i + 1) = 1$.

A weight of particular importance is $-u = v = (1, \ldots, 1)$, or more generally $-u = v = (1, \ldots, 1, 0, \ldots, 0)$. In these cases $\mathrm{gr}_{(u,v)}(D_n) \cong D_n$. On the other hand, if $u + v$ is componentwise positive, then $\mathrm{gr}_{(u,v)}(D_n)$ is commutative (compare the initial forms of $\partial_i x_i$ and $x_i \partial_i$) and isomorphic to the polynomial ring in $2n$ variables corresponding to the symbols of $x_1, \ldots, x_n, \partial_1, \ldots, \partial_n$.

If L is a left ideal in D_n we write $\mathrm{in}_{(u,v)}(L)$ for $\{\mathrm{in}_{(u,v)}(P) : P \in L\}$. This is a left ideal in $\mathrm{gr}_{(u,v)}(D_n)$. If $G \subset L$ is a finite set we call it a (u, v)-*Gröbner basis* if the left ideal of $\mathrm{gr}_{(u,v)}(D_n)$ generated by the initial forms of the elements of G agrees with $\mathrm{in}_{(u,v)}(L)$.

A *multiplicative monomial order* on D_n is a total order $\prec$ on the normally ordered monomials such that

1. $1 \prec x_i \partial_i$ for all i, and
2. $x^\alpha \partial^\beta \prec x^{\alpha'} \partial^{\beta'}$ implies $x^{\alpha+\alpha''} \partial^{\beta+\beta''} \prec x^{\alpha'+\alpha''} \partial^{\beta'+\beta''}$ for all $\alpha'', \beta'' \in \mathbb{N}^n$.

A multiplicative monomial order is a *term order* if 1 is the (unique) smallest monomial. Multiplicative monomial orders, and more specifically term orders, clearly abound.

Multiplicative monomial orders (and hence term orders) allow the construction of initial forms just like weight vectors. Now, however, the initial forms are always monomials, and always elements of $K[x, \xi]$ (due to the total order requirement on $\prec$). One defines Gröbner bases for multiplicative monomial orders analogously to the weight vector case.

For our algorithms we have need to compute weight vector Gröbner bases, and this can be done as follows. Suppose (u, v) is a weight vector on D_n and $\prec$ a term order. Define a multiplicative monomial order $\prec_{(u,v)}$ as follows:

$$x^\alpha \partial^\beta \prec_{(u,v)} x^{\alpha'} \partial^{\beta'} \Leftrightarrow [(\alpha - \alpha') \cdot u + (\beta - \beta') \cdot v < 0] \text{ or}$$
$$\left[(\alpha - \alpha') \cdot u + (\beta - \beta') \cdot v = 0 \text{ and } x^\alpha \partial^\beta \prec x^{\alpha'} \partial^{\beta'} \right].$$

Note that $\prec_{(u,v)}$ is a term order precisely when (u, v) is componentwise nonnegative.

Theorem 2.1 ([40], Theorem 1.1.6.). *Let L be a left ideal in D_n, (u, v) a weight vector for D_n, $\prec$ a term order and G a Gröbner basis for L with respect to $\prec_{(u,v)}$. Then*

1. G is a Gröbner basis for L with respect to (u, v).
2. $\text{in}_{(u,v)}(G)$ is a Gröbner basis for $\text{in}_{(u,v)}(L)$ with respect to $\prec$. □

We end this subsection with the remarks that Gröbner bases with respect to multiplicative monomial orders can be computed using the Buchberger algorithm adapted to the non-commutative situation (thus, Gröbner bases with respect to weight vectors are computable according to the theorem), and that the computation of syzygies, kernels, intersections and preimages in D_n works essentially as in the commutative algebra $K[x, \xi]$. For precise statements of the algorithms we refer the reader to [40].

2.3   D-modules

A good introduction to D-modules are the book by J.-E. Björk, [5], the nice introduction [9] by S. Coutinho, and the lecture notes by J. Bernstein [4]. In this subsection we list some properties of localizations of R_n that are important for module-finiteness over D_n. Most of this section is taken from Section 1 in [5].

Let $f \in R_n$. Then the R_n-module $R_n[f^{-1}]$ has a structure as left D_n-module via the extension of the action $\bullet$:

$$x_i \bullet \left(\frac{g}{f^k} \right) = \frac{x_i g}{f^k}, \qquad \partial_i \bullet \left(\frac{g}{f^k} \right) = \frac{\partial_i \bullet (g) f - k \partial_i \bullet (f) g}{f^{k+1}}.$$

This may be thought of as a special case of localizing a D_n-module: if M is a D_n-module and $f \in R_n$ then $M \otimes_{R_n} R_n[f^{-1}]$ becomes a D_n-module via the *product rule*

$$x_i \bullet (m \otimes \frac{g}{f^k}) = m \otimes (\frac{x_i g}{f^k}), \quad \partial_i \bullet (m \otimes \frac{g}{f^k}) = m \otimes \partial_i \bullet (\frac{g}{f^k}) + \partial_i m \otimes \frac{g}{f^k}.$$

Of particular interest are the *holonomic* modules which are those finitely generated D_n-modules M for which $\mathrm{Ext}^j_{D_n}(M, D_n)$ vanishes unless $j = n$. This innocent looking definition has surprising consequences, some of which we discuss now.

The holonomic modules form a full Abelian subcategory of the category of left D_n-modules, closed under the formation of subquotients. Our standard example of a holonomic module is

$$R_n = D_n / \Delta.$$

This equality may require some thought – it pictures R_n as a D_n-module generated by $1 \in R_n$. It is particularly noteworthy that not all elements of R_n are killed by Δ – quite impossible if D_n were commutative.

Holonomic modules are always cyclic and of finite length over D_n. These fundamental properties are consequences of the *Bernstein inequality*. To understand this inequality we associate with the D_n-module $M = D_n/L$ the Hilbert function $q_L(k)$ with values in the integers which counts for each $k \in \mathbb{N}$ the number of monomials $x^\alpha \partial^\beta$ with $|\alpha| + |\beta| \leq k$ whose cosets in M are K-linearly independent. The filtration $k \mapsto K \cdot \{x^\alpha \partial^\beta \mod L : |\alpha| + |\beta| \leq k\}$ is called the *Bernstein filtration*. The Bernstein inequality states that $q_L(k)$ is either identically zero (in which case $M = 0$) or asymptotically a polynomial in k of degree between n and $2n$. This degree is called the *dimension of M*. A holonomic module is one of dimension n, the minimal possible value for a nonzero module.

This characterization of holonomicity can be used quite easily to check with *Macaulay 2* that R_n is holonomic. Namely, let's say $n = 3$. Start a *Macaulay 2* session with

```
i1 : load "D-modules.m2"

i2 : D = QQ[x,y,z,Dx,Dy,Dz, WeylAlgebra => {x=>Dx, y=>Dy, z=>Dz}]

o2 = D

o2 : PolynomialRing

i3 : Delta = ideal(Dx,Dy,Dz)

o3 = ideal (Dx, Dy, Dz)

o3 : Ideal of D
```

The first of these commands loads the D-module library by A. Leykin, M. Stillman and H. Tsai, [23]. The second line defines the base ring $D_3 = \mathbb{Q}\langle x, y, z, \partial_x, \partial_y, \partial_z \rangle$, while the third command defines the D_3-module $D_3/\Delta \cong R_3$.

As one can see, *Macaulay 2* thinks of D as a ring of polynomials. This is using the vector space isomorphism Ψ from Subsection 2.2. Of course, two elements are multiplied according to the Leibniz rule. To see how *Macaulay 2* uses the map Ψ, we enter the following expression.

```
i4 : (Dx * x)^2

       2  2
o4 = x Dx  + 3x*Dx + 1

o4 : D
```

All Weyl algebra ideals and modules are by default left ideals and left modules in *Macaulay 2*.

If we don't explicitly specify a monomial ordering to be used in the Weyl algebra, then *Macaulay 2* uses graded reverse lex (GRevLex), as we can see by examining the options of the ring.

```
i5 : options D

o5 = OptionTable{Adjust => identity                         }
                 Degrees => {{1}, {1}, {1}, {1}, {1}, {1}}
                 Inverses => false
                 MonomialOrder => GRevLex
                 MonomialSize => 8
                 NewMonomialOrder =>
                 Repair => identity
                 SkewCommutative => false
                 VariableBaseName =>
                 VariableOrder =>
                 Variables => {x, y, z, Dx, Dy, Dz}
                 Weights => {}
                 WeylAlgebra => {x => Dx, y => Dy, z => Dz}

o5 : OptionTable
```

To compute the initial ideal of Δ with respect to the weight that associates 1 to each ∂ and to each variable, execute

```
i6 : DeltaBern = inw(Delta,{1,1,1,1,1,1})

o6 = ideal (Dz, Dy, Dx)

o6 : Ideal of QQ [x, y, z, Dx, Dy, Dz]
```

The command inw can be used with any weight vector for D_n as second argument. One notes that the output is not an ideal in a Weyl algebra any more, but in a ring of polynomials, as it should. The dimension of R_3, which is the dimension of the variety associated to DeltaBern, is computed by

```
i7 : dim DeltaBern

o7 = 3
```

As this is equal to $n = 3$, the ideal Δ is holonomic.

Let $R_n[f^{-1}, s] \otimes f^s$ be the free $R_n[f^{-1}, s]$-module generated by the symbol f^s. Using the action $\bullet$ of D_n on $R_n[f^{-1}, s]$ we define an action $\bullet$ of $D_n[s]$ on

$R_n[f^{-1}, s] \otimes f^s$ by setting

$$s \bullet \left(\frac{g(x,s)}{f^k} \otimes f^s \right) = \frac{sg(x,s)}{f^k} \otimes f^s,$$

$$x_i \bullet \left(\frac{g(x,s)}{f^k} \otimes f^s \right) = \frac{x_i g(x,s)}{f^k} \otimes f^s,$$

$$\partial_i \bullet \left(\frac{g(x,s)}{f^k} \otimes f^s \right) = \left(\partial_i \bullet \left(\frac{g(x,s)}{f^k} \right) + s\partial_i \bullet (f) \cdot \frac{g(x,s)}{f^{k+1}} \right) \otimes f^s.$$

The last rule justifies the choice for the symbol of the generator.

Writing $M = D_n/L$ and denoting by $\overline{1}$ the coset of $1 \in D_n$ in M, this action extends to an action of $D_n[s]$ on

$$\mathcal{M}_f^L = D_n[s] \bullet (\overline{1} \otimes 1 \otimes f^s) \subseteq M \otimes_{R_n} \left(R_n[f^{-1}, s] \otimes f^s \right) \qquad (2.1)$$

by the product rule for all left D_n-modules M. The interesting bit about $\mathcal{M}_f^L$ is the following fact. If $M = D_n/L$ is holonomic then there is a nonzero polynomial $b(s)$ in $K[s]$ and an operator $P(s) \in D_n[s]$ such that

$$P(s) \bullet (\overline{1} \otimes f \otimes f^s) = \overline{1} \otimes b(s) \otimes f^s \qquad (2.2)$$

in $\mathcal{M}_f^L$. This entertaining equality, often written as

$$P(s) \left(\overline{1} \otimes f^{s+1} \right) = \overline{b(s)} \otimes f^s,$$

says that $P(s)$ is roughly equal to division by f. The unique monic polynomial that divides all other polynomials $b(s)$ satisfying an identity of this type is called the *Bernstein* (or also *Bernstein-Sato*) *polynomial* of L and f and denoted by $b_f^L(s)$. Any operator $P(s)$ that satisfies (2.2) with $b(s) = b_f(s)$ we shall call a *Bernstein operator* and refer to the roots of $b_f^L(s)$ as *Bernstein roots* of f on D_n/L. It is clear from (2.2) and the definitions that $b_f^L(s)$ is the minimal polynomial of s on the quotient of $\mathcal{M}_f^L$ by $D_n[s] \bullet (\overline{1} \otimes f \otimes f^s)$.

The Bernstein roots of the polynomial f are somewhat mysterious, but related to other algebro-geometric invariants as, for example, the monodromy of f (see [29]), the Igusa zeta function (see [24]), and the log-canonical threshold (see [21]). For a long time it was also unclear how to compute $b_f(s)$ for given f. In [53] many interesting examples of Bernstein-Sato polynomials are worked out by hand, while in [1,6,28,41] algorithms were given that compute $b_f(s)$ under certain conditions on f. The general algorithm we are going to explain was given by T. Oaku. Here is a classical example.

Example 2.2. Let $f = \sum_{i=1}^n x_i^2$ and $M = R_n$ with $L = \Delta$. One can check that

$$\sum_{i=1}^n \partial_i^2 \bullet (\overline{1} \otimes 1 \otimes f^{s+1}) = \overline{1} \otimes 4(s+1)(\frac{n}{2} + s) \otimes f^s$$

and hence that $\frac{1}{4} \sum_{i=1}^n \partial_i^2$ is a Bernstein operator while the Bernstein roots of f are -1 and $-n/2$ and the Bernstein polynomial is $(s+1)(s+\frac{n}{2})$.

Example 2.3. Although in the previous example the Bernstein operator looked a lot like the polynomial f, this is not often the case and it is usually hard to guess Bernstein operators. For example, one has

$$\left(\frac{1}{27}\partial_y{}^3 + \frac{y}{6}\partial_x{}^2\partial_y + \frac{x}{8}\partial_x{}^3\right)(x^2 + y^3)^{s+1} = (s + \frac{5}{6})(s+1)(s+\frac{7}{6})(x^2 + y^3)^s.$$

In the case of non-quasi-homogeneous polynomials, there is usually no resemblance between f and any Bernstein operator.

A very important property of holonomic modules is the (somewhat counter-intuitive) fact that any localization of a holonomic module $M = D_n/L$ at a single element (and hence at any finite number of elements) of R_n is holonomic ([5], 1.5.9) and in particular cyclic over D_n, generated by $\bar{1} \otimes f^a$ for sufficiently small $a \in \mathbb{Z}$. As a special case we note that localizations of R_n are holonomic, and hence finitely generated over D_n. Coming back to the Čech complex we see that the complex $\check{C}^\bullet(M; f_1, \ldots, f_r)$ consists of holonomic D_n-modules whenever M is holonomic.

As a consequence, local cohomology modules of R_n are D_n-modules and in fact holonomic. To see this it suffices to know that the maps in the Čech complex are D_n-linear, which we will explain in Section 4. Since the category of holonomic D_n-modules and their D_n-linear maps is closed under subquotients, holonomicity of $H_f^k(R_n)$ follows.

For similar reasons, $H_\mathfrak{m}^i(H_I^j(R_n))$ is holonomic for $i, j \in \mathbb{N}$ (since $H_I^j(R_n)$ is holonomic). These modules, investigated in Subsections 4.2 and 4.3, are rather special R_n-modules and seem to carry some very interesting information about $\mathrm{Var}(I)$, see [10,52].

The fact that R_n is holonomic and every localization of a holonomic module is as well, provides motivation for us to study this class of modules. There are, however, more occasions where holonomic modules show up. One such situation arises in the study of linear partial differential equations. More specifically, the so-called GKZ-systems (which we will meet again in the final chapter) provide a very interesting class of objects with fascinating combinatorial and analytic properties [40].

3   Bernstein-Sato Polynomials and Localization

We mentioned in the introduction that for the computation of local cohomology the following is an important algorithmic problem to solve.

Problem 3.1. Given $f \in R_n$ and a left ideal $L \subseteq D_n$ such that $M = D_n/L$ is holonomic, compute the structure of the module $D_n/L \otimes R_n[f^{-1}]$ in terms of generators and relations.

This section is about solving Problem 3.1.

3.1   The Line of Attack

Recall for a given D_n-module $M = D_n/L$ the action of $D_n[s]$ on the tensor product $M \otimes_{R_n} (R_n[f^{-1}, s] \otimes f^s)$ from Subsection 2.3. We begin with defining an ideal of operators:

Definition 3.2. Let $J^L(f^s)$ stand for the ideal in $D_n[s]$ that kills $\overline{1} \otimes 1 \otimes f^s \in (D_n/L) \otimes_{R_n} R_n[f^{-1}, s] \otimes f^s$.

It turns out that it is very useful to know this ideal. If $L = \Delta$ then there are some obvious candidates for generators of $J^L(f^s)$. For example, there are $f\partial_i - \partial_i \bullet (f)s$ for all i. However, unless the affine hypersurface defined by $f = 0$ is smooth, these will not generate $J^\Delta(f^s)$. For a more general L, there is a similar set of (somewhat less) obvious candidates, but again finding all elements of $J^\Delta(f^s)$ is far from elementary, even for smooth f.

In order to find $J^L(f^s)$, we will consider the module $(D_n/L) \otimes R_n[f^{-1}, s] \otimes f^s$ over the ring $D_{n+1} = D_n\langle t, \partial_t \rangle$ by defining an appropriate action of t and ∂_t on it. It is then not hard to compute the ideal $J^L_{n+1}(f^s) \subseteq D_{n+1}$ consisting of all operators that kill $\overline{1} \otimes 1 \otimes f^s$, see Lemma 3.5. In Proposition 3.6 we will then explain how to compute $J^L(f^s)$ from $J^L_{n+1}(f^s)$.

This construction gives an answer to the question of determining a presentation of $D_n \bullet (\overline{1} \otimes f^a)$ for "most" $a \in K$, which we make precise as follows.

Definition 3.3. We say that a property depending on $a \in K^m$ *holds for a in very general position*, if there is a countable set of hypersurfaces in K^m such that the property holds for all a not on any of the exceptional hypersurfaces.

It will turn out that for $a \in K$ in very general position $J^L(f^s)$ "is" the annihilator for f^a: we shall very explicitly identify a countable number of exceptional values in K such that if a is not equal to one of them, then $J^L(f^s)$ evaluates under $s \mapsto a$ to the annihilator inside D_n of $\overline{1} \otimes f^a$.

For $a \in \mathbb{Z}$ we have of course $D_n \bullet (\overline{1} \otimes f^a) \subseteq M \otimes R_n[f^{-1}]$ but the inclusion may be strict (e.g., for $L = \Delta$ and $a = 0$). Proposition 3.11 shows how $(D_n/L) \otimes R_n[f^{-1}]$ and $J^L(f^s)$ are related.

3.2   Undetermined Exponents

Consider $D_{n+1} = D_n\langle t, \partial_t \rangle$, the Weyl algebra in $x_1, \ldots, x_n$ and the new variable t. B. Malgrange [29] has defined an action $\bullet$ of D_{n+1} on $(D_n/L) \otimes R_n[f^{-1}, s] \otimes f^s$ as follows. We require that x_i acts as multiplication on the first factor, and for the other variables we set (with $\overline{P} \in D_n/L$ and $g(x,s) \in R_n[s]$)

$$\partial_i \bullet (\overline{P} \otimes \frac{g(x,s)}{f^k} \otimes f^s) = \left(\overline{P} \otimes \left(\partial_i \bullet (\frac{g(x,s)}{f^k}) + \frac{s\partial_i \bullet (f)g(x,s)}{f^{k+1}}\right)\right.$$
$$\left. + \overline{\partial_i P} \otimes \frac{g(x,s)}{f^k}\right) \otimes f^s,$$

$$t \bullet (\overline{P} \otimes \frac{g(x,s)}{f^k} \otimes f^s) = \overline{P} \otimes \frac{g(x,s+1)f}{f^k} \otimes f^s,$$

$$\partial_t \bullet (\overline{P} \otimes \frac{g(x,s)}{f^k} \otimes f^s) = \overline{P} \otimes \frac{-sg(x,s-1)}{f^{k+1}} \otimes f^s.$$

One checks that this actually defines a left D_{n+1}-module structure (i.e., $\partial_t t$ acts like $t\partial_t + 1$) and that $-\partial_t t$ acts as multiplication by s.

Definition 3.4. We denote by $J_{n+1}^L(f^s)$ the ideal in D_{n+1} that annihilates the element $\overline{1} \otimes 1 \otimes f^s$ in $(D_n/L) \otimes R_n[f^{-1}, s] \otimes f^s$ with D_{n+1} acting as defined above. Then we have an induced morphism of D_{n+1}-modules $D_{n+1}/J_{n+1}^L(f^s) \to (D_n/L) \otimes R_n[f^{-1}, s] \otimes f^s$ sending $P + J_{n+1}^L(f^s)$ to $P \bullet (\overline{1} \otimes 1 \otimes f^s)$.

We say that an ideal $L \subseteq D_n$ is f-*saturated* if $f \cdot P \in L$ implies $P \in L$ and we say that D_n/L is f-*torsion free* if L is f-saturated. R_n and all its localizations are examples of f-torsion free modules for arbitrary f.

The following lemma is a modification of Lemma 4.1 in [29] where the special case $L = D_n \cdot (\partial_1, \ldots, \partial_n), D_n/L = R_n$ is considered (compare also [47]).

Lemma 3.5. *Suppose that* $L = D_n \cdot (P_1, \ldots, P_r)$ *is* f-*saturated. With the above definitions,* $J_{n+1}^L(f^s)$ *is the ideal generated by* $f - t$ *together with the images of the* P_j *under the automorphism* ϕ *of* D_{n+1} *induced by* $x_i \mapsto x_i$ *for all* i, *and* $t \mapsto t - f$.

Proof. The automorphism sends ∂_i to $\partial_i + \partial_i \bullet (f)\partial_t$ and ∂_t to ∂_t. So if we write P_j as a polynomial $P_j(\partial_1, \ldots, \partial_n)$ in the ∂_i with coefficients in $K[x_1, \ldots, x_n]$, then

$$\phi(P_j) = P_j(\partial_1 + \partial_1 \bullet (f)\partial_t, \ldots, \partial_n + \partial_n \bullet (f)\partial_t).$$

One checks that $(\partial_i + \partial_i \bullet (f)\partial_t) \bullet (\overline{Q} \otimes 1 \otimes f^s) = \overline{\partial_i Q} \otimes 1 \otimes f^s$ for all $Q \in D_{n+1}$, so that $\phi(P_j(\partial_1, \ldots, \partial_n)) \bullet (\overline{1} \otimes 1 \otimes f^s) = P_j(\partial_1, \ldots, \partial_n) \otimes 1 \otimes f^s = 0$. By definition, $f \bullet (\overline{1} \otimes 1 \otimes f^s) = t \bullet (\overline{1} \otimes 1 \otimes f^s)$. So $t - f \in J_{n+1}^L(f^s)$ and $\phi(P_j) \in J_{n+1}^L(f^s)$ for $j = 1, \ldots, r$.

Conversely let $P \bullet (\overline{1} \otimes 1 \otimes f^s) = 0$. The proof that $P \in \phi(J_{n+1}^L + D_{n+1} \cdot t)$ relies on an elimination idea and has some Gröbner basis flavor. We have to show that $P \in D_{n+1} \cdot (\phi(P_1), \ldots, \phi(P_r), t - f)$. We may assume, that P does not contain any power of t since we can eliminate t using $f - t$. Now rewrite P in terms of ∂_t and the $\partial_i + \partial_i \bullet (f)\partial_t$. Say, $P = \sum_{\alpha,\beta} \partial_t^\alpha x^\beta Q_{\alpha,\beta}(\partial_1 + \partial_1 \bullet$

$(f)\partial_t, \ldots, \partial_n + \partial_n \bullet (f)\partial_t)$, where the $Q_{\alpha,\beta} \in K[y_1, \ldots, y_n]$ are polynomial expressions. Then

$$P \bullet (\overline{1} \otimes 1 \otimes f^s) = \sum_{\alpha,\beta} \partial_t^\alpha \bullet (\overline{x^\beta Q_{\alpha,\beta}(\partial_1, \ldots, \partial_n)} \otimes 1 \otimes f^s).$$

Let $\overline{\alpha}$ be the largest $\alpha \in \mathbb{N}$ for which there is a nonzero $Q_{\alpha,\beta}$ occurring in $P = \sum_{\alpha,\beta} \partial_t^\alpha x^\beta Q_{\alpha,\beta}(\partial_1 + \partial_1 \bullet (f)\partial_t, \ldots, \partial_n + \partial_n \bullet (f)\partial_t)$. We show that the sum of terms that contain $\partial_t^{\overline{\alpha}}$ is in $D_{n+1} \cdot \phi(L)$ as follows. In order for $P \bullet (\overline{1} \otimes 1 \otimes f^s)$ to vanish, the sum of terms with the highest s-power, namely $s^{\overline{\alpha}}$, must vanish. Hence $\sum_\beta x^\beta Q_{\overline{\alpha},\beta}(\partial_1, \ldots, \partial_n) \otimes (-1/f)^{\overline{\alpha}} \otimes f^s \in L \otimes R_n[f^{-1}, s] \otimes f^s$ as $R_n[f^{-1}, s]$ is $R_n[s]$-flat. It follows that $\sum_\beta x^\beta Q_{\overline{\alpha},\beta}(\partial_1, \ldots, \partial_n) \in L$ (L is f-saturated!) and hence $\sum_\beta \partial_t^{\overline{\alpha}} x^\beta Q_{\overline{\alpha},\beta}(\partial_1 + \partial_1 \bullet (f)\partial_t, \ldots, \partial_n + \partial_n \bullet (f)\partial_t) \in D_{n+1} \cdot \phi(L)$ as announced.

So by the first part, $P - \sum_\beta \partial_t^{\overline{\alpha}} x^\beta Q_{\overline{\alpha},\beta}(\partial_1 + \partial_1 \bullet (f)\partial_t, \ldots, \partial_n + \partial_n \bullet (f)\partial_t)$ kills $\overline{1} \otimes 1 \otimes f^s$, but is of smaller degree in ∂_t than P was.

The claim follows by induction on $\overline{\alpha}$.   □

If we identify $D_n[-\partial_t t]$ with $D_n[s]$ then $J_{n+1}^L(f^s) \cap D_n[-\partial_t t]$ is identified with $J^L(f^s)$ since, as we observed earlier, $-\partial_t t$ multiplies by s on $\mathcal{M}_f^L$. As we pointed out in the beginning, the crux of our algorithms is to calculate $J^L(f^s) = J_{n+1}^L(f^s) \cap D_n[s]$. We shall deal with this computation now.

In Theorem 19 of [33], T. Oaku showed how to construct a generating set for $J^L(f^s)$ in the case $L = D_n \cdot (\partial_1, \ldots, \partial_n)$. Using his ideas we explain how one may calculate $J \cap D_n[-\partial_t t]$ whenever $J \subseteq D_{n+1}$ is any given ideal, and as a corollary develop an algorithm that for f-saturated D_n/L computes $J^L(f^s) = J_{n+1}^L(f^s) \cap D_n[-\partial_t t]$.

We first review some work of Oaku. On D_{n+1} we define the weight vector w by $w(t) = 1, w(\partial_t) = -1, w(x_i) = w(\partial_i) = 0$ and we extend it to $D_{n+1}[y_1, y_2]$ by $w(y_1) = -w(y_2) = 1$. If $P = \sum_i P_i \in D_{n+1}[y_1, y_2]$ and all P_i are monomials, then we will write $(P)^h$ for the operator $\sum_i P_i \cdot y_1^{d_i}$ where $d_i = \max_j(w(P_j)) - w(P_i)$ and call it the y_1-*homogenization* of P.

Note that the Buchberger algorithm preserves homogeneity in the following sense: if a set of generators for an ideal is given and these generators are homogeneous with respect to the weight above, then any new generator for the ideal constructed with the classical Buchberger algorithm will also be homogeneous. (This is a consequence of the facts that the y_i commute with all other variables and that $\partial_t t = t\partial_t + 1$ is homogeneous of weight zero.) This homogeneity is very important for the following result of Oaku:

Proposition 3.6. *Let* $J = D_{n+1} \cdot (Q_1, \ldots, Q_r)$. *Let* I *be the left ideal in* $D_{n+1}[y_1]$ *generated by the* y_1-*homogenizations* $(Q_i)^h$ *of the* Q_i, *relative to the weight* w *above, and set* $\tilde{I} = D_{n+1}[y_1, y_2] \cdot (I, 1 - y_1 y_2)$. *Let* G *be a Gröbner basis for* $\tilde{I}$ *under a monomial order that eliminates* y_1, y_2. *For each*

$P \in G \cap D_{n+1}$ set $P' = t^{-w(P)}P$ if $w(P) < 0$ and $P' = \partial_t^{w(P)}P$ if $w(P) \geq 0$.
Set $G_0 = \{P' : P \in G \cap D_{n+1}\}$. Then $G_0 \subseteq D_n[-\partial_t t]$ generates $J \cap D_n[-\partial_t t]$.

Proof. This is in essence Theorem 18 of [33]. (See the remarks in Subsection 2.2 on how to compute such Gröbner bases.) $\square$

As a corollary to this proposition we obtain an algorithm for the computation of $J^\Delta(f^s)$:

Algorithm 3.7 (Parametric Annihilator).
INPUT: $f \in R_n$; $L \subseteq D_n$ such that L is f-saturated.
OUTPUT: Generators for $J^L(f^s)$.

1. For each generator Q_i of $D_{n+1} \cdot (L, t)$ compute the image $\phi(Q_i)$ under $x_i \mapsto x_i$, $t \mapsto t - f$, $\partial_i \mapsto \partial_i + \partial_i \bullet (f)\partial_t$, $\partial_t \mapsto \partial_t$.
2. Homogenize all $\phi(Q_i)$ with respect to the new variable y_1 relative to the weight w introduced before Proposition 3.6.
3. Compute a Gröbner basis for the ideal

$$D_{n+1}[y_1, y_2] \cdot ((\phi(Q_1))^h, \ldots, (\phi(Q_r))^h, 1 - y_1 y_2)$$

   in $D_{n+1}[y_1, y_2]$ using an order that eliminates y_1, y_2.
4. Select the operators $\{P_j\}_1^b$ in this basis which do not contain y_1, y_2.
5. For each P_j, $1 \leq j \leq b$, if $w(P_j) > 0$ replace P_j by $P'_j = \partial_t^{w(P_j)}P_j$. Otherwise replace P_j by $P'_j = t^{-w(P_j)}P_j$.
6. Return the new operators $\{P'_j\}_1^b$.

End.

The output will be operators in $D_n[-\partial_t t]$ which is naturally identified with $D_n[s]$ (including the action on $\mathcal{M}_f^L$). This algorithm is in effect Proposition 7.1 of [32].

In *Macaulay 2*, one can compute the parametric annihilator ideal (for $R_n = \Delta$) by the command AnnFs:

```
i8 : D = QQ[x,y,z,w,Dx,Dy,Dz,Dw,
            WeylAlgebra => {x=>Dx, y=>Dy, z=>Dz, w=>Dw}];

i9 : f = x^2+y^2+z^2+w^2

       2   2   2   2
o9 = x + y + z + w

o9 : D

i10 : AnnFs(f)

                                                              . . .
o10 = ideal (w*Dz - z*Dw, w*Dy - y*Dw, z*Dy - y*Dz, w*Dx - x*Dw, z*Dx  · · ·
                                                              . . .

o10 : Ideal of QQ [x, y, z, w, Dx, Dy, Dz, Dw, $s, WeylAlgebra => {x = · · ·
```

If we want to compute $J^L(f^s)$ for more general L, we have to use the command AnnIFs:

```
i11 : L=ideal(x,y,Dz,Dw)

o11 = ideal (x, y, Dz, Dw)

o11 : Ideal of D

i12 : AnnIFs(L,f)

                1          1
o12 = ideal (y, x, w*Dz - z*Dw, -*z*Dz + -*w*Dw - $s)
                2          2

o12 : Ideal of QQ [x, y, z, w, Dx, Dy, Dz, Dw, $s, WeylAlgebra => {x = ···
```

It should be emphasized that saturatedness of L with respect to f is a must for AnnIFs.

3.3   The Bernstein-Sato Polynomial

Knowing $J^L(f^s)$ allows us to get our hands on the Bernstein-Sato polynomial of f on M:

Corollary 3.8. *Suppose L is a holonomic ideal in D_n (i.e., D_n/L is holonomic). The Bernstein polynomial $b_f^L(s)$ of f on (D_n/L) satisfies*

$$(b_f^L(s)) = \left(D_n[s] \cdot (J^L(f^s), f)\right) \cap K[s]. \qquad (3.1)$$

Moreover, if L is f-saturated then $b_f^L(s)$ can be computed with Gröbner basis computations.

Proof. By definition of $b_f^L(s)$ we have $(b_f^L(s) - P_f^L(s) \cdot f) \bullet (\overline{1} \otimes 1 \otimes f^s) = 0$ for a suitable $P_f^L(s) \in D_n[s]$. Hence $b_f^L(s)$ is in $K[s]$ and in $D_n[s](J^L(f^s), f)$. Conversely, if $b(s)$ is in this intersection then $b(s)$ satisfies an equality of the type of (2.2) and hence is a multiple of $b_f^L(s)$.

If we use an elimination order for which $\{x_i, \partial_i\}_1^n \gg s$ in $D_n[s]$, then if $J^L(f^s)$ is known, $b_f^L(s)$ will be (up to a scalar factor) the unique element in the reduced Gröbner basis for $D_n[s] \cdot (J^L(f^s), f)$ that contains no x_i nor ∂_i. Since we assume L to be f-saturated, $J^L(f^s)$ can be computed according to Proposition 3.6. $\qquad \Box$

We therefore arrive at the following algorithm for the Bernstein-Sato polynomial [31].

Algorithm 3.9 (Bernstein-Sato polynomial).
INPUT: $f \in R_n$; $L \subseteq D_n$ such that D_n/L is holonomic and f-torsion free.
OUTPUT: The Bernstein polynomial $b_f^L(s)$.

1. Determine $J^L(f^s)$ following Algorithm 3.7.

2. Find a reduced Gröbner basis for the ideal $J^L(f^s) + D_n[s] \cdot f$ using an elimination order for x and ∂.
3. Pick the unique element $b(s) \in K[s]$ contained in that basis and return it.

End.

We illustrate the algorithm with two examples. We first recall f which was defined at the end of the previous subsection.

```
i13 : f

        2    2    2    2
o13 = x  + y  + z  + w

o13 : D
```

Now we compute the Bernstein-Sato polynomial.

```
i14 : globalBFunction(f)

        2
o14 = $s  + 3$s + 2

o14 : QQ [$s]
```

The routine globalBFunction computes the Bernstein-Sato polynomial of f on R_n. We also take a look at the Bernstein-Sato polynomial of a cubic:

```
i15 : g=x^3+y^3+z^3+w^3

        3    3    3    3
o15 = x  + y  + z  + w

o15 : D

i16 : factorBFunction globalBFunction(g)

                  7        8             4        5
o16 = ($s + 1)($s + -)($s + -)($s + 2)($s + -)($s + -)
                  3        3             3        3

o16 : Product
```

In *Macaulay 2* one can also find $b_f^L(s)$ for more general L. We will see in the following remark what the appropriate commands are.

Remark 3.10. It is clear that $s + 1$ is always a factor of any Bernstein-Sato polynomial on R_n, but this is not necessarily the case if $L \neq \Delta$. For example, $b_f^L(s) = s$ for $n = 1$, $f = x$ and $L = x\partial_x + 1$ (in which case $D_1/L \cong R_1[x^{-1}]$, generated by $1/x$). In particular, it is not true that the roots of $b_f^L(s)$ are negative for general holonomic L.

If L is equal to Δ, and if f is nice, then the Bernstein roots are all between $-n$ and 0 [46]. But for general f very little is known besides a famous theorem of Kashiwara that states that $b_f^\Delta(s)$ factors over $\mathbb{Q}$ [19] and all roots are negative.

For L arbitrary, the situation is more complicated. The Bernstein-Sato polynomial of any polynomial f on the D_n-module generated by $\bar{1} \otimes f^a$ with

$a \in K$ is related to that of f on D_n/L by a simple shift, and so the Bernstein roots of f on the D_n-module generated by the function germ f^a, $a \in K$, are still all in K by [19]. Localizing other modules however can easily lead to nonrational roots. As an example, consider

```
i17 : D1 = QQ[x,Dx,WeylAlgebra => {x=>Dx}];

i18 : I1 = ideal((x*Dx)^2+1)

             2  2
o18 = ideal(x Dx  + x*Dx + 1)

o18 : Ideal of D1
```

This is input defined over the rationals. Even localizing D_1/I_1 at a very simple f leads to nonrational roots:

```
i19 : f1 = x;

i20 : b=globalB(I1, f1)

                               2
o20 = HashTable{Boperator => - x*Dx  + 2Dx*$s + Dx}
                               2
                Bpolynomial => $s  + 2$s + 2

o20 : HashTable
```

The routine `globalB` is to be used if a Bernstein-Sato polynomial is suspected to fail to factor over $\mathbb{Q}$. If $b_f^L(s)$ does factor over $\mathbb{Q}$, one can also use the routine `DlocalizeAll` to be discussed below. It would be very interesting to determine rules that govern the splitting field of $b_f^L(s)$ in general.

3.4 Specializing Exponents

In this subsection we investigate the result of substituting $a \in K$ for s in $J^L(f^s)$. Recall that the Bernstein polynomial $b_f^L(s)$ will exist (i.e., be nonzero) if D_n/L is holonomic. As outlined in the previous subsection, $b_f^L(s)$ can be computed if D_n/L is holonomic and f-torsion free. The following proposition (Proposition 7.3 in [32], see also Proposition 6.2 in [19]) shows that replacing s by an exponent in very general position leads to a solution of the localization problem.

Proposition 3.11. *If L is holonomic and $a \in K$ is such that no element of $\{a-1, a-2, \ldots\}$ is a Bernstein root of f on L then we have D_n-isomorphisms*

$$(D_n/L) \otimes_{R_n} \left(R_n[f^{-1}] \otimes f^a\right) \cong \left(D_n[s]/J^L(f^s)\right)|_{s=a} \cong D_n \bullet (\overline{1} \otimes 1 \otimes f^a). \tag{3.2}$$

$\square$

One notes in particular that if any $a \in \mathbb{Z}$ satisfies the conditions of the proposition, then so does every integer smaller than a. This motivates the following

Definition 3.12. The *stable integral exponent of f on L* is the smallest integral root of $b_f^L(s)$, and denoted a_f^L.

In terms of this definition,

$$\left(D_n/J^L(f^s)\right)\big|_{s=a_f^L} \cong (D_n/L) \otimes_{R_n} R_n[f^{-1}],$$

and the presentation corresponds to the generator $\bar{1} \otimes f^{a_f^L}$. If $L = \Delta$ then Kashiwara's result tells us that $b_f^L(s)$ will factor over the rationals, and thus it is very easy to find the stable integral exponent. If we localize a more general module, the roots may not even be K-rational anymore as we saw at the end of the previous subsection.

The following lemma deals with the question of finding the smallest integer root of a polynomial. We let $|s|$ denote the complex absolute value.

Lemma 3.13. *Suppose that in the situation of Corollary 3.8,*

$$b_f^L(s) = s^d + b_{d-1}s^{d-1} + \cdots + b_0,$$

and define $B = \max_i\{|b_i|^{1/(d-i)}\}$. The smallest integer root of $b_f^L(s)$ is an integer between $-2B$ and $2B$. If in particular $L = D_n \cdot (\partial_1, \ldots, \partial_n)$, it suffices to check the integers between $-b_{d-1}$ and -1.

Proof. Suppose $|s_0| = 2B\rho$ where B is as defined above and $\rho > 1$. Assume also that s_0 is a root of $b_f^L(s)$. We find

$$(2B\rho)^d = |s_0|^d = \Big| - \sum_{i=0}^{d-1} b_i s_0^i \Big| \leq \sum_{i=0}^{d-1} B^{d-i}|s_0|^i$$

$$= B^d \sum_{i=0}^{d-1} (2\rho)^i \leq B^d\big((2\rho)^d - 1\big),$$

using $\rho \geq 1$. By contradiction, s_0 is not a root.

The final claim is a consequence of Kashiwara's work [19] where he proves that if $L = D_n \cdot (\partial_1, \ldots, \partial_n)$ then all roots of $b_f^L(s)$ are rational and negative, and hence $-b_{n-1}$ is a lower bound for each single root. $\square$

Combining Proposition 3.11 with Algorithms 3.7 and 3.9 we therefore obtain

Algorithm 3.14 (Localization).
INPUT: $f \in R_n$; $L \subseteq D_n$ such that D_n/L is holonomic and f-torsion free.
OUTPUT: Generators for an ideal J such that $(D_n/L) \otimes_{R_n} R_n[f^{-1}] \cong D_n/J$.

1. Determine $J^L(f^s)$ following Algorithm 3.7.
2. Find the Bernstein polynomial $b_f^L(s)$ using Algorithm 3.9.
3. Find the smallest integer root a of $b_f^L(s)$.

4. Replace s by a in all generators for $J^L(f^s)$ and return these generators.

End.

Algorithms 3.9 and 3.14 are Theorems 6.14 and Proposition 7.3 in [32].

Example 3.15. For $f = x^2+y^2+z^2+w^2$, we found a stable integral exponent of -2 in the previous subsection. To compute the annihilator of f^{-2} using *Macaulay 2*, we use the command `Dlocalize` which automatically uses the stable integral exponent. We first change the current ring back to the ring D which we used in the previous subsection:

```
i21 : use D

o21 = D

o21 : PolynomialRing
```

Here is the module to be localized.

```
i22 : R = (D^1/ideal(Dx,Dy,Dz,Dw))

o22 = cokernel | Dx Dy Dz Dw |

                             1
o22 : D-module, quotient of D
```

The localization then is obtained by running

```
i23 : ann2 = relations Dlocalize(R,f)

o23 = | wDz-zDw wDy-yDw zDy-yDz wDx-xDw zDx-xDz yDx-xDy xDx+yDy+zDz+wD · · ·

                 1        10
o23 : Matrix D   <--- D
```

The output `ann2` is a 1×10 matrix whose entries generate $\mathrm{ann}_{D_4}(f^{-2})$.

Remark 3.16. The computation of the annihilator of f^a for values of a such that $a - k$ is a Bernstein root for some $k \in \mathbb{N}^+$ can be achieved by an appropriate syzygy computation. For example, we saw above that the Bernstein-Sato polynomial of $f = x^2 + y^2 + z^2 + w^2$ on R_4 is $(s + 1)(s + 2)$. So evaluation of $J^L(f^s)$ at -1 does not necessarily yield $\mathrm{ann}_{D_4}(f^{-1})$, as will be documented in the next remark. On the other hand, evaluation at -2 gives $\mathrm{ann}_{D_4}(f^{-2})$. It is not hard to see that $\mathrm{ann}_{D_4}(f^{-1}) = \{P \in D_n : Pf \in \mathrm{ann}_{D_n}(f^{-2})\}$ because $D_4 \bullet f^{-1} - D_4 f \bullet f^{-2} \subseteq D_4 \bullet f^{-2}$. So we set:

```
i24 : F = matrix{{f}}

o24 = | x2+y2+z2+w2 |

                 1        1
o24 : Matrix D   <--- D
```

To find $\mathrm{ann}_{D_4}(f^{-1})$, we use the command `modulo` which computes relations: `modulo(M,N)` computes for two matrices M, N the set of (vectors of) operators P such that $P \cdot M \subseteq \mathrm{im}(N)$.

```
i25 : ann1 = gb modulo(F,ann2)

o25 = {2} | wDz-zDw wDy-yDw zDy-yDz Dx^2+Dy^2+Dz^2+Dw^2 wDx-xDw zDx-xD ···

o25 : GroebnerBasis
```

The generator $\partial_2^2 + \partial_y^2 + \partial_z^2 + \partial_w^2$ is particularly interesting. To see the quotient of $D_4 \bullet f^{-2}$ by $D_4 \bullet f^{-1}$ we execute

```
i26 : gb((ideal ann2) + (ideal F))

o26 = | w z y x |

o26 : GroebnerBasis
```

which shows that $D_4 \bullet f^{-2}$ is an extension of $D_4/D_4(x, y, z, w)$ by $D_4 \bullet f^{-1}$. This is not surprising, since $(0, 0, 0, 0)$ is the only singularity of f and hence the difference between $D_4 \bullet f^{-2}$ and $D_4 \bullet f^{-1}$ must be supported at the origin.

It is perhaps interesting to note that for a more complicated (but still irreducible) polynomial f the quotient $(D_n \bullet f^a)/(D_n \bullet f^{a+1})$ can be a non-simple nonzero D_n-module. For example, let $f = x^3 + y^3 + z^3 + w^3$ and $a = a_f^{\Delta} = -2$. A computation similar to the quadric case above shows that here $(D_n \bullet f^a)/(D_n \bullet f^{a+1})$ is a (x, y, z, w)-torsion module (supported at the singular locus of f) isomorphic to $(D_4/D_4 \cdot (x, y, z, w))^6$. The socle elements of the quotient are the degree 2 polynomials in x, y, z, w.

Example 3.17. Here we show how with *Macaulay 2* one can get more information from the localization procedure.

```
i27 : D = QQ[x,y,z,Dx,Dy,Dz, WeylAlgebra => {x=>Dx, y=>Dy, z=>Dz}];

i28 : Delta = ideal(Dx,Dy,Dz);

o28 : Ideal of D
```

We now define a polynomial and compute the localization of R_3 at the polynomial.

```
i29 : f=x^3+y^3+z^3;

i30 : I1=DlocalizeAll(D^1/Delta,f,Strategy=>Oaku)

                     1        1       1          2      ···
o30 = HashTable{annFS => ideal (-*x*Dx + -*y*Dy + -*z*Dz - $s, z Dy - ···
                     3        3       3                        ···
                           2       5       4
             Bfunction => ($s + 1) ($s + -)($s + -)($s + 2)
                                   3       3
                         2      3        2      4       1   2  ···
             Boperator => --*y*z*Dx Dy*Dz - --*y*z*Dy Dz + ---*z Dx ···
                         81               81             243    ···
             GeneratorPower => -2
             LocMap => | x6+2x3y3+y6+2x3z3+2y3z3+z6 |
             LocModule => cokernel | 1/3xDx+1/3yDy+1/3zDz+2 z2Dy-y2 ···

o30 : HashTable

i31 : I2=DlocalizeAll(D^1/Delta,f)
```

```
o31 = HashTable{GeneratorPower => -2                              · · ·
                           2          2         2      1
             IntegrateBfunction => ($s) ($s + 1) ($s + -)($s + -)
                                                       3      3
             LocMap => | x6+2x3y3+y6+2x3z3+2y3z3+z6 |
             LocModule => cokernel | xDx+yDy+zDz+6 z2Dy-y2Dz z2Dx-x · · ·

o31 : HashTable
```

The last two commands both compute the localization of R_3 at f but follow different localization algorithms. The former uses our Algorithm 3.14 while the latter follows [37].

The output of the command DlocalizeAll is a hashtable, because it contains a variety of data that pertain to the map $R_n \hookrightarrow R_n[f^{-1}]$. LocMap gives the element that induces the map on the D_n-module level (by right multiplication). LocModule gives the localized module as cokernel of the displayed matrix. Bfunction is the Bernstein-Sato polynomial and annFS the generic annihilator $J^L(f^s)$. Boperator displays a Bernstein operator and the stable integral exponent is stored in GeneratorPower.

Algorithm 3.14 requires the ideal L to be f-saturated. This property is not checked by *Macaulay 2*, so the user needs to make sure it holds. For example, this is always the case if D_n/L is a localization of R_n. One can check the saturation property in *Macaulay 2*, but it is a rather involved computation. This difficulty can be circumvented by omitting the option Strategy=>Oaku, in which case the localization algorithm of [37] is used. In terms of complexity, using the Oaku strategy is much better behaved.

One can address the entries of a hashtable. For example, executing

```
i32 : I1.LocModule

o32 = cokernel | 1/3xDx+1/3yDy+1/3zDz+2 z2Dy-y2Dz z2Dx-x2Dz y2Dx-x2Dy |

                      1
o32 : D-module, quotient of D
```

one can see that $R_3[f^{-1}]$ is isomorphic to the cokernel of the LocModule entry which (for either localization method) is

$$D_3 / D_3 \cdot (x\partial_x + y\partial_y + z\partial_z + 6, \; z^2\partial_y - y^2\partial_z, \; x^3\partial_y + y^3\partial_y + y^2z\partial_z + 6y^2,$$
$$z^2\partial_x - x^2\partial_z, \; y^2\partial_x - x^2\partial_y, \; x^3\partial_z + y^3\partial_z + z^3\partial_z + 6z^2).$$

The first line of the hashtable I1 shows that $R_3[f^{-1}]$ is generated by f^{-2} over D_3, while I1.LocMap shows that the natural inclusion $D_3/\Delta = R_3 \hookrightarrow R_3[f^{-1}] = D_3/J^\Delta(f^s)|_{s=a_f^\Delta}$ is given by right multiplication by f^2, shown as the third entry of the hashtable I1. It is perhaps useful to point out that the fourth entry of hashtable I2 is a relative of the Bernstein-Sato polynomial of f, and is used for the computation of the so-called restriction functor (compare with [35,48]).

Remark 3.18. Plugging in bad values a for s (such that $a - k$ *is* a Bernstein root for some $k \in \mathbb{N}^+$) can have unexpected results. Consider the case $n = 1$,

$f = x$. Then $J^\Delta(f^s) = D_1 \cdot (s + 1 - \partial_1 x_1)$. Hence $b_f^\Delta(s) = s + 1$ and -1 is the unique Bernstein root. According to Proposition 3.11,

$$\left(D_1[s]/J^\Delta(f^s)\right)|_{s=a} \cong R_1[x_1^{-1}] \otimes x_1^{\,a} \cong D_1 \bullet x_1^{\,a}$$

for all $a \in K \setminus \mathbb{N}$. For $a \in \mathbb{N}^+$, we also have $D_1[s]/J^\Delta(f^s)|_{s=a} \cong D_1 \bullet x^a$, but this is of course not $R_1[x_1^{-1}]$ but just R_1.

For $a = 0$ however, $\left(D_1[s]/J^\Delta(f^s)\right)|_{s=a}$ has x_1-torsion! It equals in fact what is called the Fourier transform of $R_1[x_1^{-1}]$ and fits into an exact sequence

$$0 \to H^1_{x_1}(R_1) \to \mathcal{F}(R_1[x_1^{-1}]) \to R_1 \to 0.$$

Remark 3.19. If D_n/L is holonomic but has f-torsion, then $(D_n/L) \otimes R_n[f^{-1}]$ and $((D_n/L)/H^0_{(f)}(D_n/L)) \otimes R_n[f^{-1}]$ are of course isomorphic. So if we knew how to find $M/H^0_f(M)$ for holonomic modules M, our localization algorithm could be generalized to all holonomic modules. There are two different approaches to the problem of f-torsion, presented in [35] and in [43,44]. The former is based on homological methods and restriction to the diagonal while the latter aims at direct computation of those $P \in D_n$ for which $f^k P \in L$ for some k.

There is also another direct method for localizing $M = D_n/L$ at f that works in the situation where the nonholonomic locus of M is contained in the variety of f (irrespective of torsion). It was proved by Kashiwara, that $M[f^{-1}]$ is then holonomic, and in [37] an algorithm based on integration is given that computes a presentation for it.

4   Local Cohomology Computations

The purpose of this section is to present algorithms that compute for given $i, j, k \in \mathbb{N}, I \subseteq R_n$ the structure of the local cohomology modules $H^k_I(R_n)$ and $H^i_\mathfrak{m}(H^j_I(R_n))$, and the invariants $\lambda_{i,j}(R_n/I)$ associated to I. In particular, the algorithms detect the vanishing of local cohomology modules.

4.1   Local Cohomology

We will first describe an algorithm that takes a finite set of polynomials $\{f_1, \ldots, f_r\} \subset R_n$ and returns a presentation of $H^k_I(R_n)$ where $I = R_n \cdot (f_1, \ldots, f_r)$. In particular, if $H^k_I(R_n)$ is zero, then the algorithm will return the zero presentation.

Definition 4.1. Let Θ^r_k be the set of k-element subsets of $\{1, \ldots, r\}$ and for $\theta \in \Theta^r_k$ write F_θ for the product $\prod_{i \in \theta} f_i$.

Consider the Čech complex $\check{C}^\bullet = \check{C}^\bullet(f_1, \ldots, f_r)$ associated to $f_1, \ldots, f_r$ in R_n,

$$0 \to R_n \to \bigoplus_{\theta \in \Theta_1^r} R_n[F_\theta^{-1}] \to \bigoplus_{\theta \in \Theta_2^r} R_n[F_\theta^{-1}] \to \cdots \to R_n[(f_1 \cdots f_r)^{-1}] \to 0.$$

$$(4.1)$$

Its k-th cohomology group is $H_I^k(R_n)$. The map

$$M_k : \left(\check{C}^k = \bigoplus_{\theta \in \Theta_k^r} R_n[F_\theta^{-1}] \right) \to \left(\bigoplus_{\theta' \in \Theta_{k+1}^r} R_n[F_{\theta'}^{-1}] = \check{C}^{k+1} \right) \qquad (4.2)$$

is the sum of maps

$$R_n[(f_{i_1} \cdots f_{i_k})^{-1}] \to R_n[(f_{j_1} \cdots f_{j_{k+1}})^{-1}] \qquad (4.3)$$

which are zero if $\{i_1, \ldots, i_k\} \not\subseteq \{j_1, \ldots, j_{k+1}\}$, or send $\frac{1}{1}$ to $\frac{1}{1}$ (up to sign). With $D_n/\Delta \cong R_n$, identify $R_n[(f_{i_1} \cdots f_{i_k})^{-1}]$ with $D_n/J^\Delta((f_{i_1} \cdots f_{i_k})^s)|_{s=a}$ and $R_n[(f_{j_1} \cdots f_{j_{k+1}})^{-1}]$ with $D_n/J^\Delta((f_{j_1} \cdots f_{j_{k+1}})^s)|_{s=a'}$ where a, a' are sufficiently small integers. By Proposition 3.11 we may assume that $a = a' \leq 0$. Then the map (4.3) is in the nonzero case multiplication from the right by $(f_l)^{-a}$ where $l = \{j_1, \ldots, j_{k+1}\} \setminus \{i_1, \ldots, i_k\}$, again up to sign. For example, consider the inclusion

$$D_2/D_2 \cdot (\partial_x x, \partial_y) = R_2[x^{-1}] \hookrightarrow R_2[(xy)^{-1}] = D_2/D_2 \cdot (\partial_x x, \partial_y y).$$

Since $\frac{1}{x} = \frac{y}{xy}$, the inclusion on the level of D_2-modules maps $P + \mathrm{ann}(x^{-1})$ to $Py + \mathrm{ann}((xy)^{-1})$.

It follows that the matrix representing the map $\check{C}^k \to \check{C}^{k+1}$ in terms of D_n-modules is very easy to write down once the annihilator ideals and Bernstein polynomials for all k- and $(k+1)$-fold products of the f_i are known: the entries are 0 or $\pm f_l^{-a}$ where f_l is the new factor. These considerations give the following

Algorithm 4.2 (Local cohomology).
INPUT: $f_1, \ldots, f_r \in R_n; k \in \mathbb{N}$.
OUTPUT: $H_I^k(R_n)$ in terms of generators and relations as finitely generated D_n-module where $I = R_n \cdot (f_1, \ldots, f_r)$.

1. Compute the annihilator ideal $J^\Delta((F_\theta)^s)$ and the Bernstein polynomial $b_{F_\theta}^\Delta(s)$ for all $(k-1)$-, k- and $(k+1)$-fold products F_θ of $f_1, \ldots, f_r$ following Algorithms 3.7 and 3.9 (so θ runs through $\Theta_{k-1}^r \cup \Theta_k^r \cup \Theta_{k+1}^r$).
2. Compute the stable integral exponents $a_{F_\theta}^\Delta$, let a be their minimum and replace s by a in all the annihilator ideals.
3. Compute the two matrices M_{k-1}, M_k representing the D_n-linear maps $\check{C}^{k-1} \to \check{C}^k$ and $\check{C}^k \to \check{C}^{k+1}$ as explained above.

4. Compute a Gröbner basis G for the kernel of the composition

$$\bigoplus_{\theta \in \Theta_k^r} D_n \twoheadrightarrow \bigoplus_{\theta \in \Theta_k^r} D_n/J^\Delta(F_\theta{}^s)|_{s=a} \xrightarrow{M_k} \bigoplus_{\theta' \in \Theta_{k+1}^r} D_n/J^\Delta(F_{\theta'}{}^s)|_{s=a}.$$

5. Compute a Gröbner basis G_0 for the preimage in $\bigoplus_{\theta \in \Theta_k^r} D_n$ of the module

$$\text{im}(M_{k-1}) \subseteq \bigoplus_{\theta \in \Theta_k^r} D_n/J^\Delta((F_\theta)^s)|_{s=a} \twoheadleftarrow \bigoplus_{\theta \in \Theta_k^r} D_n$$

under the indicated projection.
6. Compute the remainders of all elements of G with respect to G_0.
7. Return these remainders and G_0.

End.

The nonzero elements of G generate the quotient $G/G_0 \cong H_I^k(R_n)$ so that in particular $H_I^k(R_n) = 0$ if and only if all returned remainders are zero.

Example 4.3. Let I be the ideal in $R_6 = K[x, y, z, u, v, w]$ that is generated by the 2×2 minors f, g, h of the matrix $\begin{pmatrix} x & y & z \\ u & v & w \end{pmatrix}$. Then $H_I^i(R_6) = 0$ for $i < 2$ and $H_I^2(R_6) \neq 0$ because I is a height 2 prime, and $H_I^i(R_6) = 0$ for $i > 3$ because I is 3-generated, so the only open case is $H_I^3(R_6)$. This module in fact does not vanish, and our algorithm provides a proof of this fact by direct calculation. The *Macaulay 2* commands are as follows.

```
i33 : D= QQ[x,y,z,u,v,w,Dx,Dy,Dz,Du,Dv,Dw, WeylAlgebra =>
              {x=>Dx, y=>Dy, z=>Dz, u=>Du, v=>Dv, w=>Dw}];

i34 : Delta=ideal(Dx,Dy,Dz,Du,Dv,Dw);

o34 : Ideal of D

i35 : R=D^1/Delta;

i36 : f=x*v-u*y;

i37 : g=x*w-u*z;

i38 : h=y*w-v*z;
```

These commands define the relevant rings and polynomials. The following three compute the localization of R_6 at f:

```
i39 : Rf=DlocalizeAll(R,f,Strategy => Oaku)

o39 = HashTable{annFS => ideal (Dw, Dz, x*Du + y*Dv, y*Dy - u*Du, x*Dy ...
                Bfunction => ($s + 1)($s + 2)
                Boperator => - Dy*Du + Dx*Dv
                GeneratorPower => -2
                LocMap => | y2u2-2xyuv+x2v2 |
                LocModule => cokernel | Dw Dz xDu+yDv yDy-uDu xDy+uDv ...

o39 : HashTable
```

of $R_6[f^{-1}]$ at g:

```
i40 : Rfg=DlocalizeAll(Rf.LocModule,g, Strategy => Oaku)
```

```
                                                             . . .
o40 = HashTable{annFS => ideal (Dz*Dv - Dy*Dw, x*Du + y*Dv + z*Dw, z*D ···
                Bfunction => ($s + 1)($s)
                Boperator => - Dz*Du + Dx*Dw
                GeneratorPower => -1
                LocMap => | -zu+xw |
                LocModule => cokernel | DzDv-DyDw xDu+yDv+zDw zDz-uDu- ···
```

```
o40 : HashTable
```

and of $R_6[(fg)^{-1}]$ at h:

```
i41 : Rfgh=DlocalizeAll(Rfg.LocModule,h, Strategy => Oaku)
```

```
                                                             . . .
o41 = HashTable{annFS => ideal (x*Du + y*Dv + z*Dw, z*Dz - u*Du - v*Dv ···
                Bfunction => ($s - 1)($s + 1)
                Boperator => - Dz*Dv + Dy*Dw
                GeneratorPower => -1
                LocMap => | -zv+yw |
                LocModule => cokernel | xDu+yDv+zDw zDz-uDu-vDv-2 yDy- ···
```

```
o41 : HashTable
```

From the output of these commands one sees that $R_6[(fgh)^{-1}]$ is generated by $1/f^2gh$. This follows from considering the stable integral exponents of the three localization procedures, encoded in the hashtable entry stored under the key GeneratorPower: for example,

```
i42 : Rf.GeneratorPower
```

```
o42 = -2
```

shows that the generator for $R_6[f^{-1}]$ is f^{-2}. Now we compute the annihilator of $H_I^3(R_6)$. From the Čech complex it follows that $H_I^3(R_6)$ is the quotient of the output of Rfgh.LocModule (isomorphic to $R_6[(fgh)^{-1}]$) by the submodules generated by f^2, g and h. (These submodules represent $R_6[(gh)^{-1}]$, $R_6[(fh)^{-1}]$ and $R_6[(fg)^{-1}]$ respectively.)

```
i43 : Jfgh=ideal relations Rfgh.LocModule;
```

```
o43 : Ideal of D
```

```
i44 : JH3=Jfgh+ideal(f^2,g,h);
```

```
o44 : Ideal of D
```

```
i45 : JH3gb=gb JH3
```

```
o45 = | w z uDu+vDv+wDw+4 xDu+yDv+zDw yDy-uDu-wDw-1 xDy+uDv uDx+vDy+wD ···
```

```
o45 : GroebnerBasis
```

So JH3 is the ideal of D_3 generated by

$$w, \ z, \ u\partial_u + v\partial_v + w\partial_w + 4, \ x\partial_u + y\partial_v + z\partial_w, \ y\partial_y - u\partial_u - w\partial_w - 1,$$
$$x\partial_y + u\partial_v, \ u\partial_x + v\partial_y + w\partial_z, \ y\partial_x + v\partial_u, \ x\partial_x - v\partial_v - w\partial_w - 1,$$
$$v^2, \ uv, \ yv, \ u^2, \ yu + xv, \ xu, \ y^2, \ xy,$$
$$x^2, \ xv\partial_v + 2x, \ v\partial_y\partial_u + w\partial_z\partial_u - v\partial_x\partial_v - w\partial_x\partial_w - 3\partial_x$$

which form a Gröbner basis. This proves that $H_I^3(R) \neq 0$, because 1 is not in the Gröbner basis of JH3. (There are also algebraic and topological proofs to this account. Due to Hochster, and Bruns and Schwänzl, they are quite ingenious and work only in rather special situations.)

From our output one can see that $H_I^3(R_6)$ is (x, y, z, u, v, w)-torsion as JH3 contains $(x, y, z, u, v, w)^2$. The following sequence of commands defines a procedure testmTorsion which as the name suggests tests a module D_n/L for being m-torsion. We first replace the generators of L with a Gröbner basis. Then we pick the elements of the Gröbner basis not using any ∂_i. If now the left over polynomials define an ideal of dimension 0 in R_n, the ideal was m-torsion and otherwise not.

```
i46 : testmTorsion = method();
```

```
i47 : testmTorsion Ideal := (L) -> (
         LL = ideal generators gb L;
         n = numgens (ring (LL)) // 2;
         LLL = ideal select(first entries gens LL, f->(
                 l = apply(listForm f, t->drop(t#0,n));
                 all(l, t->t==toList(n:0))
                 ));
         if dim inw(LLL,toList(apply(1..2*n,t -> 1))) == n
         then true
         else false);
```

If we apply testmTorsion to JH3 we obtain

```
i48 : testmTorsion(JH3)
```

```
o48 = true
```

Further inspection shows that the ideal JH3 is in fact the annihilator of the fraction $f/(wzx^2y^2u^2v^2)$ in $R_6[(xyzuvw)^{-1}]/R_6 \cong D_6/D_6 \cdot (x, y, z, u, v, w)$, and that the fraction generates $D_6/D_6 \cdot (x_1, \ldots, x_6)$. Since $D_6/D_6 \cdot (x_1, \ldots, x_6)$ is isomorphic to $E_{R_6}(R_6/R_6 \cdot (x_1, \ldots, x_6))$, the injective hull of $R_6/R_6 \cdot (x_1, \ldots, x_6) = K$ in the category of R_6-modules, we conclude that $H_I^3(R_6) \cong E_{R_6}(K)$. (In the next subsection we will display a way to use Macaulay 2 to find the length of an m-torsion module.)

In contrast, let I be defined as generated by the three minors, but this time over a field of finite characteristic. Then $H_I^3(R_6)$ is zero because Peskine and Szpiro proved using the Frobenius functor [39] that R_6/I Cohen-Macaulay implies that $H_I^k(R_6)$ is nonzero only if $k = \text{codim}(I)$.

Also opposite to the above example, but in any characteristic, is the following calculation. Let I be the ideal in $K[x, y, z, w]$ describing the twisted cubic: $I = R_4 \cdot (f, g, h)$ with $f = xz - y^2$, $g = yw - z^2$, $h = xw - yz$. The projective variety V_2 defined by I is isomorphic to the projective line. It is of interest to determine whether V_2 and other Veronese embeddings of the projective line are complete intersections. The set-theoretic complete intersection property can occasionally be ruled out with local cohomology techniques: if V is of codimension c in the affine variety X and $H_{I(V)}^{c+k}(O(X)) \neq 0$ for any positive k then V cannot be a set-theoretic complete intersection. In the case of the twisted cubic, it turns out hat $H_I^3(R_4) = 0$ as can be seen from the following computation:

```
i49 : D=QQ[x,y,z,w,Dx,Dy,Dz,Dw,WeylAlgebra => {x=>Dx, y=>Dy, z=>Dz,
      w=>Dw}];

i50 : f=y^2-x*z;

i51 : g=z^2-y*w;

i52 : h=x*w-y*z;

i53 : Delta=ideal(Dx,Dy,Dz,Dw);

o53 : Ideal of D

i54 : R=D^1/Delta;

i55 : Rf=DlocalizeAll(R,f,Strategy => Oaku)
```

$$o55 = \text{HashTable}\{\text{annFS} => \text{ideal (Dw, x*Dy + 2y*Dz, y*Dx + } \tfrac{1}{2}\text{*z*Dy, x*Dx} \cdots$$

$$\text{Bfunction} => (\$s + \tfrac{3}{2})(\$s + 1)$$

$$\text{Boperator} => \tfrac{1}{4}\text{*Dy}^2 - \text{Dx*Dz}$$

```
                  GeneratorPower => -1
                  LocMap => | y2-xz |
                  LocModule => cokernel | Dw xDy+2yDz yDx+1/2zDy xDx-zDz ···
```

```
o55 : HashTable

i56 : Rfg=DlocalizeAll(Rf.LocModule,g, Strategy => Oaku);

i57 : Rfgh=DlocalizeAll(Rfg.LocModule,h, Strategy => Oaku);

i58 : Ifgh=ideal relations Rfgh.LocModule;

o58 : Ideal of D

i59 : IH3=Ifgh+ideal(f,g,h);

o59 : Ideal of D

i60 : IH3gb=gb IH3

o60 = | 1 |

o60 : GroebnerBasis
```

It follows that we cannot conclude from local cohomological considerations that V_2 is not a set-theoretic complete intersection. This is not an accident but typical, as the second vanishing theorem of Hartshorne, Speiser,

Huneke and Lyubeznik shows [14,15,18]: if a homogeneous ideal $I \subseteq R_n$ describes an geometrically connected projective variety of positive dimension then $H_I^{n-1}(R_n) = H_I^n(R_n) = 0$.

4.2   Iterated Local Cohomology

Recall that $\mathfrak{m} = R_n \cdot (x_1, \ldots, x_n)$. As a second application of Gröbner basis computations over the Weyl algebra we show now how to compute the $\mathfrak{m}$-torsion modules $H_\mathfrak{m}^i(H_I^j(R_n))$. Note that we cannot apply Lemma 3.5 to $D_n/L = H_I^j(R_n)$ since $H_I^j(R_n)$ may well contain some torsion.

$\check{C}^j(R_n; f_1, \ldots, f_r)$ denotes the j-th module in the Čech complex to R_n and $\{f_1, \ldots, f_r\}$. Let $\check{C}^{\bullet,\bullet}$ be the double complex

$$\check{C}^{i,j} = \check{C}^i(R_n; x_1, \ldots, x_n) \otimes_{R_n} \check{C}^j(R_n; f_1, \ldots, f_r),$$

with vertical maps $\phi^{\bullet,\bullet}$ induced by the identity on the first factor and the usual Čech maps on the second, and horizontal maps $\xi^{\bullet,\bullet}$ induced by the Čech maps on the first factor and the identity on the second. Now $\check{C}^{i,j}$ is a direct sum of modules $R_n[g^{-1}]$ where $g = x_{\alpha_1} \cdots x_{\alpha_i} \cdot f_{\beta_1} \cdots f_{\beta_j}$. So the whole double complex can be rewritten in terms of D_n-modules and D_n-linear maps using Algorithm 3.14:

$$
\begin{array}{ccccc}
\check{C}^{i-1,j+1} & \xrightarrow{\xi^{i-1,j+1}} & \check{C}^{i,j+1} & \xrightarrow{\xi^{i,j+1}} & \check{C}^{i+1,j+1} \\
\big\uparrow{\scriptstyle\phi^{i-1,j}} & & \big\uparrow{\scriptstyle\phi^{i,j}} & & \big\uparrow{\scriptstyle\phi^{i+1,j}} \\
\check{C}^{i-1,j} & \xrightarrow{\xi^{i-1,j}} & \check{C}^{i,j} & \xrightarrow{\xi^{i,j}} & \check{C}^{i+1,j} \\
\big\uparrow{\scriptstyle\phi^{i-1,j-1}} & & \big\uparrow{\scriptstyle\phi^{i,j-1}} & & \big\uparrow{\scriptstyle\phi^{i+1,j-1}} \\
\check{C}^{i-1,j-1} & \xrightarrow{\xi^{i-1,j-1}} & \check{C}^{i,j-1} & \xrightarrow{\xi^{i,j-1}} & \check{C}^{i+1,j-1}
\end{array}
$$

Since $\check{C}^i(R_n; x_1, \ldots, x_n)$ is R_n-flat, the column cohomology of $\check{C}^{\bullet,\bullet}$ at (i,j) is $\check{C}^i(R_n; x_1, \ldots, x_n) \otimes_{R_n} H_I^j(R_n)$ and the induced horizontal maps in the j-th row are simply the maps in the Čech complex $\check{C}^\bullet(H_I^j(R_n); x_1, \ldots, x_n)$. It follows that the row cohomology of the column cohomology at (i_0, j_0) is $H_\mathfrak{m}^{i_0}(H_I^{j_0}(R_n))$, the object of our interest.

We have, denoting by $X_{\theta'}$ in analogy to F_θ the product $\prod_{i \in \theta'} x_i$, the following

Algorithm 4.4 (Iterated local cohomology).
INPUT: $f_1, \ldots, f_r \in R_n; i_0, j_0 \in \mathbb{N}$.
OUTPUT: $H_\mathfrak{m}^{i_0}(H_I^{j_0}(R_n))$ in terms of generators and relations as finitely generated D_n-module where $I = R_n \cdot (f_1, \ldots, f_r)$.

1. For $i = i_0 - 1, i_0, i_0 + 1$ and $j = j_0 - 1, j_0, j_0 + 1$ compute the annihilators $J^\Delta((F_\theta \cdot X_{\theta'})^s)$, Bernstein polynomials $b_{F_\theta \cdot X_{\theta'}}^\Delta(s)$, and stable integral exponents $a_{F_\theta \cdot X_{\theta'}}^\Delta$ of $F_\theta \cdot X_{\theta'}$ where $\theta \in \Theta_j^r, \theta' \in \Theta_i^n$.

2. Let a be the minimum of all $a_{F_\theta \cdot X_{\theta'}}^{\Delta}$ and replace s by a in all the annihilators computed in the previous step.
3. Compute the matrices to the D_n-linear maps $\phi^{i,j} : \check{C}^{i,j} \to \check{C}^{i,j+1}$ and $\xi^{k,l} : \check{C}^{k,l} \to \check{C}^{k+1,l}$, for $(i,j) \in \{(i_0, j_0), (i_0 + 1, j_0 - 1), (i_0, j_0 - 1), (i_0 - 1, j_0)\}$ and $(k,l) \in \{(i_0, j_0), (i_0 - 1, j_0)\}$.
4. Compute a Gröbner basis G for the module

$$D_n \cdot G = \ker(\phi^{i_0,j_0}) \cap \left[(\xi^{i_0,j_0})^{-1} (\mathrm{im}(\phi^{i_0+1,j_0-1})) \right] + \mathrm{im}(\phi^{i_0,j_0-1})$$

and a Gröbner basis G_0 for the module

$$D_n \cdot G_0 = \xi^{i_0-1,j_0} (\ker(\phi^{i_0-1,j_0})) + \mathrm{im}(\phi^{i_0,j_0-1}).$$

5. Compute the remainders of all elements of G with respect to G_0.
6. Return these remainders together with G_0.

End.

Note that $(D_n \cdot G)/(D_n \cdot G_0)$ is isomorphic to

$$\ker \frac{\ker \left(\dfrac{\ker(\phi^{i_0,j_0})}{\mathrm{im}(\phi^{i_0,j_0-1})} \xrightarrow{\xi^{i_0,j_0}} \dfrac{\ker(\phi^{i_0+1,j_0})}{\mathrm{im}(\phi^{i_0+1,j_0-1})} \right)}{\xi^{i_0-1,j_0} \left(\dfrac{\ker(\phi^{i_0-1,j_0})}{\mathrm{im}(\phi^{i_0-1,j_0-1})} \right)} \cong H_{\mathfrak{m}}^{i_0}(H_I^{j_0}(R_n)).$$

The elements of G will be generators for $H_{\mathfrak{m}}^{i_0}(H_I^{j_0}(R_n))$ and the elements of G_0 generate the extra relations that are not syzygies.

The algorithm can of course be modified to compute any iterated local cohomology group $H_J^j(H_I^i(R_n))$ for $J \supseteq I$ by replacing the generators $x_1, \ldots, x_n$ for $\mathfrak{m}$ by those for J. Moreover, the iteration depth can also be increased by considering "tricomplexes" etc. instead of bicomplexes.

Again we would like to point out that with the methods of [35] or [37] one could actually compute first $H_I^i(R_n)$ and from that $H_J^j(H_I^i(R_n))$, but probably that is quite a bit more complex a computation.

4.3 Computation of Lyubeznik Numbers

G. Lyubeznik proved in [25] that if K is a field, $R = K[x_1, \ldots, x_n]$, $I \subseteq R$, $\mathfrak{m} = R \cdot (x_1, \ldots, x_n)$ and $A = R/I$ then $\lambda_{i,j}(A) = \dim_K \mathrm{soc}_R H_{\mathfrak{m}}^i(H_I^{n-j}(R))$ is invariant under change of presentation of A. In other words, it only depends on A and i, j but not the projection $R \twoheadrightarrow A$. Lyubeznik proved that $H_{\mathfrak{m}}^i(H_I^j(R_n))$ is in fact an injective $\mathfrak{m}$-torsion R_n-module of finite socle dimension $\lambda_{i,n-j}(A)$ and so isomorphic to $(E_{R_n}(K))^{\lambda_{i,n-j}(A)}$ where $E_{R_n}(K)$ is the injective hull of K over R_n. We are now in a position to compute these invariants of R_n/I in characteristic zero..

Algorithm 4.5 (Lyubeznik numbers).
INPUT: $f_1, \ldots, f_r \in R_n; i, j \in \mathbb{N}$.
OUTPUT: $\lambda_{i,n-j}(R_n/R_n \cdot (f_1, \ldots, f_r))$.

1. Using Algorithm 4.4 find $g_1, \ldots, g_l \in D_n{}^d$ and $h_1, \ldots, h_e \in D_n{}^d$ such that $H^i_{\mathfrak{m}}(H^j_I(R_n))$ is isomorphic to $D_n \cdot (g_1, \ldots, g_l)$ modulo $H = D_n \cdot (h_1, \ldots, h_e)$.
2. Assume that after a suitable renumeration g_1 is not in H. If such a g_1 cannot be chosen, quit.
3. Find a monomial $m \in R_n$ such that $m \cdot g_1 \notin H$ but $x_i m g_1 \in H$ for all x_i.
4. Replace H by $D_n m g_1 + H$ and reenter at Step 2.
5. Return $\lambda_{i,n-j}(R_n/I)$, the number of times Step 3 was executed.

End.

The reason that this works is as follows. We know that $(D_n \cdot g_1 + H)/H$ is $\mathfrak{m}$-torsion (as $H^i_{\mathfrak{m}}(H^j_I(R_n))$ is) and so it is possible (with trial and error, or a suitable syzygy computation) to find the monomial m in Step 3. The element $m g_1 \mod H \in D_n/H$ has annihilator equal to $\mathfrak{m}$ over R_n and therefore generates a D_n-module isomorphic to $D_n/D_n \cdot \mathfrak{m} \cong E_{R_n}(K)$. The injection

$$(D_n \cdot m g_1 + H)/H \hookrightarrow (D_n \cdot (g_1, \ldots, g_l) + H)/H$$

splits as map of R_n-modules because $E_{R_n}(K)$ is injective and so the cokernel $D_n \cdot (g_1, \ldots, g_l)/D_n \cdot (m g_1, h_1, \ldots, h_e)$ is isomorphic to $(E_{R_n}(K))^{\lambda_{i,n-j}(A)-1}$.

Reduction of the g_i with respect to a Gröbner basis of the new relation module and repetition will lead to the determination of $\lambda_{i,n-j}(A)$.

Assume that D_n/L is an $\mathfrak{m}$-torsion module. For example, we could have $D_n/L \cong H^i_{\mathfrak{m}}(H^j_I(R_n))$. Here is a procedure that finds by trial and error the monomial socle element m of Step 3 in Algorithm 4.4.

```
i61 : findSocle = method();

i62 : findSocle(Ideal, RingElement):= (L,P) -> (
            createDpairs(ring(L));
            v=(ring L).dpairVars#0;
            myflag = true;
            while myflag do (
                w = apply(v,temp -> temp*P % L);
                if all(w,temp -> temp == 0) then myflag = false
                else (
                    p = position(w, temp -> temp != 0);
                    P = v#p * P;)
            );
        P);
```

For example, if we want to apply this socle search to the ideal JH3 describing $H^3_I(R_6)$ of Example 4.3 we do

```
i63 : D = ring JH3

o63 = D

o63 : PolynomialRing
```

(as D was most recently the differential operators on $\mathbb{Q}[x, y, z, w]$)

```
i64 : findSocle(JH3,1_D)

o64 = x*v

o64 : D
```

One can then repeat the socle search and kill the newly found element as suggested in the explanation above:

```
i65 : findLength = method();

i66 : findLength Ideal := (I) -> (
           l = 0;
           while I != ideal 1_(ring I) do (
               l = l + 1;
               s = findSocle(I,1_(ring I));
               I = I + ideal s;);
           l);
```

Applied to JH3 of the previous subsection this yields

```
i67 : findLength JH3

o67 = 1
```

and hence JH3 does indeed describe a module isomorphic to $E_{R_6}(K)$.

5   Implementation, Examples, Questions

5.1   Implementations and Optimizing

The Algorithms 3.7, 3.9 and 3.14 have first been implemented by T. Oaku and N. Takayama using the package Kan [42] which is a postscript language for computations in the Weyl algebra and in polynomial rings. In *Macaulay 2* Algorithms 3.7, 3.9 and 3.14 as well as Algorithm 4.2 have been implemented by A. Leykin, M. Stillman and H. Tsai. They additionally implemented a wealth of *D*-module routines that relate to topics which we cannot all cover in this chapter. These include homomorphisms between holonomic modules and extension functors, restriction functors to linear subspaces, integration (de Rham) functors to quotient spaces and others. For further theoretical information the reader is referred to [35,34,36,40,45,48,50,51].

Computation of Gröbner bases in many variables is in general a time and space consuming enterprise. In commutative polynomial rings the worst case performance for the number of elements in reduced Gröbner bases is doubly exponential in the number of variables and the degrees of the generators. In the (relatively) small Example 4.3 above R_6 is of dimension 6, so that the intermediate ring $D_{n+1}[y_1, y_2]$ contains 16 variables. In view of these facts the following idea has proved useful.

The general context in which Lemma 3.5 and Proposition 3.11 were stated allows successive localization of $R_n[(fg)^{-1}]$ in the following way. First one computes $R_n[f^{-1}]$ according to Algorithm 3.14 as quotient $D_n/J^{\Delta}(f^s)|_{s=a}$, $\mathbb{Z} \ni a \ll 0$. Then $R_n[(fg)^{-1}]$ may be computed using Algorithm 3.14 again since $R_n[(fg)^{-1}] \cong R_n[g^{-1}] \otimes_{R_n} D_n/J^\Delta(f^s)|_{s=a}$. (Note that all localizations of R_n are automatically f-torsion free for $f \in R_n$ so that Algorithm 3.14 can be used.) This process may be iterated for products with any finite number of factors. Of course the exponents for the various factors might be different. This requires some care when setting up the Čech complex. In particular one needs to make sure that the maps $\check{C}^k \to \check{C}^{k+1}$ can be made explicit using the f_i. (In our Example 4.3, this is precisely how we proceeded when we found Rfgh.)

Remark 5.1. One might hope that for all holonomic fg-torsion free modules $M = D_n/L$ we have (with $M \otimes R_n[g^{-1}] \cong D_n/L'$):

$$a_f^L = \min\{s \in \mathbb{Z} : b_f^L(s) = 0\} \leq \min\{s \in \mathbb{Z} : b_f^{L'}(s) = 0\} = a_f^{L'}. \quad (5.1)$$

This hope is unfounded. Let $R_5 = K[x_1, \ldots, x_5]$, $f = x_1^2 + x_2^2 + x_3^2 + x_4^2 + x_5^2$. One may check that then $b_f^\Delta(s) = (s+1)(s+5/2)$. Hence $R_5[f^{-1}] = D_5 \bullet f^{-1}$, let $L = \ker(D_5 \to D_5 \bullet f^{-1})$. Set $g = x_1$. Then $b_g^\Delta(s) = s + 1$, let $L' = \ker(D_5 \to D_5 \bullet g^{-1})$.

Then $b_f^{L'}(s) = (s+1)(s+2)(s+5/2)$ and $b_g^L(s) = (s+1)(s+3)$ because of the following computations.

```
i68 : erase symbol x; erase symbol Dx;
```

These two commands essentially clear the history of the variables x and Dx and make them available for future computations.

```
i70 : D = QQ[x_1..x_5, Dx_1..Dx_5, WeylAlgebra =>
            apply(toList(1..5), i -> x_i => Dx_i)];
```

```
i71 : f = x_1^2 + x_2^2 + x_3^2 + x_4^2 +x_5^2;
```

```
i72 : g = x_1;
```

```
i73 : R = D^1/ideal(Dx_1,Dx_2,Dx_3,Dx_4,Dx_5);
```

As usual, these commands defined the base ring, two polynomials and the D_5-module R_5. Now we compute the respective localizations.

```
i74 : Rf =DlocalizeAll(R,f,Strategy => Oaku);
```

```
i75 : Bf = Rf.Bfunction
```

```
              5
o75 = ($s + -)($s + 1)
              2
```

```
o75 : Product
```

```
i76 : Rfg = DlocalizeAll(Rf.LocModule,g,Strategy => Oaku);
```

```
i77 : Bfg = Rfg.Bfunction
```

```
o77 = ($s + 1)($s + 3)
```

```
o77 : Product
```

```
i78 : Rg = DlocalizeAll(R,g,Strategy => Oaku);
```

```
i79 : Bg = Rg.Bfunction

o79 = ($s + 1)

o79 : Product

i80 : Rgf = DlocalizeAll(Rg.LocModule,f,Strategy => Oaku);

i81 : Bgf = Rgf.Bfunction

                     5
o81 = ($s + 2)($s + 1)($s + -)
                     2

o81 : Product
```

The output shows that $R_n[(fg)^{-1}]$ is generated by $f^{-2}g^{-1}$ or $f^{-1}g^{-3}$ but not by $f^{-1}g^{-2}$ and in particular not by $f^{-1}g^{-1}$. This can be seen from the various Bernstein-Sato polynomials: as for example the smallest integral root of Bf is -1 and that of Bfg is -3, $R_3[f^{-1}]$ is generated by f^{-1} and $R_3[(fg)^{-1}]$ by $f^{-1}g^{-3}$. This example not only disproves the above inequality (5.1) but also shows the inequality to be wrong if $\mathbb{Z}$ is replaced by $\mathbb{R}$ (as $-3 < \min(-5/2, -1)$).

Nonetheless, localizing $R_n[(fg)^{-1}]$ as $(R_n[f^{-1}])[g^{-1}]$ is heuristically advantageous, apparently for two reasons. For one, it allows the exponents of the various factors to be distinct which is useful for the subsequent cohomology computation: it helps to keep the degrees of the maps small. So in Example 4.3 we can write $R_6[(fg)^{-1}]$ as $D_6 \bullet (f^{-1}g^{-2})$ instead of $D_6 \bullet (fg)^{-2}$. Secondly, since the computation of Gröbner bases is potentially doubly exponential it seems to be advantageous to break a big problem (localization at a product) into several "easy" problems (successive localization).

An interesting case of this behavior is our Example 4.3. If we compute $R_n[(fgh)^{-1}]$ as $((R_n[f^{-1}])[g^{-1}])[h^{-1}]$, the calculation uses approximately 6MB and lasts a few seconds using *Macaulay 2*. If one tries to localize R_n at the product of the three generators at once, *Macaulay 2* runs out of memory on all machines the author has tried this computation on.

5.2   Projects for the Future

This is a list of theoretical and implementational questions that the author finds important and interesting.

Prime Characteristic. In [26], G. Lyubeznik gave an algorithm for deciding whether or not $H^i_I(R) = 0$ for any given $I \subseteq R = K[x_1, \ldots, x_n]$ where K is a computable field of positive characteristic. His algorithm is built on entirely different methods than the ones used in this chapter and relies on the Frobenius functor. The implementation of this algorithm would be quite worthwhile.

Ambient Spaces Different from $\mathbb{A}_K^n$. If A equals $K[x_1, \ldots, x_n]$, $I \subseteq A$, $X = \mathrm{Spec}(A)$ and $Y = \mathrm{Spec}(A/I)$, knowledge of $H_I^i(A)$ for all $i \in \mathbb{N}$ answers of course the question about the local cohomological dimension of Y in X. If $W \subseteq X$ is a smooth variety containing Y then Algorithm 4.2 for the computation of $H_I^i(A)$ also leads to a determination of the local cohomological dimension of Y in W. Namely, if J stands for the defining ideal of W in X so that $R = A/J$ is the affine coordinate ring of W and if we set $c = \mathrm{ht}(J)$, then it can be shown that $H_I^{i-c}(R) = \mathrm{Hom}_A(R, H_I^i(A))$ for all $i \in \mathbb{N}$. As $H_I^i(A)$ is I-torsion (and hence J-torsion), $\mathrm{Hom}_A(R, H_I^i(A))$ is zero if and only if $H_I^i(A) = 0$. It follows that the local cohomological dimension of Y in W equals $\mathrm{cd}(A, I) - c$ and in fact $\{i \in \mathbb{N} : H_I^i(A) \neq 0\} = \{i \in \mathbb{N} : H_I^{i-c}(R) \neq 0\}$.

If however $W = \mathrm{Spec}(R)$ is not smooth, no algorithms for the computation of either $H_I^i(R)$ or $\mathrm{cd}(R, I)$ are known, irrespective of the characteristic of the base field. It would be very interesting to have even partial ideas for computations in that case.

De Rham Cohomology. In [35,48] algorithms are given to compute de Rham (in this case equal to singular) cohomology of complements of complex affine hypersurfaces and more general varieties. In [51] an algorithm is given to compute the multiplicative (cup product) structure, and in [50] the computation of the de Rham cohomology of open and closed sets in projective space is explained. Some of these algorithms have been implemented while others are still waiting.

For example, de Rham cohomology of complements of hypersurfaces, and partially the cup product routine, are implemented.

Example 5.2. Let $f = x^3 + y^3 + z^3$ in R_3. One can compute with *Macaulay 2* the de Rham cohomology of the complement of $\mathrm{Var}(f)$, and it turns out that the cohomology in degrees 0 and 1 is 1-dimensional, in degrees 3 and 4 2-dimensional and zero otherwise – here is the input:

```
i82 : erase symbol x;
```

Once x gets used as a subscripted variable, it's hard to use it as a nonsubscripted variable. So let's just erase it.

```
i83 : R = QQ[x,y,z];
```

```
i84 : f=x^3+y^3+z^3;
```

```
i85 : H=deRhamAll(f);
```

H is a hashtable with the entries Bfunction, LocalizeMap, VResolution, TransferCycles, PreCycles, OmegaRes and CohomologyGroups. For example, we have

```
i86 : H.CohomologyGroups

                     1
o86 = HashTable{0 => QQ }
                     1
                1 => QQ
```

```
                   2
          2 => QQ
                   2
          3 => QQ

o86 : HashTable
```

showing that the dimensions are as claimed above. One can also extract information on the generator of $R_3[f^{-1}]$ used to represent the cohomology classes by

```
i87 : H.LocalizeMap

o87 = | $x_1^6+2$x_1^3$x_2^3+$x_2^6+2$x_1^3$x_3^3+2$x_2^3$x_3^3+$x_3^6 |

o87 : Matrix
```

which proves that the generator in question is f^{-2}. The cohomology classes that *Macaulay 2* computes are differential forms:

```
i88 : H.TransferCycles

o88 = HashTable{0 => | -1/12$x_1^4$x_2^3$D_1-1/3$x_1$x_2^6$D_1-1/12$x_  ···
                1 => | 2/3$x_1^5+2/3$x_1^3$x_2^3+2/3$x_1^2$x_3^3  |
                     | -2/3$x_1^3$x_2^2-2/3$x_2^5-2/3$x_2^2$x_3^3 |
                     | 2/3$x_1^3$x_3^2+2/3$x_2^3$x_3^2+2/3$x_3^5  |
                2 => | 48$x_1$x_2$x_3^2 600$x_3^4      |
                     | 48$x_1$x_2^2$x_3 600$x_2$x_3^3 |
                     | 48$x_1^2$x_2$x_3 600$x_1$x_3^3 |
                3 => | -$x_1$x_2$x_3 -$x_3^3 |

o88 : HashTable
```

So, for example, the left column of the three rows that correspond to $H^2_{\mathrm{dR}}(\mathbb{C}^3 \setminus \mathrm{Var}(f), \mathbb{C})$ represent the form $(xyz(48z\,dx\,dy + 48y\,dz\,dx + 48x\,dy\,dz))/f^2$. If we apply the above elements of D_3 to f^{-2} and equip the results with appropriate differentials one arrives (after dropping unnecessary integral factors) at the results displayed in the table in Figure 1. In terms of de Rham cohomology, the cup product is given by the wedge product of differential forms. So from the table one reads off the cup product relations $o \cup t_1 = d_1$, $o \cup t_2 = d_2$, $o \cup o = 0$, and that e operates as the identity.

For more involved examples, and the algorithms in [50], an actual implementation would be necessary since paper and pen are insufficient tools then.

Remark 5.3. The reader should be warned: if f^{-a} generates $R_n[f^{-1}]$ over D_n, then it is not necessarily the case that each de Rham cohomology class of $U = \mathbb{C}^n \setminus \mathrm{Var}(f)$ can be written as a form with a pole of order at most a. A counterexample is given by $f = (x^3 + y^3 + xy)xy$, where $H^1_{\mathrm{dR}}(U; \mathbb{C})$ has a class that requires a third order pole, although -2 is the smallest integral Bernstein root of f on R_2.

Hom and Ext. In [36,44,45] algorithms are explained that compute homomorphisms between holonomic systems. In particular, rational and polynomial solutions can be found because, for example, a polynomial solution to the

Group	Dimension	Generators
H^0_{dR}	1	$e := \dfrac{f^2}{f^2}$
H^1_{dR}	1	$o := \dfrac{(x^2 dx - y^2 dy + z^2 dz)f}{f^2}$
H^2_{dR}	2	$t_1 := \dfrac{xyz(zdxdy + ydzdx + xdydz)}{f^2}$
		$t_2 := \dfrac{(zdxdy + ydzdx + xdydz)z^3}{f^2}$
H^3_{dR}	2	$d_1 := \dfrac{xyzdxdydz}{f^2}$
		$d_2 := \dfrac{z^3 dxdydz}{f^2}$

Fig. 1.

system $\{P_1, \ldots, P_r\} \in D_n$ corresponds to an element of $\mathrm{Hom}_{D_n}(D_n/I, R_n)$ where $I = D_n \cdot (P_1, \ldots, P_r)$.

Example 5.4. Consider the GKZ system in 2 variables associated to the matrix $(1, 2) \in \mathbb{Z}^{1 \times 2}$ and the parameter vector $(5) \in \mathbb{C}^1$. Named after Gelfand-Kapranov-Zelevinski [11], this is the following system of differential equations:

$$(x\partial_x + y\partial_y) \bullet f = 5f,$$
$$(\partial_x^2 - \partial_y) \bullet f = 0.$$

With *Macaulay 2* one can solve systems of this sort as follows:

```
i89 : I = gkz(matrix{{1,2}}, {5})

             2
o89 = ideal (D  - D , x D  + 2x D  - 5)
             1    2   1 1    2 2

o89 : Ideal of QQ [x , x , D , D , WeylAlgebra => {x  => D , x  => D }]
                   1   2   1   2                   1     1   2     2
```

This is a simple command to set up the GKZ-ideal associated to a matrix and a parameter vector. The polynomial solutions are obtained by

```
i90 : PolySols I

        5       3       2
o90 = {x  + 20x x  + 60x x }
       1      1 2      1 2

o90 : List
```

This means that there is exactly one polynomial solution to the given GKZ-system, and it is

$$x^5 + 20x^3y + 60xy^2.$$

The algorithm for $\mathrm{Hom}_{D_n}(M, N)$ is implemented and can be used to check whether two given D-modules are isomorphic. Moreover, there are algorithms (not implemented yet) to compute the ring structure of $\mathrm{End}_D(M)$ for a given D-module M of finite holonomic rank which can be used to split a given holonomic module into its direct summands. Perhaps an adaptation of these methods can be used to construct Jordan-Hölder sequences for holonomic D-modules.

Finiteness and Stratifications. Lyubeznik pointed out in [27] the following curious fact.

Theorem 5.5. *Let $P(n, d; K)$ denote the set of polynomials of degree at most d in at most n variables over the field K of characteristic zero. Let $B(n, d; K)$ denote the set of Bernstein-Sato polynomials*

$$B(n, d; K) = \{b_f(s) : f \in P(n, d; K)\}.$$

Then $B(n, d; K)$ is finite. □

So $P(n, d; K)$ has a finite decomposition into strata with constant Bernstein-Sato polynomial. A. Leykin proved in [22] that this decomposition is independent of K and computable in the sense that membership in each stratum can be tested by the vanishing of a finite set of algorithmically computable polynomials over $\mathbb{Q}$ in the coefficients of the given polynomial in $P(n, d; K)$. In particular, the stratification is algebraic and for each K induced by base change from $\mathbb{Q}$ to K. It makes thus sense to define $B(n, d)$ which is the finite set of Bernstein polynomials that can occur for $f \in P(n, d; K)$ (where K is in fact irrelevant).

Example 5.6. Consider $P(2, 2; K)$, the set of quadratic binary forms over K. With *Macaulay 2*, Leykin showed that there are precisely 4 different Bernstein polynomials possible:

- $b_f(s) = 1$ iff $f \in V_1 = V_1' \setminus V_1''$, where $V_1' = V(a_{1,1}, a_{0,1}, a_{0,2}, a_{1,0}, a_{2,0})$, while $V_1'' = V(a_{0,0})$
- $b_f(s) = s + 1$ iff $f \in V_2 = (V_2' \setminus V_2'') \cup (V_3' \setminus V_3'')$, where $V_2' = V(0)$, $V_2'' = V(\gamma_1)$, $V_3' = V(\gamma_2, \gamma_3, \gamma_4)$, $V_3'' = V(\gamma_3, \gamma_4, \gamma_5, \gamma_6, \gamma_7, \gamma_8)$,
- $b_f(s) = (s + 1)^2$ iff $f \in V_4' \setminus V_4''$, where $V_4' = V(\gamma_1)$, $V_4'' = V(\gamma_2, \gamma_3, \gamma_4)$,
- $b_f(s) = (s + 1)(s + \frac{1}{2})$ iff $f \in V_5' \setminus V_5''$, where $V_5' = V(\gamma_3, \gamma_4, \gamma_5, \gamma_6, \gamma_7, \gamma_8)$, $V_5'' = V(a_{1,1}, a_{0,1}, a_{0,2}, a_{1,0}, a_{2,0})$.

Here we have used the abbreviations

- $\gamma_1 = a_{0,2}a_{1,0}^2 - a_{0,1}a_{1,0}a_{1,1} + a_{0,0}a_{1,1}^2 + a_{0,1}^2 a_{2,0} - 4a_{0,0}a_{0,2}a_{2,0}$,
- $\gamma_2 = 2a_{0,2}a_{1,0} - a_{0,1}a_{1,1}$,
- $\gamma_3 = a_{1,0}a_{1,1} - 2a_{0,1}a_{2,0}$,
- $\gamma_4 = a_{1,1}^2 - 4a_{0,2}a_{2,0}$,

- $\gamma_5 = 2a_{0,2}a_{1,0} - a_{0,1}a_{1,1}$,
- $\gamma_6 = a_{0,1}^2 - 4a_{0,0}a_{0,2}$,
- $\gamma_7 = a_{0,1}a_{1,0} - 2a_{0,0}a_{1,1}$,
- $\gamma_8 = a_{1,0}^2 - 4a_{0,0}a_{2,0}$.

Similarly, Leykin shows that there are 9 possible Bernstein polynomials for $f \in B(2,3;K)$:

$$B(2,3) = \left\{ (s+1)^2(s+\frac{2}{3})(s+\frac{4}{3}), \ (s+1)^2(s+\frac{1}{2}), \ (s+1), \ 1, \right.$$

$$(s+1)(s+\frac{2}{3})(s+\frac{1}{3}), \ (s+1)^2, \ (s+1)(s+\frac{1}{2}),$$

$$\left. (s+1)(s+\frac{7}{6})(s+\frac{5}{6}), \ (s+1)^2(s+\frac{3}{4})(s+\frac{5}{4}) \right\}.$$

It would be very interesting to study the nature of the stratification in larger cases, and its restriction to hyperplane arrangements.

A generalization of this stratification result is obtained in [49]. There it is shown that there is an algorithm to give $P(n, d; K)$ an algebraic stratification defined over $\mathbb{Q}$ such that the algebraic de Rham cohomology groups of the complement of $\mathrm{Var}(f)$ do not vary on the stratum in a rather strong sense. Again, the study and explicit computation of this stratification should be very interesting.

Hodge Numbers. If Y is a projective variety in $\mathbb{P}_{\mathbb{C}}^n$ then algorithms outlined in [50] show how to compute the dimensions not only of the de Rham cohomology groups of $\mathbb{P}_{\mathbb{C}}^n \setminus Y$ but also of Y itself. Suppose now that Y is in fact a smooth projective variety. An interesting set of invariants are the Hodge numbers, defined by $h^{p,q} = \dim H^p(Y, \Omega^q)$, where Ω^q denotes the sheaf of $\mathbb{C}$-linear differential q-forms with coefficients in $\mathcal{O}_Y$. At present we do not know how to compute them. Of course there is a spectral sequence $H^p(Y, \Omega^q) \Rightarrow H_{\mathrm{dR}}^{p+q}(Y, \mathbb{C})$ and we know the abutment (or at least its dimensions), but the technique for computing the abutment does not seem to be usable to compute the E^1 term because on an affine patch $H^p(Y, \Omega^q)$ is either zero or an infinite dimensional vector space.

Hodge structures and Bernstein-Sato polynomials are related as is for example shown in [46].

5.3   Epilogue

In this chapter we have only touched a few highlights of the theory of computations in D-modules, most of them related to homology and topology. Despite this we hardly touched on the topics of integration and restriction, which are the D-module versions of a pushforward and pullback, [20,30,35,48].

A very different aspect of *D*-modules is discussed in [40] where at the center of investigations is the combinatorics of solutions of hypergeometric differential equations. The combinatorial structure is used to find series solutions for the differential equations which are polynomial in certain logarithmic functions and power series with respect to the variables.

Combinatorial elements can also be found in the work of Assi, Castro and Granger, see [2,3], on Gröbner fans in rings of differential operators. An important (open) question in this direction is the determination of the set of ideals in D_n that are initial ideals under *some* weight.

Algorithmic *D*-module theory promises to be an active area of research for many years to come, and to have interesting applications to various other parts of mathematics.

References

1. A.G. Aleksandrov and V.L. Kistlerov: Computer method in calculating *b*-functions of non-isolated singularities. *Contemp. Math.*, 131:319–335, 1992.
2. A. Assi, F.J. Castro-Jiménez, and M. Granger: How to calculate the slopes of a *D*-module.. *Compositio Math.*, 104(2):107–123, 1996.
3. A. Assi, F.J. Castro-Jiménez, and M. Granger: The Gröbner fan of an A_n-module.. *J. Pure Appl. Algebra*, 150(1):27–39, 2000.
4. J. Bernstein: Lecture notes on the theory of *D*-modules. Unpublished.
5. J.-E. Björk: *Rings of Differential Operators*. North Holland, New York, 1979.
6. J. Briançon, M. Granger, Ph. Maisonobe, and M. Miniconi: Algorithme de calcul du polynôme de Bernstein: cas nondégénéré. *Ann. Inst. Fourier*, 39:553–610, 1989.
7. M. Brodmann and J. Rung: Local cohomology and the connectedness dimension in algebraic varieties. *Comment. Math. Helvetici*, 61:481–490, 1986.
8. M.P. Brodmann and R.Y. Sharp: *Local Cohomology: an algebraic introduction with geometric applications.*. Cambridge Studies in Advanced Mathematics, 60. Cambridge University Press, 1998.
9. S.C. Coutinho: *A primer of algebraic D-modules*. London Mathematical Society Student Texts 33. Cambridge University Press, Cambridge, 1995.
10. R. Garcia and C. Sabbah: Topological computation of local cohomology multiplicities. *Coll. Math.*, XLIX(2-3):317–324, 1998.
11. I. Gelfand, M. Kapranov, and A. Zelevinski: *Discriminants, resultants, and multidimensional determinants*. Mathematics: Theory & Applications. Birkhäuser Boston, Inc., Boston, MA, 1994.
12. P. Griffiths and J. Harris: *Principles of Algebraic Geometry*. A Wiley Interscience Publication. WILEY and Sons, 1978.
13. R. Hartshorne: *Local Cohomology. A seminar given by A. Grothendieck*, volume 41 of *Lecture Notes in Mathematics*. Springer Verlag, 1967.
14. R. Hartshorne: Cohomological Dimension of Algebraic Varieties. *Ann. Math.*, 88:403–450, 1968.
15. R. Hartshorne and R. Speiser: Local cohomological dimension in characteristic *p*. *Ann. Math.(2)*, 105:45–79, 1977.
16. Robin Hartshorne: On the De Rham cohomology of algebraic varieties. *Inst. Hautes Études Sci. Publ. Math.*, 45:5–99, 1975.

17. C. Huneke: Problems on local cohomology. *Res. Notes Math.*, 2:93–108, 1992.
18. C. Huneke and G. Lyubeznik: On the vanishing of local cohomology modules. *Invent. Math.*, 102:73–93, 1990.
19. M. Kashiwara: *B*-functions and holonomic systems, rationality of *B*-functions.. *Invent. Math.*, 38:33–53, 1976.
20. M. Kashiwara: On the holonomic systems of linear partial differential equations, II. *Invent. Math.*, 49:121–135, 1978.
21. J. Kollár: Singularities of pairs, *in* Algebraic Geometry, Santa Cruz, 1995. *Proc. Symp. Pure Math. Amer. Math. Soc.*, 62:221–287, 1997.
22. A. Leykin: Towards a definitive computation of Bernstein-Sato polynomials. *math.AG/0003155*[1], 2000.
23. A. Leykin, M. Stillman, and H. Tsai: The *D*-module package for *Macaulay 2*. http://www.math.umn.edu/~leykin/.
24. François Loeser: Fonctions d'Igusa *p*-adiques, polynômes de Bernstein, et polyèdres de Newton. *J. Reine Angew. Math.*, 412:75–96, 1990.
25. G. Lyubeznik: Finiteness Properties of Local Cohomology Modules: an Application of *D*-modules to Commutative Algebra. *Invent. Math.*, 113:41–55, 1993.
26. G. Lyubeznik: *F*-modules: applications to local cohomology and *D*-modules in characteristic $p > 0$. *Journal für die Reine und Angewandte Mathematik*, 491:65–130, 1997.
27. G. Lyubeznik: On Bernstein-Sato polynomials. *Proc. Amer. Math. Soc.*, 125(7):1941–1944, 1997.
28. Ph. Maisonobe: *D*-modules: an overview towards effectivity, in: E. Tournier, ed., Computer Algebra and Differential Equations. *London Math. Soc. Lecture Note Ser., 193, Cambridge Univ. Press, Cambridge*, pages 21–55, 1994.
29. B. Malgrange: Le polynôme de Bernstein d'une singularité isolée. *Lecture Notes in Mathematics, Springer Verlag*, 459:98–119, 1975.
30. Z. Mebkhout: Le formalisme des six opérations de Grothendieck pour les D_X-modules cohérents.. Travaux en Cours. Hermann, Paris, 1989.
31. T. Oaku: An algorithm for computing *B*-functions. *Duke Math. Journal*, 87:115–132, 1997.
32. T. Oaku: Algorithms for *b*-functions, restrictions, and algebraic local cohomology groups of *D*-modules. *Advances in Appl. Math.*, 19:61–105, 1997.
33. T. Oaku: Algorithms for the *b*-function and *D*-modules associated with a polynomial. *J. Pure Appl. Algebra*, 117-118:495–518, 1997.
34. T. Oaku and N. Takayama: An algorithm for de Rham cohomology groups of the complement of an affine variety via *D*-module computation. *J. Pure Appl. Algebra*, 139:201–233, 1999.
35. T. Oaku and N. Takayama: Algorithms for *D*-modules – restriction, tensor product, localization and local cohomology groups. *Journal of Pure and Applied Algebra*, 156(2-3):267–308, February 2001.
36. T. Oaku, N. Takayama, and H. Tsai: Polynomial and rational solutions of holonomic systems. *math.AG/0001064*, 1999.
37. T. Oaku, N. Takayama, and U. Walther: A localization algorithm for D-modules. *J. Symb. Comp.*, 29(4/5):721–728, May 2000.

---

[1] The webpage http://xxx.lanl.gov/ is a page designed for the storage of preprints, and allows posting and downloading free of charge.

38. A. Ogus: Local cohomological dimension of algebraic varieties. *Ann. Math. (2)*, 98:327–365, 1973.
39. C. Peskine and L. Szpiro: Dimension projective finie et cohomologie locale. *Inst. Hautes Études Sci. Publ. Math.*, 42:47–119, 1973.
40. M. Saito, B. Sturmfels, and N. Takayama: *Gröbner deformations of hypergeometric differential equations*. Algorithms and Computation in Mathematics, 6. Springer Verlag, 1999.
41. M. Sato, M. Kashiwara, T. Kimura, and T. Oshima: Micro-local analysis of prehomogeneous vector spaces. *Inv. Math.*, 62:117–179, 1980.
42. N. Takayama: Kan: A system for computation in algebraic analysis. Source code available at www.math.kobe-u.ac.jp/KAN. *Version 1 (1991), Version 2 (1994)*, the latest Version is 2.990914 (1999).
43. H. Tsai: Weyl closure, torsion, and local cohomology of *D*-modules. Preprint, 1999.
44. H. Tsai: Algorithms for algebraic analysis. *Thesis, University of California at Berkeley*, 2000.
45. H. Tsai and U. Walther: Computing homomorphisms between holonomic *D*-modules. *math.RA/0007139*, 2000.
46. A. Varchenko: Asymptotic Hodge structure in the vanishing cohomology. *Math. USSR Izvestija*, 18(3):469–512, 1982.
47. U. Walther: Algorithmic Computation of Local Cohomology Modules and the Local Cohomological Dimension of Algebraic Varieties. *J. Pure Appl. Algebra*, 139:303–321, 1999.
48. U. Walther: Algorithmic Computation of de Rham Cohomology of Complements of Complex Affine Varieties. *J. Symb. Comp.*, 29(4/5):795–839, May 2000.
49. U. Walther: Homotopy Type, Stratifications and Gröbner bases. In preparation, 2000.
50. U. Walther: Cohomology with Rational Coefficients of Complex Varieties. *Contemp. Math.*, to appear.
51. U. Walther: The Cup Product Structure for Complements of Complex Affine Varieties. *J. Pure Appl. Algebra*, to appear.
52. U. Walther: On the Lyubeznik numbers of a local ring. *Proc. Amer. Math. Soc.*, to appear.
53. T. Yano: On the theory of *b*-functions. *Publ. RIMS, Kyoto Univ.*, 14:111–202, 1978.

Index

Druck: Strauss Offsetdruck, Mörlenbach
Verarbeitung: Schäffer, Grünstadt